ADVANCE JAVA

By
Kanika Lakhani Chaudhary
Ph.d (Pursuing), M.Tech. (CSE)
Assistant Professor
Manav Rachna College of Engineering,
Faridabad

S.K. KATARIA & SONS®
Educational Publishers
4885/109, Prakash Mahal, Dr. Subhash Bhargav Lane,
Opposite Delhi Medical Association, Daryaganj, New Delhi-110002 (INDIA)
Phone : +91-11-43551243, Mobile: +91-9871775858, 7678425209
e-mail: katariabook@gmail.com, katariabooks@yahoo.com
Website: www.skkatariaandsons.com
(e-books, subscriptions and bulk buying also available on our website)

Published by:

S.K. Kataria & Sons®

4885/109, Prakash Mahal, Dr. Subhash Bhargav Lane,
Opposite Delhi Medical Association, Daryaganj, New Delhi-110002 (INDIA)
Phone: +91-11-43551243, Mobile: +91-9871775858, 7678425209
e-mail: katariabook@gmail.com, katariabooks@yahoo.com
Website: www.skkatariaandsons.com

First Edition: 2010, Third Edition: 2012, Reprint: 2025

Laser Typesetting by :
AISA Enterprises, Delhi-110092

Printed at:
L.B. Enterprises

PREFACE

The book Advance Java is meant for the engineering students of various universities of Haryana and U.P. Technical University. This book has been written in accordance with the recent syllabus of the various universities for B.Tech, B.E. and MCA courses, with illustrative figures, tables and examples. The aim behind writing this book is to provide an up-to-date account of the relevant topics, with problems and solutions and particularly emphasizing on the examples. Due consideration has been taken to make the language easy to understand. To facilitate easy understanding, the text and problems have been put in simple form. The emphasis is on simpler set of examples relating to every concept in each chapter. This will definitely help the students to understand the subject in a better way.

AUTHOR

ACKNOWLEDGEMENT

I am thankful to Almighty God for blessing me with the strength to face any situation with courage and patience. I extend my thanks to my dear husband who encouraged me finish the task successfully on time. I also thank my parents for their inspiration and blessings. I convey my thanks to Mr. Sanjeev Kataria, the publisher of this book, Mr. Sharda Singh and his entire team for their continuous interaction and suggestions for improvement in the manuscript and bringing out this book in time.

Kanika Lakhani

FORWARD

The topic **Java** is an important and mandatory subject in B.Tech./M.Tech. Course curriculum in Computer Science & Engineering, IT, Software Engineering and in all the allied areas like Electronics, Management, Electrical Engineering, etc. in India as well as in any renowned universities in the world. Unfortunately, student friendly types of books are not available in the market. Those available books are highly text in nature, and consequently many of them can not reach completely the students mind.

This book has been the hard work of the authors who have worked on their toes to come up with this product in a record time. The book has been designed in such a way that it provides an easier perception of the advanced programming in Java and helps in acquiring an in-depth knowledge of the subject. The contents have been structured such that it would be easier and simpler for the students to understand the subject smoothly.

This creation of Kanika Lakhani and Ashish Kumar is appreciable and it will definitely benefit the readers in all spheres.

Dr. Ranjit Biswas
Professor & HOD
Department of Computer Engineering
National Institute of Technology, Agartala.

March'2010

Dedicated to

My Parents

CONTENTS

CHAPTER 1

Introduction to Java

INSIDE THIS CHAPTER

Java is a general-purpose Object-Oriented programming language which was developed by Sun Microsystems, USA in 1991.It was originally called Oak by James Gosling. There are several features of Java that distinguishes it from other programming languages. Java's support to GUI makes it user-friendly. Other than GUI feature, Java utilizes Internet and provides client-server solutions. Java supports all platforms such as Mac OS, UNIX, and WINDOWS. Before discussing the features and structure of Java, the concept of OOP has to be understood. The chapter discusses about object oriented programming with its benefits and applications. Then, the structure of Java program is discussed in detail.

1.1. BASIC CONCEPT OF OOPS

Object-Oriented Programming is an advancement towards organizing programs that works towards the reduction in flaws caused by traditional programming methods. OOP provides several new concepts and features that result into a powerful platform for programming languages. OOP was actually developed to overcome the difficulties caused due to structured programming.

1.1.1. Problems with Structured Programming

Structured programming emphasis on structured approach for problem-solving. It considers data and functions as different units. It divides the program structure into sub-structures where each sub-structure provides a different understanding of the program. It is generally considered top-down approach where there is a single entry point and a single exit point. Structured programming works with any programming language. In this method, the procedures are independent of each other. It supports modular programming. Maintaining large systems is a complex job in structured programming. Examples of structured programming languages are Ada, Pascal, ALGOL and PL/I. To overcome these drawbacks of structured programming; object-oriented programming was developed.

1.1.2. Object-Oriented Programming

OOP concentrates more on data rather the procedure to be followed or the structure of the program. It considers data as a crucial element and work towards its security. It ties data and functions together under a single entity called *class* and prevents any modification to the data by providing access specifiers to the class. Some of the features of OOP are explained below.

Class

It provides a picture of a set of objects that share same attributes, operations and relationships. A *class* is a model for creating and managing objects. When a program is written in an object-oriented language like Java, the classes are defined which are used to create objects. Classes are considered as user-defined data types. A class can be defined using the keyword '*class*'. For example, a class named *Employee* can be defined as:

```
class Employee
```

Object

It is a logical entity that contains data and code to manipulate the data. An object to a class is chosen such that it matches real-world entities. An object may represent a person, place, thing, or even a bank account. Any problem related to programming can easily be analyzed using objects. An object takes some space in the memory and has an associated address like structures in C. The data elements are hidden from the code that is using the data. During program execution, objects communicate by sending messages to each other. For example, the objects to the class *Employee* can be created as:

```
Employee emp1, emp2;
```

where *emp1, emp2* are the objects of class *Employee.*

Abstraction

It is one of the chief features of OOP that hides the implementation details from their specifications. In simple words, it is the blend of encapsulation and information hiding. It encapsulates all the essential features of the objects of a class. The attributes of the objects are called *data members* and the functions that operate on that data are called *member functions* or *methods.* Since Java works with classes, that feature abstraction, so, the classes are also called *abstract data types.*

Encapsulation

It is the wrapping up of data and functions under single unit called class. It is one of the most important features of OOP. The data is not accessible to the outside world. It provides a most striking feature called *information hiding* which means hiding data from direct access by the outside world. The example below shows encapsulation:

```
class Employee
{
        public static void main(String args[])
    {
        int emp_id, salary;
        char emp_name[10];
        void emp_detail()
      {
    ...Code....
      } } }
```

Inheritance

Another feature of OOP is inheritance. Inheritance allows programmers to create new classes from existing ones. The main class is called *base class* or *parent class* and the class derived from the base class is called *derived class* or *child class*. A child class inherits its properties and attributes from its parents. There are various methods of achieving inheritance such as *single inheritance, multilevel inheritance and hybrid inheritance*. We will discuss inheritance in detail in forthcoming chapters.

Polymorphism

It provides the ability to take more than one form. Functions as well as operators can take more than one form. When an operator exhibits different forms or behavior, it is called *operator overloading*. When a function exhibits more than one form or behavior, it is called *function overloading*. Some complexity is generated through operator overloading, so this feature has been removed in Java programming

1.2. BENEFITS OF OOP

OOP provides several benefits to the programmers. It provides solution to many problems related to the development of programs. Its advantages are:

- It makes the maintenance of application easier.
- Encapsulation and abstraction enhances security.
- Abstraction provides the flexibility in approach.
- It provides the modular approach to the problems.
- Polymorphism helps in the reusability of code.
- It helps in achieving robustness.
- Object-Oriented systems can easily be upgraded.
- Software complexity is manageable.
- Information hiding and data abstraction increases reliability
- Dynamic binding increases flexibility

These are the few benefits of object-oriented programming that makes it much user-friendly. It is easier to understand and implement for the users and programmers. Object-oriented systems can easily be handled and modified as and when required. The use of objects and classes provide a descriptive view of the problem and provides an easy to understand solution to the problem. Thus, OOP can be easily and efficiently implemented in Java technology. In the following section we will be discussing about the application of object-oriented features in Java technology.

1.3. APPLICATIONS OF OOP

There are several areas for OOP applications of which the most important function of OOP is in the area of user interface design. OOP is beneficial in the solution of complex problems. Its most promising areas are:

- Neural networks
- Parallel Programming
- Office Automation system
- Simulation and Modeling
- Object-Oriented databases
- Hypertext and Hypermedia
- AI and Expert System
- Computer-Aided Engineering

- Data System
- Internet Applications
- Operating System design
- Multimedia applications
- CAD/CAM Systems
- Image Processing
- Web Processing
- E-Commerce

1.4. THE CREATION OF JAVA

Java was developed by James Gosling, Patrick Naughton, Chris Warth, Ed Frank and Mike Sheridan in 1991 at Sun Microsystems. It was initially called 'oak' but was renamed as Java in 1995. The original motivation for Java was the need for a platform-independent, architecture-neutral language that could be used to create software for various consumer electronic devices, such as microwave ovens and remote controls. The problem with other programming languages such as C, C++ was that they were designed to achieve some specific targets and their compilers were expensive and time-consuming to create. An easier and more cost-efficient solution was needed. In the course of finding a solution to this problem, Gosling and others began work on a portable, platform-independent language that could produce code that would run in differing environments. This effort ultimately led to the creation of Java. The popularity of Java was due to the evolution of web. If World Wide Web would not have developed at that time, it would have been difficult for Java to gain such popularity.

1.5. IMPORTANCE OF JAVA FOR THE INTERNET

The Internet has helped projecting Java to the front of programming and in turn Java has an intense effect on the Internet. The reason for this is that Java provides vast range of objects that can move freely in cyberspace. Java can be used to create two types of programs: applications and applets.

1.5.1. Java Applets and Applications

An *application* is a program that runs on your computer. Java is not much different from any other computer language when used to create applications. It is Java's ability to create applets that makes it important.

An *applet* is an application designed to be transmitted over the Internet and executed by a Java-compatible Web browser. An applet is a tiny Java program that is dynamically downloaded across the network like an image, sound file, or video clip. An applet is an *intelligent program*, not just an animation file that can react to user input and act accordingly.

1.5.2. Security

Every time that we download a normal program, we are risking a viral infection to our computer system. Before the development of Java, users were not downloading executable programs frequently and those who were downloading scanned the programs for viruses before executing. Not only viruses but another type of malicious program exists that must be guarded against. This type of program can gather private information, such as credit card numbers, bank account balances, and passwords, by searching the contents of your computer's local file system. Java provides a *firewall* between the network applications and your computer system. When we are using a Java-compatible Web browser, we can safely download Java applets without the fear of viral infection or malicious activity. Java achieves this protection by confining a Java program to the Java execution environment and not allowing it access to other parts of the computer.

1.5.3. Portability

As many types of computers and operating systems are in use throughout the world and many of them are connected to the Internet; for programs to be dynamically downloaded to various types of platforms connected to the Internet, some means of generating portable executable code is needed. The same mechanism that helps ensuring security also helps creating portability. Java's solution to these two problems is both well-designed and proficient.

1.6. JAVA VIRTUAL MACHINE

Java technology has been developed to support diverse environments. It supports applications that can easily be executed on a variety of platforms. Java Virtual Machine (JVM) is a run-time environment that is capable of executing Java applications and applets. JVM is architectural-neutral, portable language platform for Java technology that understands the file format of the particular platform and implementation independent class file produced by the Java compiler. JVM acts as an interface between the program code and the platform at which we are trying to execute the code. Source code of the Java program is converted to *Bytecode* by the *Java Compiler* which in turn is converted to machine code by *Java Interpreter.* This process is depicted in the figure below:

Fig. 1.1. Process to convert Source code to Machine code

Further, Java programs are written such that they can be easily downloaded over the network. It causes a major security concern. There may be a program that may damage the entire file system on your computer. The *bytecode* interpreter acts as a buffer between the programmer and the program you download. The interpreter protects our system from dangerous actions on the part of that program.

1.7. JAVA'S MAGIC: THE BYTE CODE

The means that allows Java to solve both the security and the portability problems is that the output of a Java compiler is not executable code but it is the bytecode. *Bytecode* is a highly optimized set of instructions designed to be executed by the Java run-time system, which is called the *Java Virtual Machine (JVM)*. JVM is an *interpreter for bytecode.* The fact that a Java program is executed by JVM helps solve the major problems associated with downloading programs over the Internet. Translating a Java program into bytecode helps makes it much easier to run a program in a wide variety of environments. This is because only the JVM needs to be implemented for each platform.

Once the run-time package exists for a given system, any Java program can run on it. If a Java program is compiled to its subject code, then different versions of the same program would exist for each type of CPU connected to the Internet. This is not a practical solution. Thus, the understanding of bytecode is the most effective way to create strictly portable programs. Now, if a Java program is interpreted, it also helps to make it secure. Since the execution of every Java program is under the control of the JVM, the JVM can contain the program and prevent it from generating side effects outside of the system. The use of bytecode enables the Java run-time system to execute programs much faster.

> **DO YOU KNOW**
>
> Just-in-time approach yields a considerable advancement in performance.

Sun provides a facility called *Just In Time (JIT) compiler* for bytecode. It is not possible to compile an entire Java program into executable code all at once, because Java performs various run-time checks that can be done only at run time. JIT compiles the code as and when needed.

1.8. FEATURES OF JAVA

In addition to object-oriented features, Java comprises some other features that make it much simpler and easy to understand. One of the most striking features of Java, as already discussed, is that it is architectural-neutral. Moreover, it is more secure and portable. We will discuss all the features of Java in detail. Its various features are:

Compiled and Interpreted

Java interpreter can execute Java bytecode directly on any machine to which the interpreter and run-time systems have been ported. Java enables the creation of multi-platform programs. The bytecode can be interpreted on any system that provides a Java Virtual Machine. Previously, many attempts have been made for multi-platform systems but at the cost of hampering the performance. Java was designed to perform well on very low-power CPUs. The Java bytecode was carefully designed so that it can easily be translated into native machine code for high performance by using a just-in-time compiler.

Robust and Secure

Java provides extensive compile-time checking followed by run-time checking. Memory management process is quite secure. The objects are created with a ***new*** operator. There is neither a pointer-type data type, nor the automatic garbage collection. Java technology is designed to operate in distributed environments, so, the security is a prime concern. The ability to create robust programs was given a high priority in the design of Java. Java provides reliability and to achieve this, it restricts the programmers in some key areas to force them to find mistakes in program development. Also, Java frees programmers from worrying about the most common causes of programming errors. Java checks your code at compile time. It also checks the code at run time. Java considers two main reasons for program failure: *memory management mistakes* and *run-time errors.* Memory management can be a difficult, tedious job in traditional programming environments. Run-time errors in traditional environments arise in situations such as division by zero or "file not found". Java helps in this area by providing object-oriented exception handling.

Platform Independent and Portable

The main issue with Java developers was the need for a platform-independent, portable language. One of the main problems that the programmers were facing was that there was no guarantee that if you write a program today, it will run tomorrow or even on the same machine. Operating system upgrades, processor upgrades, and changes in core system resources sometimes combine to cause a program breakdown. The Java designers made several tough decisions in the development of Java language and Java Virtual Machine to overcome these issues. Their goal was "write once; run anywhere, anytime, forever." This goal was accomplished to a great extend. The figure below shows platform-independent feature of Java.

Distributed

Java handles TCP/IP protocols, which make it easier to use with Internet. Java provides Remote Method Invocation (RMI) that brings an unparalleled level of abstraction to client/server programming.

Simple

Simplicity is the primary feature of Java technology as the programs are simple to run without much training. The concepts of Java technology can be grasped easily and the programmers can begin with a productive form of programming. If one is clear about the primary concept of OOP, then working with java will be much easier for the user.

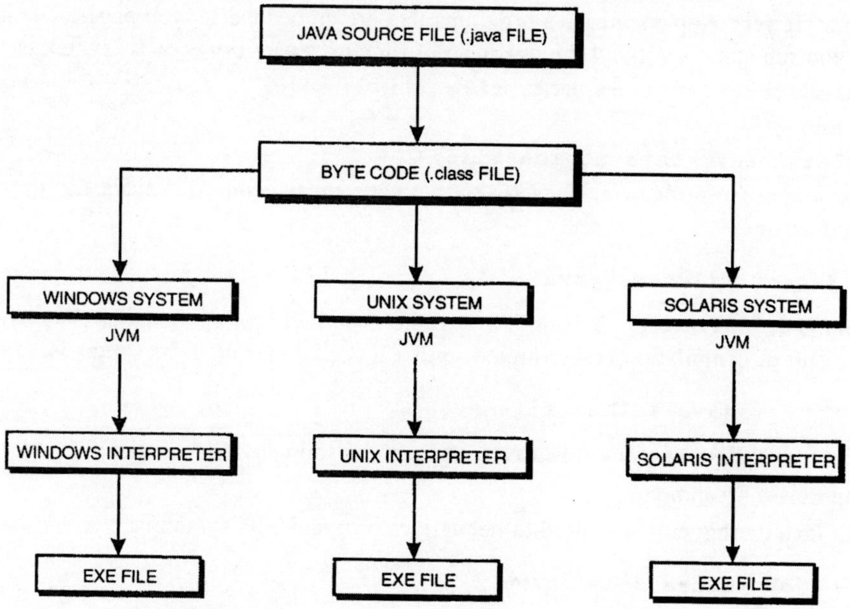

Fig. 1.2. Platform-independence in Java.

Multithreaded

Java supports multithreading and provides library for the Thread class. Java was designed to meet the real-world requirement of creating interactive network programs. It allows the programmers to write programs that perform many tasks concurrently. Java run-time system comes with a refined solution for multi-process synchronization that enables us to construct efficient and simple programs.

High Performance

Java technology achieves high performance by adopting a scheme by which the interpreter can run at full speed without needing to check the run-time environment. Automatic garbage collection runs as a low-priority background thread, ensuring the high availability of memory, thus leading to better performance.

Case Sensitive

Java is a case sensitive programming language. '*Program*' and '*program*' provide different meanings in Java.

Object-Oriented

Java programming language is designed to be object-oriented. Java includes all the object-oriented features of OOP such as abstraction, encapsulation, inheritance, polymorphism etc. this provides a clean, usable, practical approach to objects.

1.9. JAVA ENVIRONMENT

Java environment include a large number of development tools and various classes and methods. The development tools are part of the system known as *Java Development Kit* (JDK) and the classes and methods are part of *Java Standard Library* (JSL), also known as the *Application Programming Interface* (API). The Java Development Kit comes with a collection of tools that are used to develop and run Java programs. These may include the following

- **Appletviewer:** Applets are java programs that are embedded in web pages. The applet viewer lets you run applets without the overhead of running a web browser. Its syntax is:

```
appletviewer <URL of HTML file >
    and
appletviewer <Java Source File>
```

- **javac (Java Compiler)**: It converts source code to bytecode, i.e. .java file to .class file. Its syntax is:

```
javac <filename.java>
```

- **java (Java Interpreter)**: It is used to execute compiled java applications. The bytecode that is the result of compilation is interpreted so that it can be executed. Its syntax is:

```
java <class filename>
```

- **javap (Java dissembler) –** It is used to recover the source code from the bytecode. Its syntax is: javap <class filename>
- **jdb (Java debugger)-** It is used to debug java programs. Its syntax is:

```
jdb <class filename>
```

- **javadoc (HTML documents)**: It is the document generator that creates HTML documentation for the classes. Its syntax is:

```
javadoc <filename.java>
```

- **javah (C header file)**: It creates header and stub files that let you extend Java code with C language. Its syntax is:

```
javah <class filename>
```

1.10. JAVA PROGRAM STRUCTURE

As discussed earlier, Java is an object-oriented programming language that contains all the features of OOP. This means, Java comprises of classes and objects. The methods are defined in the main class. Figure 1.3. depicts the structure of Java program:

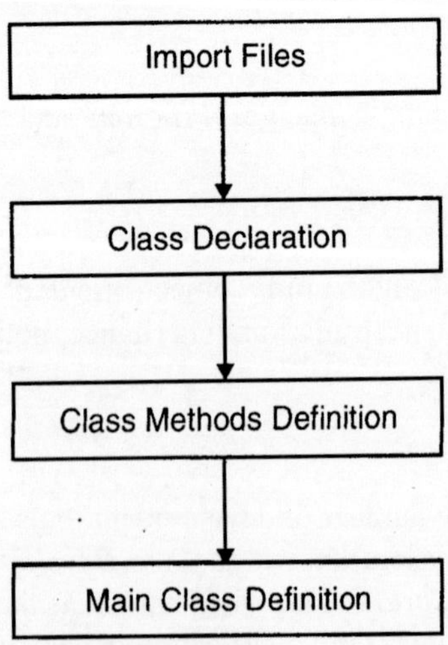

Fig. 1.3. Structure of Java Program

The sections below discuss about the data types, variables and various operators used in Java and finally various control statements.

1.11. DATA TYPES AND OPERATORS

Java has eight built-in data types namely int, char, long, short, byte, float, double and Boolean. The table below shows the size in bytes of all the built-in data types used in Java.

TABLE 1.1. Data Types in Java

Data Type	Size (bytes)
int	2
char	1
long	4
short	2
byte	8 bits
float	4
double	8
boolean	1 bit

Java deals with different kinds of operators categorized as arithmetic operators, relational operators and logical operators. The tables below depict the three types of operators used in Java with their operator symbols and functions:

TABLE 1.2. Arithmetic Operators

Operator	Function
+	Addition
-	Subtraction
*	Multiplication
/	Division
%	Modulus

TABLE 1.3. Relational Operators

Operator	Function
==	Equality
!=	Not Equal
<	Less Than
>	Greater Than
<=	Less than Equal to
>=	Greater than Equal to

TABLE 1.4. Logical Operators

Operator	Function
&&	AND
||	OR
!	NOT

POINTS TO REMEMBER

Java is strongly type-checked. This means you cannot assign a float value to an integer as allowed in C++ as C++ does not provide strong type-checking.

1.12. CONTROL AND ITERATIVE STATEMENTS

Java program is a set of statements that either executes sequentially or in the order in which they appear. There may arise situations in which a program breaks the normal sequential flow and jumps to another part of the code. This condition is called branching. Branching can either be conditional or unconditional. Conditional branching occurs on a particular situation or a statement in the program code while unconditional branching occurs without any particular decision. Java deals with different control and iterative statements listed below which will be discussed in detail in preceding sections.

- if-else
- switch statement
- while loop
- do-while loop
- for loop

if-else

It is a powerful decision making státement that is used to control the flow of execution of statements. It can be implemented in different ways which are discussed in this section in detail with their syntax and examples. The syntax for *if* is:

```
if(expression)
```

Various forms of *if-else* are discussed below.

- Simple *if* statement

Its syntax is:

```
if(expression)
{
    Block of statements;
}
```

This can be explained with the help of example below:

"If employee salary is greater than 28000, then his income is taxable". This statement can be explained with the help of simple if loop as:

```
if(emp_salary>28000)
{
        System.out.println("Taxable Income");
}
```

- *if...else* statement

Its syntax is:

```
if(expression)
{
    Block of statements;
}
```

```
        else
        {
            Another block of statements;
        }
```

This can be explained with the help of example below:

"If employee salary is greater than 28000, then his income is taxable else his income in non-taxable". This statement can be explained with the help of simple if loop as:

```
if(emp_salary>28000)
{
        System.out.println("Taxable Income");
}
else
{
    System.out.println("Non-taxable Income");
}
```

➠ nested *if...else* statement

Its syntax is:

```
if(expression1)
{
    if(expression2)
    {
        First block of statements;
    }
    else
    {
        Second block of statements;
    }
    Third block of statements;
}
else
{
    Next block of statements;
}
```

This can be explained with the help of example below:

"If employee salary is greater than 28000 and if employee is a male candidate, then his income is taxable". This statement can be explained with the help of simple if loop as:

```
if(emp_salary>28000)
{
        if(emp_sex=='M')
                {
            System.out.println("Taxable Income");
        }
        else
        {
            System.out.println("Non-taxable Income");
```

```
        }
    }
    else
    {
        System.out.println("Condition does not satisfy");
    }
```

➡ *else...if* statement

Its syntax is:

```
if(expression1)
    statement-1;
  else if(expression2)
      statement-2;
   else if(expression3)
      statement-3;
      ..................
    else if(expressionN)
        statement-N;
   else
       Default statement;
                           statement-X;
```

This can be explained with the help of example below:

"If employee has an experience of 28 years, he is designated as 'General Manager'. If his experience is 20 years, he is designated as 'Asstt. General Manager'. If his experience is less than 20 years but more than 15 years, he is designated as 'Chief Manager'. If his experience is 15 years, he is designated as 'Manager'. If his experience is less than 15 years, he is designated as 'Supervisor'. Otherwise he is a worker". This statement can be expiained with the help of simple if loop as:

```
if(emp_exp==28)
 designation="General Manager";
   else if (emp_exp==20)
      designation="Asstt. General Manager";
        else if(15<emp_exp<20)
 designation="Chief Manager";
                  else if(emp_exp==15)
                      designation="Manager";
      else
         designation="Worker";
```

System.out.println("Enter Correct Experience for Employee");

switch statement

switch is an another decision-making statement that can be used as an alternative to ***if*** statement.

It is a special multi-way decision maker that checks whether an expression matches one of the given values. Its syntax is:

```
switch(expression)
{
   case value1:
    statement11;
```

```
        statement12;
        break;
    case value2:
        statement 21;
    statement22;
        break;
        .
        .
        .
    case valuen:
        statement n;
        break;
    default:
        statement;
}
```

This can be explained with the help of example below:

"To check vowels and consonants." The code for this using switch statement can be given as:

```
char ch='A';
switch(ch)
{
  case 'a':
  case 'e':
  case 'i':
  case 'o':
  case 'u':
  case 'A':
  case 'E':
  case 'I':
  case 'O':
  case 'U':
      System.out.println("Character is vowel.");
      break;
  default:
      System.out.println("Character is Consonant.");
}
```

while loop

***while** loop* is used when we are not sure if a loop will execute or not. If the condition is true, then the loop will continue until the condition satisfies. Its syntax is:

```
while(expression)
{
    //Block of statements;
}
```

This can be explained with the help of example below:

"To print an integer in reverse order". The code using *while loop* for this is:

```
while(n!=0)
{
    System.out.print(n%10);
    n=n/10;
}
```

do-while Loop

do-while loop is a repetitive loop supported by Java. It's an exit condition loop i.e. the set of statement will execute first, and then at the exit time the condition will be checked for repetition. If the condition return true then the loop will re-execute the statements otherwise it will terminate. Its syntax is:

```
do
{
    Block of statements
    ........
} while(expression);
```

This can be explained with the help of example below:

```
do
{
    System.out.println(i*count);
    count++;
}while(count<=10);
```

for loop

for loop is the simplest loop in Java. It contains three expressions:

- initial expression: used to initialize the counter of the loop
- termination expression: used to give a termination condition on the loop
- steps expression: used to either increase or decrease the value of the counter.

All above three expressions are optional, according to need of the logic, we will add in for loop. The syntax for *for loop* is:

```
for(initial expression; termination expression; steps expression)
{
    Body of the loop;
}
```

This can be explained with the help of example below:

"To find the Armstrong number between 1 and 300". The code for the same using *for loop* is:

```
for(i=0;i<=300;i++)
{
    n=i;
    x=n%10;
    n=n/10;
    y=n%10;
    n=n/10;
    sum=(a*a*a)+(b*b*b)+(c*c*c);
    if(sum==i)
```

```
    System.out.println("Armstrong No. is:" =sum);
}
```

1.13. BUILDING AND RUNNING JAVA APPLICATION PROGRAM

A simple Java program is defined using *class*. As discussed earlier, a *class* is a collection of *data types* and member functions called *methods*. We consider a simple example and create a class in Java for *Employee*. The class is created using keyword class followed by the class name.

class Employee

First line in the class begins with the statement below:

```
public static void main(String args[])
```

In this statement, the keyword ***public*** is an access specifier which can either be public, private or protected that allows the programmer to control the visibility of class members. When a class member is preceded by ***public***, then that member may be accessed outside the class in which it is declared. ***main*** specifies the main function which must be declared public since the code has to be accessed by the class. ***static*** allows ***main( )*** to be called without having to instantiate a particular instance of the class. ***void*** specifies the return type. ***String args[ ]*** declares a parameter named ***args***, which is an array of instances of the class ***String***. Objects of type ***String*** store character strings. ***args*** receives any command-line arguments present when the program is executed.

A very important point to be remembered is that since java is case-sensitive, so the syntax has to be implemented carefully.

A simple variable is declared as:

```
return-type variable_name;
```

For example,

```
int num;
```

To print value of the variable *num* in Java, the statement below is used:

```
System.out.println("The value is" +num);
```

In the *println* method, the arguments are passed in the form of String but in the above example, there are two arguments, *"The value is"* and *num*, of type *String* and *int* respectively. Here, first *num* will be converted to *String* implicitly by the Java compiler and then concatenated with *String* object through + operator.

A simple program in Java defining the value of ð would look like:

```
class Pie
{
  public static void main(String args[])
  {
    float pi=3.14f;
    System.out.println("The value of Pie is : " +pi);
  }
}
```

To write the code in Java, follow the steps below:

1. Write the source code in any editor (e.g.Notepad).
2. Save the file as per the rule e.g. Pie.java

3. to compile the file, follow the following steps:
   - Start → Run → type cmd
   - *javac filename.java* (*e.g.* javac Pie.java)

   It will create a class file of the corresponding source code (e.g. Pie.class)
4. To run the file, write the command below:

   `java classfilename (e.g. java Pie)`

The point to be noticed here is that never give the extension .class with the *classfilename*. JVM will search implicitly the class file by concatenating the .class extension.

In the screenshot below, the point to be observed here is that, there was no file in the ***JAVA_prg*** directory except ***Pie.java***. After compiling it using the command ***"javac Pie.java"***, it generates the file name ***"Pie.class"***. Now, the directory contains two files, ***Pie.java*** (source code) and ***Pie.class*** (bytecode). The file is then executed using the command ***"java Pie"***.

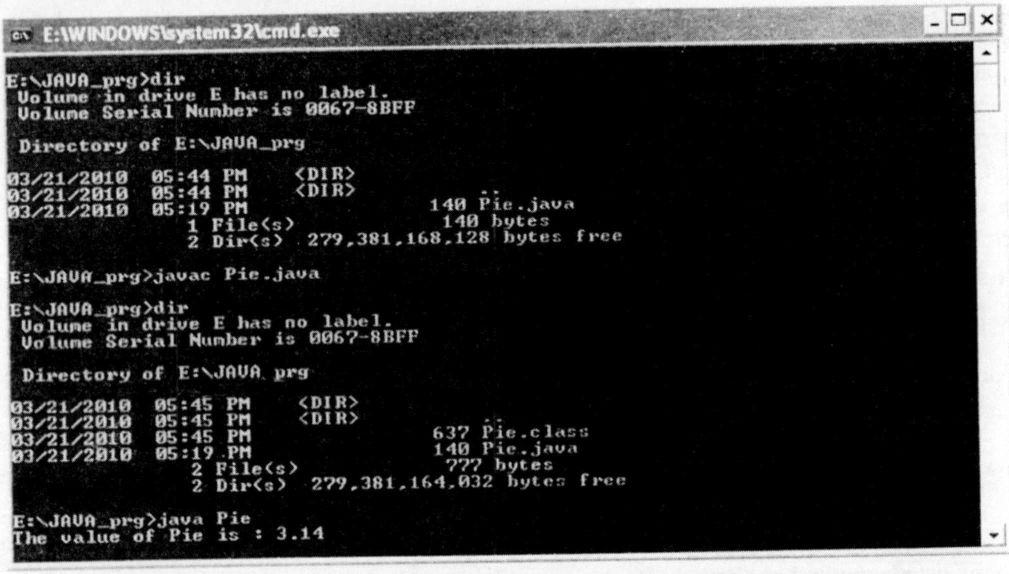

1.14. ARRAYS

An *Array* is a (fixed length) data structures used to store multiple data elements of same type. The memory allocated to an array is consecutive. All the elements of an array are under one variable name. Every data element of an array is accessed by its position and that position is indicated by an integer value that start from 0 (zero) called *index*. The array size is fixed and cannot be changed at runtime of the program.

Java supports two types of arrays:

- One Dimensional Array
- Multiple Dimensional Array

In Java, arrays can be of either types, *primitive data types* or *references types*. In *primitive data type array*, all elements are of a definite primitive data type. In *references type array* all elements are of definite reference type.

Declaration of Array Variables

The syntax to declare either type of array, i.e. one-dimensional or multiple dimensional array is:

```
        <data-type>[] <array-name>;           //One Dimensional Array
OR
        <data-type> <array-name>[];           //One Dimensional Array
```

For Example:

```
int an_Array[];
Employee[] emp;
```

```
        <data-type>[][]...[] <array-name>;       //Multidimensional Array
OR
        <data-type> <array-name>[][]...[];       //Multidimensional Array
```

For Example:

```
int an_Array[][];
char[] char_Array[];
Employee[][] emp;
```

> **DO YOU KNOW**
>
> Arrays in Java work differently than in any other programming language.

Dimensions of an array are omitted in the declaration variable of an array. *<data-type>* can be of any primitive type or reference type. Very important to notice here that declaration does not create an array, it will just create or declares the references of an array object.

Array Construction

As defined earlier array length is fixed at its creation time and cannot be changed, i.e. using *new* operator an array can be created for a definite number of elements of a homogeneous type. The constructing statement of array return reference value of resulting array and that will be assigned to corresponding type array variable.

Syntax to create an array:

```
<array-name> =  new <data-type> [<array-size>];  //One Dimensional Array
```

For Example:

```
an_Array = new int[10];          //One dimensional array of 10 integers
emp = new Employee[50];          //One dimensional array of 50 Employees

an_Array = new int[5][5];        //Two dimensional array of 5 * 5
                                 integers
char_Array =  new char[5][30]; //Two dimensional array of 5 * 30
                                 characters
```

Declaration and Construction of an array can be combined as the following manner:

```
int userList[] = new int[12];   //One dimensional array of 12 integers
int matrix[][] = new int[4][5]; // Two dimensional array of 4 * 5 integers
```

As we construct an array of either type (primitive or reference type), its elements are initialized implicitly with their default values.

```
int an_Array = new int[10];       //Default value = 0 of each element
Employee emp[] = new Employee[10]; //Default value = null of each element
Employee salariedEmp[] = new SalariedEmp[5];
                                  //Default value = null of each element
```

<array-size> can be minimum 0, because Java support zero-length arrays constructions. But it can never be negative, otherwise it will throw *NegativeArraySizeException*. The length of constructed array will be the *<array-size>* given at the creation time. Later you can find the length of an array using *<array-name>.length* .

For Example:

```
int userList[] = new int[12];
int x = userList.length;        //Return the length of userList = 12 in x
```

Java gives flexibility in multidimensional array, that there is no need to have same length, called *ragged arrays*. At the time of constructing multidimensional array the deeply nested arrays length may be omitted.

For Example:

```
int matrix[][] = new int[5][];              //Just 5 rows given
```

The above declarations create an array of 5 rows. Later we can add different number of columns in each row separately.

```
matrix[0] = new int[5];         //The row contains 5 columns
matrix[1] = new int[2];         //The row contains 2 columns
matrix[2] = new int[4]; //The row contains 4 columns
matrix[3] = new int[3]; //The row contains 3 columns
matrix[4] = new int[5]; //The row contains 5 columns
```

Initialization of Array

Initialization means to assign the value to the array at the time of constructing and declaring in one statement simultaneously.

```
<data-type>[] <array-name> = { <initial value list> };
```

Arrays are initialized by sandwiching the values between braces. As you initialize an array at the time of declaration, there is no need to create an array using 'new' explicitly.

In the initialization expression the *<initial value list>* are evaluated from left to right, and list must be terminated by right curly braces.

For Example:

```
int arr[] = {1, 2, 3, 4};                    //One dimensional array
char chr[] = {'a', 'b', 'c', 'd'};
String[] flower =  {"Rose", "Lotus", "Lily", "SunFlower"};
Object obj[] = { new Employee(), new Employee(), new Employee(), null0;
int matrix[][] ={ {1, 2, 3}, {4, 5, 6}, {7, 8, 9}};
                                                   //Two dimensional array
```

An array that does not have a name is called an *anonymous array*. An anonymous array is declared in the manner given below.

```
new <data-type>[] { <initial value list> };
```

For Example:

```
new int[] {1, 2, 3, 4};
```

Array Access

To access elements of an array there is one array expression with a certain format that contains an array reference followed by an index number. An index number can be calculated by an expression, that must lie between integers 0 (zero) to n-1 (n is length of array).

At run time array accesses are checked which is an attempt to check whether an index is less than zero or greater than or equal to n (n is length of array) then *ArrayIndexOutOfBoundsException* will be thrown.

Array index must be *int* value, but it can be *short, byte*, or *char* because they are implicitly converted to unary numeric value and become *int* values. A *long* index value generates a compile-time error.

For Example:

```
int an_Array = {1, 2, 3, 4, 5, 6, 7, 8 ,9}; //One dimensional array
  for(int i=0; i<10;i++)
   System.out.println(anArray[i]);       //Accessing ith index value.

int matrix[][]={ {1, 2, 3}, {4, 5, 6}, {7, 8, 9}};
                                          //Two dimensional array
for(int i=0; i<3; i++)
   {
          for(int j=0; j<3;, j++)
               System.out.print(matrix[i][j]);
   System.out.println();
   }
```

Arrays of Characters

Java support *char array*, and are initialized and access in the same manner as given in earlier examples.

For Example:

```
char chars[] = {'a', 'b', 'c'};
char chars[] = {'a', 'b', 'c', 'd', 'e' };

class  CharacterArrayDemo{
       public static void main(String args[]) {
char c[] = {'I','', 'a','m','', 'l','e', 'a', 'r', 'n', 'i', 'n', 'g',
'J', 'a', 'v', 'a'};
for(int i=0;i<c.length;i++)
System.out.print(c[i]);
 }
       }
```

Output:

```
I am learning Java
```

POINTS TO REMEMBER

In languages like C, C++, Strings are implemented as array of characters. But this is not the case in Java. Strings are actually object types in Java.

1.15. STRING HANDLING IN JAVA

The string is a collection of characters with blank spaces. Java supports handling sequences of character called string through three final classes – *String*, *StringBuilder*, and *StringBuffer*. In this section we'll discuss about *String* class. *String* object has a full description that requires numerous object related features. Neither, it is simple type nor an array of characters. String is defined as in-built standard object. An object of *String* type contains set of immutable character, called string, which is read-only, after the object has been created and initialized. While *StringBuilder* creates dynamic character string, and *StringBuffer* class is a *thread-safe* class.

There are three ways to create String implicitly:

- Using a string literals e.g. "Rosy"
- Using '+'or '+=' operators on at least two *String* object
- Using *String* constructor

Constructors

You can construct *String* objects explicitly using the new mechanism. The *String* class supports the following simple constructors.

public String()

The constructor creates a *String* object with " "(empty) value.

```
String str = new String();
```

public String(String stringObject)

The constructor creates a *String* object, which is a copy of *String* object *stringObject.* It is a copy constructor, and is rarely used. It will be in use when there is need of an explicit copy of original *String* object, since Strings are immutable.

Parameters:

```
stringObject - A String
```

```
String str1 = new String ("Blossom");
String str2 = new String (str1);
```

In the above code *str* will be "Blossom".

```
public String(char chars[ ])
```

The string is a combination of set of characters, so, it can be mapped to an array of *char* and vice versa. The constructor creates a *String* object, which is initialized by an array of characters. The constructor copies the contents of the *char* array to *String* object. The modifications in char array do not affect the *String* object created by it.

Parameters:

Chars: The original value of the string

```
char ch[] = {'F','l','o','w','e','r'};
String str = new String (ch);
```

In the above code *str* will be "Flower".

```
public String(char chars[ ], int startIndex, int numChars)
```

The constructor constructs an object of *String*, which contains characters from an array from *startIndex* up to a maximum *numChars* number of characters. The constructor copies the sub array contents of the *chars[]* to *String* object. The modifications in *char* array do not affect the *String* object created by it.

Parameters:

chars: Array that is the source of characters

startIndex: The starting index number, from where the characters have to copy.

numChars: The number of characters to be copied.

But if *startIndex* and *numChars* index outside the bounds of the value array then *IndexOutOfBoundsException* will be thrown.

```
char ch[] = {'F','l','o','w','e','r'};
String str = new String (ch,0,4));
```

In the above code *str* will be "Flow".

```
public String(byte asciiChars[ ])
```

Some time there is needs to convert an array of raw 8-bit character to 16-bit Unicode and vice versa. There is some encoding for such conversion. This constructor creates a *String* object by decoding the *byte* array.

Parameters:

asciiChars: The *byte* array to be decoded into characters

```
byte asciiChars[] = {97, 98,99,100,101,102};
String str = new String(asciiChars);
```

In the above code the value of *str* will be "abcdef".

```
public String(byte asciiChars[ ], int startIndex, int numChars)
```

The constructor constructs an object of *String*, which contains characters by converting the bytes from index *startIndex* up to maximum *numChars* number bytes.

Parameters:

asciiChars: The byte array to be decoded into characters

startIndex: The starting index from where bytes to be decoded

numChars: The number of bytes to decode

But if *startIndex* and *numChars* index outside the bounds of the value array then *IndexOutOfBoundsException* will be throws.

```
byte asciiChars[] = {97, 98,99,100,101,102};
String str = new String(asciiChars,2,3);
```

In the above code the value of *str* will be "cde".

public String(StringBuffer bufferObject)

The constructor creates an object of *String* that contains the characters sequence from *StringBuffer* object. It will copy content of *StringBuffer* to String object and any change in *StringBuffer* will not affect the String object value.

Parameters:

bufferObject - A *StringBuffer* object of which value will be copied

```
StringBuffer sb = new StringBuffer("Hello! You are on earth.");
String str = new String(sb);
```

In the above code the value of *str* will be "Hello! You are on earth."

```
public String(StringBuilder builderObject)
```

The constructor creates an object of *String* that contains the characters sequence from *StringBuilder* object. It will copy content of *StringBuilder* to String object and any change in *StringBuilder* will not affect the String object value.

Parameters:

builderObject: A StringBuilder object of which value will be copied

```
StringBuilder sb = new StringBuilder();
sb.append("You are perfect.");
String str=new String(sb);
```

In the above code the value of *str* will be "You are Perfect".

Operations on String Handling

There are a bunch of operations on String can be possible. But before we move to more discussion on operations let's have some look on the String class functions.

Method: *public String toString( )*

Description: Method is defined by Object class. When you want to give your own string representation of your class, then override this method. Return a String object that contains some human readable string that describes your class object.

Method: *public int length()*

Description: Method returns the length of this string. The length is equal to the count of 16-bit Unicode characters in the string.

Method: *public boolean isEmpty()*

Description: Method returns true if, and only if, length() is 0, otherwise false.

Method: *public byte[ ] getBytes( )*

Description: Method encodes the contents of the String object to the byte array provided by the platform.

Method: *public char[ ] toCharArray( )*

Description: Method converts the contents to String object into a character array. It returns an array of characters.

(i) Reading Characters from a String

Method: ***public char charAt(int where)***

Description: Method returns a character at the given index *where* from the content of *String* object. The value of index lies between 0 to *length()* – 1.

Example:

```
String str = new String("Rose is beautiful");
System.out.println(str.charAt(9));
```

Output:

```
e
```

Method: *public void getChars(int sourceStart, int sourceEnd, char target[ ], int targetStart)*

Description: Method copies sequence of characters starting from *sourceStart* index upto *sourceEnd* index, from the contents of source *String* object to destination character array *target[]* at *targetStart* index location.

Example:

```
String str = new String("Rose is beautiful");
String str2 = new String("Lotus");
char target[]=new char[21];
str2.getChars(0,str2.length(),target,0);
str.getChars(4, str.length() ,target,5);
System.out.println(target);
```

Output:

```
Lotus is beautiful.
```

(ii) Comparing Strings

Method: *public boolean equals(Object str)*

Description: Method will compare two String object on the base of their same sequence of characters. It return *true* in two conditions only, if and only if the argument is not null and both objects have the same character sequence.

Example:

```
String s1 = "Hello";
String s2 = new String(s1);
String s3 = "Hello";
System.out.println(s1 + " equals " + s2 + " -> " +
s1.equals(s2));
System.out.println(s1 + " == " + s2 + " -> " + (s1 == s2));
System.out.println(s1 + " == " + s3 + " -> " + (s1 == s3));
```

Output:

```
Hello equals Hello -> true
Hello == Hello -> false
Hello == Hello -> true
```

Method:

```
public int compareTo(String str)
```

Description: Method compares two String object on the base of lexicography. The comparison is based on the Unicode value at each index. It return positive, negative and zero on the base of comparison. If String object lexicographically precedes the argument string, then it returns negative value. If String object lexicographically follows the argument string, then it returns positive value. If the strings are equals, then it returns zero.

Example:

```
String s1 = "Hello";
String s2 = new String(s1);
String s3 = "HelLo";
String s4 = "hello";
System.out.println(s1 + " equals " + s2 + " -> " +
s1.compareTo(s2));
System.out.println(s1 + " equals " + s3 + " -> " +
                        s1.compareTo(s3));
System.out.println(s1 + " equals " + s4 + " -> " +
                           s1.compareTo(s4));
```

Output:

```
Hello equals Hello -> 0
Hello equals HelLo -> 32
Hello equals hello -> -32
```

Method: *public boolean equalsIgnoreCase(String str)*

Description: Method is used to compare one String object to another String object with ignoring the case. Two strings are equals, if both are of same length and having same set of character with considering of cases.

Two characters c1 and c2 are considered the same ignoring case if at least one of the following is true:

- The two characters are the same (compared by the == operator)
- By applying the method Character.toUpperCase(char) to each character produces the same result.
- By applying the method Character.toLowerCase(char) to each character produces the same result.

Example:

```
String s1 = "Hello";
String s2 = new String(s1);
String s3 = "HELLO";
String s4 = "hello";
String s5 = "HeLlO";
System.out.println(s1 + " equals " + s2 + " -> " +
                    s1.equalsIgnoreCase(s2));
System.out.println(s1 + " equals " + s3 + " -> " +
                          s1.equalsIgnoreCase(s3));
System.out.println(s1 + " equals " + s4 + " -> " +
                          s1.equalsIgnoreCase(s4));
System.out.println(s1 + " equals " + s5 + " -> " +
                          s1.equalsIgnoreCase(s5));
```

Output:

```
Hello equals Hello -> true
Hello equals HELLO -> true
Hello equals hello -> true
Hello equals HeLlO -> true
```

(iii) Character Case in a String

Method: *public String toUpperCase()*

Description: Method converts all the characters of the String to upper case, and is equivalent to toUpperCase(Locale.getDefault()).

The method produces the unexpected result, so that, to obtain correct results for locale insensitive string, use toUpperCase(Locale.ENGLISH).

Example:

```
String str = new String("Rose is beautiful");
System.out.println(str.toUpperCase());
```

Output:

```
ROSE IS BEAUTIFUL
```

Method: *public String toUpperCase(Locale locale)*

Description: Method converts all the characters of the String to upper case according the passed Locale value.

Example:

```
String str = new String("Rose is beautiful");
System.out.println(str.toUpperCase( java.util.Locale.ENGLISH));
```

Output:

```
ROSE IS BEAUTIFUL
```

Method: *public String toLowerCase()*

Description: Method converts all the characters of the String to upper case, and is equivalent to toLowerCase(Locale.getDefault()).

The method produces the unexpected result, so that, to obtain correct results for locale insensitive string, use toLowerCase(Locale.ENGLISH).

Example:

```
String str = new String("ROSE IS BEAUTIFUL");
System.out.println(str.toLowerCase());
```

Output:

```
Rose is beautiful
```

Method: *public String toLowerCase(Locale locale)*

Description: Method converts all the characters of the String to upper case according the passed Local value.

Example:

```
String str = new String("ROSE IS BEAUTIFUL");
System.out.println(str.toLowerCase(
                                    java.util.Locale.ENGLISH));
```

Output:

```
Rose is beautiful
```

(iv) Concatenation of Strings

Method: *public String concat(String str)*

Description: Method returns the concatenation of the contents of two String objects. If the length of the argument string is 0 , this String object will return. Otherwise, new String object is created and represents a sequence of characters, which is concatenation of the content of this String object and followed by argument String object content.

Example:

```
String str1 = new String("Hello!");
String str2 = new String(" How are you?");
String str3 =str1.concat(str2);
System.out.println(str3);
```

Output:

```
Hello! How are you?
```

(v) Searching for Characters and Substrings

Method: *public int indexOf(char ch)*

Description: Method return the first occurrence index of given character as argument in this String. In either case, if no such character occurs in this string, then -1 is returned.

Example:

```
String str= new String("This is a test string. Java is a language");
System.out.println(str.indexOf('i'));
```

Output:

```
2
```

Method: *public int indexOf(char ch, int start)*

Description: Method will return the first occurrence of given character argument, *ch,* start the searching after the given index *start.*

Example:

```
String str= new String("This is a test string. Java is a language.");
    System.out.println(str.indexOf('i',4));
```

Output:

```
5
```

Method: *public int indexOf(String str)*

Description: Method return first occurrence index of given string as argument in this String.

Example:

```
String str= new String("This is a test string. Java is a
                                                    language.");

System.out.println(str.indexOf("is"));
```

Output:

```
2
```

Method: *public int indexOf(String str, int start)*

Description: Method return the first occurrence of the sub string passed as argument in this String, searching will start from given index *start.* If no such value exists, then -1 is returned.

Example:

```
String str= new String("This is a test string. Java is a
                                                    language.");

System.out.println(str.indexOf("is",10));
```

Output:

```
28
```

Method: *public int lastIndexOf(char ch)*

Description: Method return the last occurrence index of given character as argument in this String. In either case, if no such character occurs in this string, then -1 is returned.

Example:

```
String str= new String("This is a test string. Java is a
                                                    language.");

System.out.println(str.lastIndexOf('i'));
```

Output:

```
28
```

Method: *public int lastIndexOf(char ch, int start)*

Description: Method will return the last occurrence of given character argument, ch, start the searching backward at the given index *start.*

Example:

```
String str= new String("This is a test string. Java is a
                                                    language.");

System.out.println(str.lastIndexOf('i',10));
```

Output:

```
5
```

Method: *public int lastIndexOf(String str)*

Description: Method return last occurrence index of given string as argument in this String.

Example:

```
String str= new String("This is a test string. Java is a
                                              language.");
System.out.println(str.lastIndexOf("is"));
```

Output:

```
28
```

Method: *public int lastIndexOf(String str, int start)*

Description: Method return the first occurrence of the sub string passed as argument in this String, searching backward at the given index *start*. If no such value exists, then -1 is returned.

Example:

```
String str= new String("This is a test string. Java is a
                                              language.");
System.out.println(str.lastIndexOf("is",10));
```

Output:

```
5
```

Method: *public String replace(char oldCharacter, char newCharacter)*

Description: Method returns the new *String*, result of replacing *oldCharacter* with *newCharacter* in this *String*. If the *oldCharacter* is not exist in this String, then this String object will return.

Example:

```
String str= new String("This is a test string. Java is a
                                              language.");
   String str2 = str.replace('s', 'x');
   System.out.println(str2);
```

Output:

```
Thix is a text xtring. Java ix a language.
```

Method: *public String replace(CharSequence targetSequence, CharSequence replacementSequence)*

Description: Method replaces the sequence of characters, *targetSequence* with another sequence of characters, *replacementSequence*. The replacement starts form the beginning of the string to the end of string. For example, replacing "xx" with "y" in the string "xxx" will result in "yx" rather than "xy".

Example:

```
String str1= new String("Rose is beautiful");
String str2 = new String("Lotus is a flower");
CharSequence cs1 = str1.subSequence(0,4);
CharSequence cs2 = str2.subSequence(0,5);

String str3=str1.replace(cs1,cs2);
System.out.println(str3);
```

Output:

Lotus is beautiful

Method: *public booleans contains(CharSequence cs)*

Description: Method returns *true* if and only if the given sequence of character as an argument are exists in this String values.

Example:

```
String str= new String("Rose is beautiful");
CharSequence cs = str.subSequence(0,4);
System.out.println(str.contains(cs));
```

Output:

```
true
```

Operations on StringBuffer

(i) Append to StringBuffer

The append operations is similar to concatenation of the string operation. There are different version of append() method, and each call String.valueOf() for each type of parameter to get the string representation. The version of the append() method are as follows:

StringBuffer append(boolean b)

Method appends the string representation of the boolean argument to the sequence of StringBuffer.

StringBuffer append(char c)

Method appends the string representation of the char argument to this sequence of StringBuffer.

StringBuffer append(char[] charsStr)

Method appends the string representation of the char array argument to this sequence of StringBuffer.

StringBuffer append(char[] str, int startIndex, int len)

Method appends the string representation of a subarray of the char array argument to this sequence of StringBuffer.

StringBuffer append(int i)

Method appends the string representation of the int argument to this sequence of StringBuffer.

StringBuffer append(Object obj)

Method appends the string representation of the Object argument to this sequence of StringBuffer.

Example:

```
String s="Hello!";
int a = 72;
StringBuffer sb = new StringBuffer(s);
sb.append(a).toString();
System.out.println(sb);
```

Output:

```
Hello!72
```

Similarly any type of data can be appended to the

```
StringBuffer object.
```

(ii) Insert in StringBuffer

The insert() method is used to insert any type of data at any position of SringBuffer object content. There are varies version of insert() methods are as follows:

StringBuffer insert(int offset, boolean b)

Method inserts the string representation of the boolean argument into this sequence of StringBuffer.

StringBuffer insert(int offset, char c)

Method inserts the string representation of the char argument into this sequence of StringBuffer.

StringBuffer insert(int offset, char[] str)

Method inserts the string representation of the char array argument into this sequence of StringBuffer.

StringBuffer insert(int index, char[] str, int offset, int len)

Method inserts the string representation of a subarray of the str array argumnt into this sequence of StringBuffer.

StringBuffer insert(int targetOffset, CharSequence cs)

Method inserts the specified CharSequence into this sequence of StringBuffer.

StringBuffer insert(int offset, Object obj)

Method inserts the string representation of the Object argument into this character sequence of StringBuffer.

StringBuffer insert(int offset, String str)

Method inserts the string into this character sequence of StringBuffer.

Example:

```
StringBuffer sb = new StringBuffer("Lotus a flower");
                                            sb.insert(6, "is ");
          System.out.println(sb);
```

Output:

```
Lotus is a flower
```

(iii) Delete from StringBuffer

The method StringBuffer the allows to delete characters using the methods delete() methods. There are following two version:

StringBuffer delete(int startIndex, int endIndex)

Method removes the characters in a substring of this sequence of StringBuffer

StringBuffer deleteCharAt(int index)

Method removes the char at the specified position in this sequence of StringBuffer.

Example:

```
StringBuffer sb = new StringBuffer("Lotus is a flower.");
```

```
sb.delete(5, 7);
System.out.println("After delete: " + sb);
sb.deleteCharAt(9);
System.out.println("After deleteCharAt: " + sb);
```

Output: After delete: Lotuss a flower.

After deleteCharAt: Lotuss a lower.

(iv) Replace

The method replace a set of characters sequence with another set of character inside a StringBuffer object.

StringBuffer replace(int startIndex, int endIndex, String str)

Replaces the characters in a substring of this sequence with characters in the specified String.

Example:

```
StringBuffer sb = new StringBuffer("Lotus is a flower");
sb.replace(0, 5, "Rose");
System.out.println("After replace: " + sb);
```

Output:

```
After replace: Rose is a flower
```

(v) Extracting Substring

The methods are used to extract a substring from the StringBuffer.

CharSequence subSequence(int startIndex, int endIndex)

Method returns a new character sequence that is a subsequence of this sequence of StringBuffer.

String substring(int startIndex)

Method returns a new String that contains a subsequence of characters currently contained in this character sequence of StringBuffer.

String substring(int startIndex, int endIndex)

Method returns a new String that contains a subsequence of characters currently contained in this sequence of StringBuffer.

Example:

```
StringBuffer sb = new StringBuffer("Rose is a flower");
String str = sb.substring(5,8);
System.out.println(str);
```

Output:

```
is
```

(vi) Searching Substring

For searching the string or a character StringBuffer also provides some methods as follows:

```
void getChars(int srcBeginIndex, int srcEndindex, char[] target,
              int targetBeginIndex)
```

Method use to copy character from this sequence into the destination character array *target*.

int indexOf(String str)

Method returns the index within this string of the first occurrence of the specified substring.

int indexOf(String str, int startIndex)

Method returns the index within this string of the first occurrence of the specified substring, starting at the specified index.

int lastIndexOf(String str)

Method returns the index within this string of the rightmost occurrence of the specified substring.

int lastIndexOf(String str, int startIndex)

Method returns the index within this string of the last occurrence of the specified substring.

These methods are used in the same manner as the method of String class were used. Just change the object type String to StringBuffer in the examples.

(vii) Others Methods

int length()

Method returns the length (character count).

StringBuffer reverse()

Method replaced the original sequence by the reverse of the sequence of StringBuffer..

void setCharAt(int index, char ch)

Method replace character at the specified index is set to ch.

void ensureCapacity(int minimumCapacity)

Method ensures that the capacity is at least equal to the specified minimum.

int capacity()

Method returns the current capacity of StringBuffer.

void setLength(int newLength)

Method sets the length of the character sequence of StringBuffer.

1.16. CLASSES AND METHODS

A class is the core construct of Java programming upon which entire Java is based. A class defines the shape or architecture of the class. A class is a user defined data type that acts as a template or blueprint for creating multiple objects. A class encapsulates the data and methods. A class behaves like an in-built type. The class is declared with the keyword 'class'. A class in Java is a collection of data types and member functions, called methods. The syntax for class declaration is:

```
class classname
{
    type variable1;
    type variable2;
```

```
  // ...
  type variableN;

type methodname1(parameter-list)
{
  // body of method
}

type methodname2(parameter-list)
{
  // body of method
}

  // ...

type methodnameN(parameter-list)
{
  // body of method
}
}
```

A simple class in Java looks like:

```
class Employee
{
  int emp_id, salary;           // Variable Declaration
  char emp_name[10];            // Variable Declaration
  void emp_data();              // Method Declaration
  void emp_income(){}           // Method Declaration
}
```

This *Employee* class creates as many objects as required. The objects can be created as:

```
Employee emp1, emp2;
```

emp1, emp2 are actually not the objects but are the variables that are used to create objects using *new* operator. The objects are created as:

```
emp1= new Employee ();
```

Methods define the interface to most classes that allows the class implementor to hide the specific layout of internal data structures. Methods are the executable set of statements that provide a structured approach to programming. A program can have any number of methods. We can access data types with the help of methods. Methods work like functions, means that, if the same set of code is to be repeated several times in the program, then the code does not require to be rewritten. It will simply be called again and again.

The general syntax for of a method is:

```
type methodname(parameter-list)
{
  // body of method
}
```

A simple method looks like:

```
public static void main (String args[ ])
{
  System.out.println ("Employee Details");
}
```

We can also have multiple methods in a class, as in the example below:

```
class Emp_Details
{

  void emp_Data()
   {
      System.out.println("This method provides Employee Data");
   }

  void emp_Income()
   {
      System.out.println("This method provides details about Employee
Income");
    }

}

class Organization
{
  public static void main(String args[ ])
  {
    Emp_Details e1= new Emp_Details();
     e1. emp_Data();
     e1. emp_Income();
     System.out.println("This is a MAIN method");
  }
}
```

The methods can also be parameterized that can operate on variety of data. In that case, the method name contains some arguments or parameters to be used in the program code. The advantages of using methods are:

- The debugging becomes easier. While debugging, a debugger can directly jump to a particular method and can make necessary alterations.
- A program can be logically divided.
- Methods give access to the data of the class. Only the member functions or methods of the same class can access the variables of that class.

Access Specifiers

Java controls access to variables and methods through the use of access specifiers or, sometimes called, access modifiers. There are three types of access specifiers: **private, public and protected. Private** is the most restrictive access specifier. A variable or a method that has been declared private cannot be accessed by any other class. The default access specifier is **public**. If the variables and methods of a class have been declared public, then they can be viewed and used by any other class. If a variable or a method is declared **protected**, it can be seen from any subclass of that class.

Method Overloading

It is possible to define two or more methods within the same class that share the same name but with different parameter list in most object-oriented programming languages. These methods are called *overloaded,* and the process is called *method overloading.* Method overloading is one of the means to implement polymorphism in Java. When an overloaded method is invoked, Java uses the type and number of arguments as its guide to determine which overloaded method was actually called. Thus, overloaded methods must differ in the type and number of their parameters. When Java encounters a call to an overloaded method, it simply executes the method whose parameters match with the arguments in the method call. The program below explains method overloading in java:

```
class Test_Overlaod
{
  void overload_test()
  {
    System.out.println("No parameters are available");
  }

 void overload_test(int a)
 {
   System.out.println("The value of a is" + a);
 }

 void overload_test(int a, int b)
 {
   System.out.println("The values of a and b is " + a + " ," + b
"respectively");
 }

 double overload_test(double a)
 {
   System.out.println("The value of double a is " + a);
   return a*a;
  }
 }

 class OverloadMain
 {
   public static void main(String args[])
 {
   Test_Overload to = new Test_Overload();
   double result;

   to.overload_test();
   to.overload_test(10);
```

```
        to.overload_test(10, 20);
        result = to.overload_test(12.24);
        System.out.println("Result of to.overload_test(12.24): " + result);
    }
}
```

This program generates the following output:

```
No parameters are available
The value of a is 10
The values of a and b are 10, 20 respectively
The value of double a is 12.24
Result of to.overload_test(12.24): 149.8176
```

overload_test( ) is overloaded four times. The first version takes no parameters, the second takes one integer parameter, the third takes two integer parameters, and the fourth takes one **double** parameter. When an overloaded method is called, Java looks for a match in between the arguments that are used to call the method and its parameters.

Method overloading supports polymorphism because it is one way that Java implements the "one interface, multiple methods" paradigm. Through the application of polymorphism, several names have been reduced to one. When you overload a method, each type of that method can perform any activity that you desire. There is no rule stating that overloaded methods must relate to one another. In practice, the closely related operations must not be overloaded.

Static members

There are cases when we want to define a class member to be used independently of any object of that class. To create such a member, its declaration is preceded with the keyword **static**. When a member is declared **static**, it can be accessed before any objects of its class are created, and without reference to any object. Both methods and variables can be declared **static**. The most common example of a **static** member is **main( )**. **main( )** is declared as **static** because it must be called before any objects exist. Instance variables declared as **static** are the global variables. When objects of its class are declared, no copy of a **static** variable is made. Rather all instances of the class share the same **static** variable.

Methods declared as **static** have several restrictions:

- They can only call other **static** methods.
- They must only access **static** data.
- They cannot refer to **this** or **super** in any way.

If you need to do computation in order to initialize your **static** variables, you can declare a **static** block which gets executed exactly once, when the class is first loaded. The example below shows a class that has a **static** method, some **static** variables, and a **static** initialization block:

```
class Static_Test
{
        static int x = 5;
        static int y;
        static void static_method(int a)
        {
          System.out.println("x = " + x);
          System.out.println("y = " + y);
          System.out.println("a = " + a);
        }
```

```
   static
   {
      System.out.println("Static block initialized.");
      y = x * 2;
   }
   public static void main(String args[])
   {
      static_method(20);
   }
}
```

As soon as the **Static_Test** class is loaded, all of the **static** statements are run. First, **x** is set to 5, then the **static** block executes and then y is initialized to **x * 2** or **10**. Then **main( )** is called, which further calls static_**method( ) and** passes **20** to a. The three **println( )** statements refer to the two **static** variables **x** and y, as well as to the local variable a.

Here is the output of the program:

```
Static block initialized.
x = 5
y = 10
a = 20
```

static methods and variables can be used independently of any object. For this, we only need to specify the name of their class followed by the dot operator. For example, if you wish to call a **static** method from outside its class, you can do so using the following general form:

```
classname.method( )
```

Here, *classname* is the name of the class in which the **static** method is declared. A **static** variable can be accessed by the use of the dot operator on the name of the class.

1.17. CONSTRUCTORS

Constructors of a class are used to initialize data members. They bear the same name as the class name and do not have a return type. Constructors cannot be inherited. Java automatically calls constructors when its object is created. There can be more than one constructor with different signature and it is called constructor overloading. It is the constructor's job to initialize the internal state of an object so that the code creating an instance will have a fully initialized, usable object immediately. A simple class with a constructor is:

```
class Emp_Details
{

Emp_Details()
   {
      System.out.println("This is Java Constructor");
   }
}


class Organization
{
   public static void main(String args[ ])
```

```
        {
         Emp_Details e1= new Emp_Details();
          System.out.println("This is MAIN Program");
        }
  }
```

When an object e1 is created for the class Organization, the storage is allocated with the 'new' operator and a call is made to the constructor.

Java also provides parameterized constructor. A program depicting parameterized constructor is as follows:

```
class Emp_Details
{
  Emp_Details(int x)
  {
    System.out.println("This is Java Constructor" +x);
  }
}

class Employee
{
   public static void main(String args[ ])
     {
        for( int i=0; i<=5;i++)
         new  Emp_Details(i);
        }
  }
```

Copy Constructor

A copy constructor is the one that replicates or duplicates an existing object. It takes the object of the class as a reference to the parameters. This means that whenever we initialize an instance using value of another instance of same type, a copy constructor is used. For example,

```
testClass t1 = new testClass( );    //default constructor

testClass t2 = new testClass (t1);   //copy constructor
```

In the example above, testClass is the user-defined class name. We create the object **t1** of the class testClass without any arguments so, it becomes a default constructor. In the second statement, the compiler copies the object **t1** to **t2** member. It we have not defined a copy constructor; the compiler will automatically invoke it and makes it public.

Constructor Overloading

In addition to overloading normal methods, you can also overload constructor methods. Consider the example below for the nesting of constructor overloading:

```
class Const_Overload
{
    int x;
    int y;
    int z;
```

```
// This is the constructor for Const_Overload.

Const_Overload(int a, int b, int c)
{
  x=a;
  y=b;
  z=c;
 }
Const_Overload()
{
  x = 10;
  y = 5;
  z = 15;
 }

Const_Overload(int m)
{
  x = y = z = m;
}
// compute and return volume

int result()
{
  return x * y * z;
 }
}

class OverloadConsMain
{
 public static void main(String args[])
 {
      Const_Overload co1 = new Const_Overload(10, 20, 15);
      Const_Overload co2 = new Const_Overload( );
      Const_Overload co3 = new Const_Overload (10);

      int calculate;

      calculate = co1.result( );
      System.out.println("The value 1 is " + calculate);

      calculate = co2.result( );
      System.out.println("The value 2 is " + calculate);
      calculate = co3.result( );
      System.out.println("The value 3 is " + calculate);
    }
 }
```

The Const_Overload( ) constructor requires three parameters. This means that the all declarations of Const_Overload objects must pass three arguments to the Const_Overload( ) constructon Since Const_Overload( ) requires three arguments, it's an error to call it without them.

The output produced by this program is:

```
The value of 1 is 3000
The value of 2 is 750
The value of 3 is 1000
```

1.18. CREATING OBJECTS OF A CLASS

This *Employee* class creates as many objects as required. The objects can be created as:

```
Employee emp1, emp2;
```

emp1, emp2 are actually not the objects but are the variables that are used to create objects using *new* operator. The objects are created as:

```
emp1= new Employee ();
emp2=emp1;
```

The object declaration is actually a two-step process; first, to declare a variable of that class, second, is to acquire the actual physical copy of the object and assign that copy to the variable. The new operator dynamically allocates memory for an object and returns a reference to it. This reference is then stored in the variable. It is worth noting that since new allocates memory at runtime; we are at an advantage of creating as many objects as required during the execution of the program. Since, the memory is finite; there may be a case when we have created a lot of objects and are not able to create objects further. In this case, a runtime exception will occur.

1.19. REFERENCE VARIABLES

Object reference variables work differently in Java. When we are assigning one object reference variable to another object reference variable, we are not creating a copy of the object; we are only making a copy of the reference. Consider the example in the section above. It looks like that emp2 is being assigned a reference to a copy of the object referred to by emp1. This means, we might think that emp1 and emp2 refer to separate and distinct objects. However, this is completely wrong. Actually, as soon as this fragment executes, emp1 and emp2 will both refer to the same object. The assignment of emp1 and emp2 did not allocate any memory or copy any part of the original object. It simply makes emp2 refer to the same object as emp1. Thus, any changes made to the object through emp2 will affect the object to which emp1 is referring. The figure below explains this.

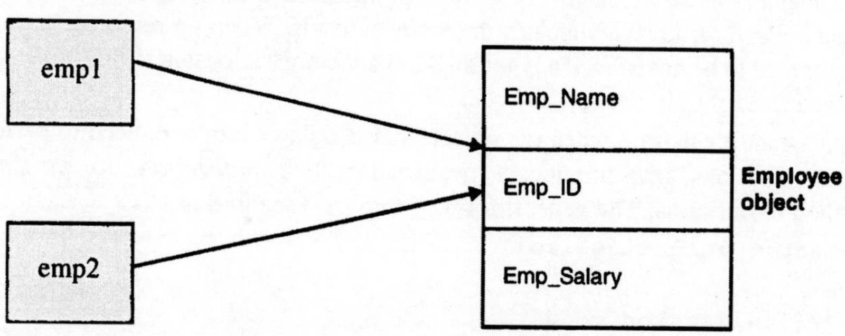

Fig. 1.4. Referencing variables

Although emp1 and emp2 refer to the same object, they are not linked in any way.

1.20. WHAT IS "THIS"

'this' pointer is used when a method has to refer to an object that has invoked it. **'this' pointer** can be used inside any method to refer to the *current* object. This means that **this** is always used as a reference to the object on which the method was invoked. We can use **this** anywhere as reference to an object of the current class type is permitted. Since, it is illegal in Java to declare two local variables with the same name inside the same or enclosing scopes; we can have local variables, which include formal parameters to methods, which may overlap with the names of the class instance variables. However, when a local variable has the same name as an instance variable, the local variable ***hides*** the instance variable.

emp_id and emp_salary have been used as the names of the parameters to the Employee() constructor inside the **Employee** class. If they had been, then **emp_id** would have referred to a formal parameter, hiding the instance variable **emp_id**. While it is usually easier to simply use different names, there is another way around this situation. Because **this** lets you refer directly to the object, you can use it to resolve any name space collisions that might occur between instance variables and local variables. For example, here is another version of **Employee( )**, which uses **emp_id and emp_salary** for parameter names and then uses **this** to access the instance variables by the same name:

```
Employee(double emp_id, double emp_salary)
{
  this.emp_id = emp_id;
  this.emp_salary = emp_salary;
}
```

The use of **this** in such a context can sometimes be confusing, and some programmers are careful not to use local variables and formal parameter names that hide instance variables. But other programmers believe that it is a good convention to use the same names for clarity, and use **this** to overcome the instance variable hiding. It is a matter of taste which approach you adopt. Although **this** is of no significant value in the examples just shown, it is very useful in certain situations.

1.21. AUTOMATIC GARBAGE COLLECTION

It is quite a tedious job to keep track of all the objects that have been created during the program development. It is tougher to keep track of the objects that are not being used now or their reference have been removed or the objects that have gone out of scope. An object is eligible for garbage collection when there are no more references to that object. As soon as the variable goes out of scope, the references in the variable are automatically dropped. The references can also be explicitly dropped by setting the variable value to null. In Java, the memory is allocated to the objects using 'new' operator. In languages like C++, the memory is deallocated using 'delete' operator. This process of deallocating memory is called *garbage collection*. Java deallocates the memory itself. When no reference to an object is found, the object is assumed to be not needed any more. So, the memory allocated to that object can be reclaimed. There is no need to explicitly destroy objects.

There may occur a situation when the object, that is being destroyed, need to perform some action. To handle such situations, Java provides a mechanism called *finalization*. So, we simply need to add *finalize()* method to our class. The general form of *finalize()* method is:

```
protected void finalize()
{
    //finalization code
}
```

We need not worry about the use of finalize() in our programs since it is used in very rare cases where the object is not under the control of the garbage collector.

REVIEW QUESTIONS

Q. 1. What is a Default Constructor? How are an object instance variables initialized if a class has only a class constructor? **[MDU, May 2011]**

Ans. When you create a new instance (a new object) of a class using the new keyword, a constructor for that class is called. Constructors are used to initialize the instance variables (fields) of an object. Constructors are similar to methods, but with some important differences. If you don't define a constructor for a class, a default parameterless constructor is automatically created by the compiler. The default constructor calls the default parent constructor (super()) and initializes all instance variables to default value (zero for numeric types, null for object references, and false for booleans). When a class has only the default constructor, its instance variables are initialized to their default values. Variables that contain numbers are initialized to 0, variables of type bool are initialized to false, and reference-type variables are initialized to null.

Q. 2. How the platform independent feature of Java implemented? Explain the role of main() method in the execution of Java program. **[MDU, May 2008]**

Ans. The main issue with Java developers was the need for a platform-independent a portable language. One of the main problems that the programmers were facing was that there was no guarantee exists that if you write a program today, it will run tomorrow or even on the same machine. Operating system upgrades, processor upgrades, and changes in core system resources sometimes combine to cause a program breakdown.

Q. 3. What do you mean by bytecode? Explain its working. **[MDU, Dec. 2008]**

Ans. *Bytecode* is a highly optimized set of instructions designed to be executed by the Java run-time system, which is called the *Java Virtual Machine (JVM)*.

Q. 4. Explain the concept of arrays in Java with suitable example. **[MDU, May 2009]**

Ans. An *Array* is a (fixed length) data structures used to store multiple data elements of same type. The memory allocated to an array is consecutive. All the elements of an array are under one variable name. Every data element of an array is accessed by its position and that position is indicated by an integer value that start from 0 (zero) called *index*.

Q. 5. Why is Java more suitable than other languages? **[MDU, Dec. 2008]**

Ans. Java is more suitable as compared to other languages because of the following features:

- Platform independent and portable
- Object Oriented
- Robust and Secure
- Distributed
- Familiar, Simple and small
- Multithreaded and Interactive
- High Performance
- Dynamic and Extensible
- Compiled and Interpreted

❑❑❑

2

CHAPTER

Inheritance and Interfaces

A very important characteristic of the object oriented methodology is *Inheritance*, in which a class is derived by another class, mean to say, inheriting fields and methods from the classes. The concept supports the reusability of the classes. New class is defined with some additional features. An abstract class is a collection of abstract and non abstract methods, an abstract class can be fully abstract, but there is a limitation of a class, that more than one class can be inherited. Similar limitation is with abstract class. So, instead of defining a fully abstract class, and give the flexibility of multiple inheritance, Java provides a significant concept, called *interface*. We will have a detailed discussion about inheritance and interface in this chapter.

INSIDE THIS CHAPTER

2.1. CLASS INHERITANCE

Inheritance is one of the most dominant and vital feature of the object oriented programming because it supports the hierarchical classifications, reusability of class; and defined to specialization. Java is a language that supports inheritance but with some additional advantages and features.

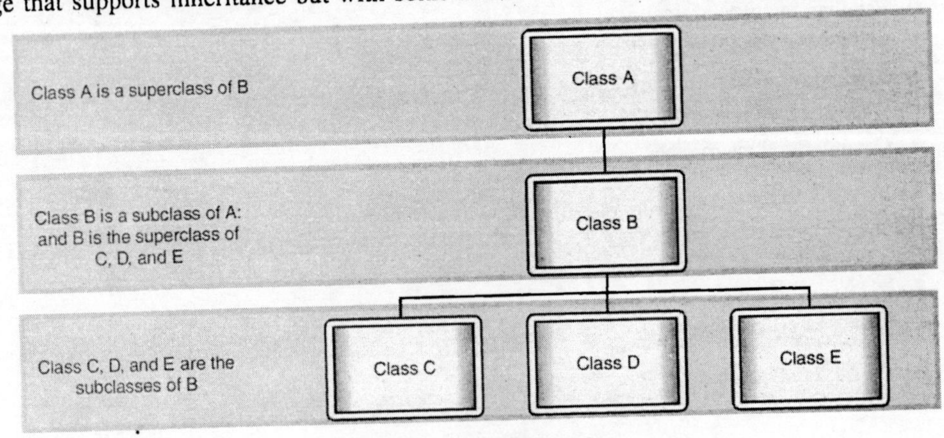

Fig. 2.1. Class Hierarchy

Here we create a generalized class that have some set of related items. A class that is derived from any another class (may be generalized class) is called a *subclass* or *derived* class or *extended* class or *child* class. The class from which a subclass is derived is called *superclass* or *base* class or *parent* class. A subclass is a specialized version of the it's parent class, and it inherits all the members

Each class has a superclass, and each class can have one or more subclasses. There is a hierarchical relation between the classes. Java supports single and multilevel inheritance.

The class Object is at the top of the Java class hierarchy. It is the most general class, which defines the specific behavior required by all object. If you are not extending any class then that class will automatically extends the Object class by the compiler. Excepting Object class, every class has one and only one direct superclass (single inheritance). Only Object class does not have any superclass. In the absence of any other explicit superclass, every class is implicitly a subclass of Object.

Whenever you create a new class, you define all the related information like fields, methods, and nested class. But in case of inheritance, when we create a new class and if there is an existing class having some of the code that you want, you can derive your new class from the existing class with some additional code (includes instance fields and methods).

The class is a collection or a wrapper of the methods and fields. The question arise here that how these members (fields and methods) are accessible from outside the class. The description of how these members are predictable to behave is called class *contract*. To define the class's contract is the responsibility of the class designer. Extending of class provides two types of inheritance as given below:

- *Inheritance of type or contract:* The subclass obtains the type of the superclass.
- *Inheritance of implementation:* The subclass obtains the implementation of accessible fields and methods of superclass.

The class can be extended for a number of purposes. The extended class (subclass) is called the specialized version of the superclass; this subclass defines new behavior. Yet you can change the implementation of some inherited method, but keep in mind, we can change it within in a limit. The changes in implementation must not violate that contract. The access control mechanism work when interacting or accessing the members. Each class provides two different contracts – one is for the users of the class and other is for the extenders of the class.

2.2. REUSABILITY OF EXISTING CLASSES

For the reusability of existing classes, inheritance came into use, and java supports only two types of inheritance, in which a class extends another class.

- Single Inheritance
- Multilevel Inheritance

In Java, a class can be derived from another class using the general form:

```
access_specifier class Subclass extends Superclass
{
 //Derived class data and methods
}
```

Various types of inheritance are discussed in the section below.

Single Inheritance

In this type of inheritance, a single sub class is derived from a single base class. This can be depicted in the Fig. 2.2.

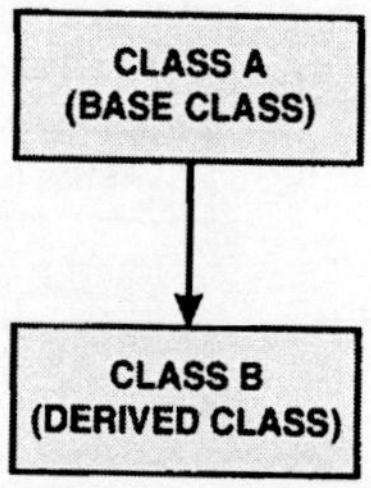

Fig. 2.2. ***Single Inheritance***

Multiple Inheritance

In multiple inheritance, there is more than one base class from which the child class is derived. This results in the duplication of data with the derived class. This is a major problem with OOP and a mechanism called *Virtual Base Class* is provided for this, in which, one of the base classes is made virtual to avoid the duplicate inheritance of data. Multiple Inheritance can be depicted as:

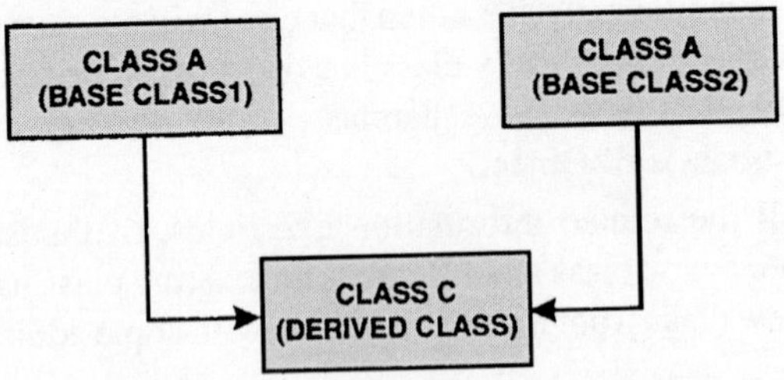

Fig. 2.3. Multiple Inheritance

> **DO YOU KNOW**
>
> Java does not directly support multiple inheritance.

> **DO YOU KNOW**
>
> Java only supports multiple inheritance when used with Interfaces.

Java does not directly support multiple inheritance. This is because different classes may have different variables with same name that may be contradicted and can cause confusions in JVM, thus resulting in errors. Java only supports multiple inheritance when used with Interfaces. We can extend one class only to avoid ambiguity problem with JVM. In interface we have to define the functions. So there is no ambiguity. In C++, it is big problem but in JAVA this issue has been improved by introducing Interfaces.

Multilevel Inheritance

In multilevel inheritance, a class derives another class, and from the derived class, another sub class is derived. This can be depicted as:

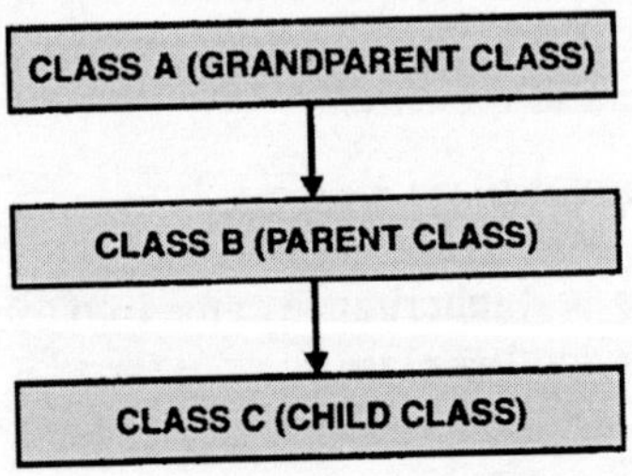

Fig. 2.4. *Multilevel Inheritance*

Hybrid Inheritance

There may occur situations where we need to apply two or more types of inheritance in the design of a program. For such cases, hybrid inheritance is used. This can be depicted as:

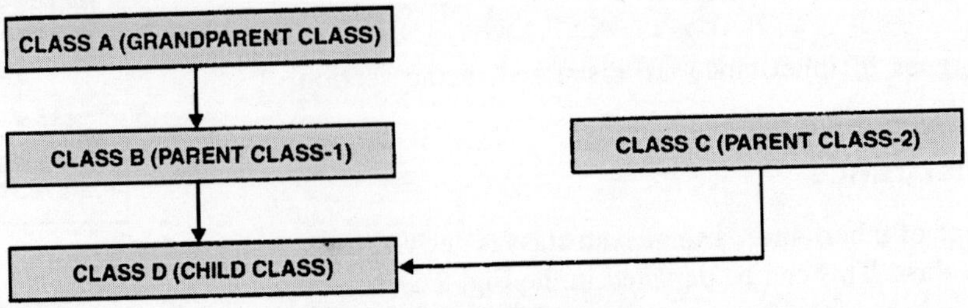

Fig. 2.5. Hybrid Inheritance

Multiple inheritances cannot be possible by extending more than one class, it can be possible with one another and important features of the Java, called *interface*.

To inherit a class into another class *extends* keyword is used.

```
class subclassName extends superclassName
{
//body of subclass
}
```

The most important point in inheritance is that, the subclass object will be created and all the accessible members of superclass and its own are accessed by subclass object.

Example:

```
//Create a superclass
class SuperClass
{
int a,b;
public void storeab()
{
a = 41;
b = 41;
}
public void showab()
{
System.out.println("SuperClass contents :: a = "+a+" b = "+b);
}
}
//Create a sub class by extending class SuperClass
class SubClass extends SuperClass
{
int s;
public void sumab()
{
s=a+b;
}
public void shows()
{
System.out.println("SubClass contents:: s = "+s);
}
}
public class SingleInheritanceDemo
{
public static void main(String ar[])
{
SubClass ob = new SubClass();     //Create object of subclass
ob.storeab();                     //Call super class method
ob.sumab();                            //Call own method
ob.showab();                      //Call super class method
ob.shows();                            //Call own method
}
}
```

Output:

```
SuperClass contents :: a = 41 b = 41
SubClass contents:: s = 82
```

In the give example, the object of the class *SubClass*, ob, is used to call the public member of it's superclass as well as it's own. So that we can observe, that the public member of *SuperClass* are inherited in the *SubClass*, i.e. become the part of it, so that using ob, we're able to call them or access them. In the method *sumab*() of *SubClass*, we are directly using the a,b of *SuperClass* without using any instance of *SuperClass*, because these are inherited in the *SubClass*. The fields are without any access modifier, count under the category of default access modifier, and methods are defined as *public*.

2.3. ACCESS ATTRIBUTES

The class member attributes (fields) and methods are bounded with some accessibility modifier, which defines the access scope of the member. In this section, we'll study how to access the data members (attributes) of the class.

All the public and protected members of the parent of subclass always inherit, doesn't matter what package the subclass is in.

- Just like any other fields of the subclass, the inherited fields can be used.
- If you declare a field in the subclass, with the same name as in superclass, then the subclass field will hide the superclass inherited filed.
- The inherited method can be used as same as the subclass method are used.
- You can define a new instance method in the subclass, that has the same signature as one of the superclass method, called overriding.
- You can define a new static method (class member) for the subclass, that has the same signature as one of the superclass method, called hiding.
- The subclass constructor invokes the superclass constructor either implicitly if the superclass constructor is default or by using the keyword super.

The attributes can be with any access modifier like private, protected, public, and default. There is fixed scope of accessibility or extendibility of the attribute in subclass. The attributes will be inherited or not, it totally depends on its access modifier. There are four access modifiers as follows with its scope:

- private : never inherit
- protected : only in its subclass
- public : inherit
- default : inherit

Example:

```
//Create a superclass
class SuperClass
{
private int a;
protected int b;
public int c;
int d;
public void storeab()
{
a = 41;
```

```
b = 41;
c = 41;
d = 41;
}
public void showab()
{
System.out.println("SuperClass contents :: a = "+a+" b = "+b+" c =
    "+c+" d = "+d);
}
}
//Create a sub class by extending class SuperClass
class SubClass extends SuperClass
{
int s;
public void sumab()
{
s=a+b+c+d;          //ERROR! a is private, not be inherited
s = b+c+d;
}
public void shows()
{
System.out.println("SubClass contents:: s = "
}
}
public class SingleInheritanceDemo
{
public static void main(String ar[])
{
SubClass ob = new SubClass();          //Create object of subclass
ob.storeab();                          //Call super class method
ob.sumab();                            //Call own method
ob.showab();                           //Call super class method
ob.shows();                            //Call own method
} }
```

When you compile this code you'll find the following message:

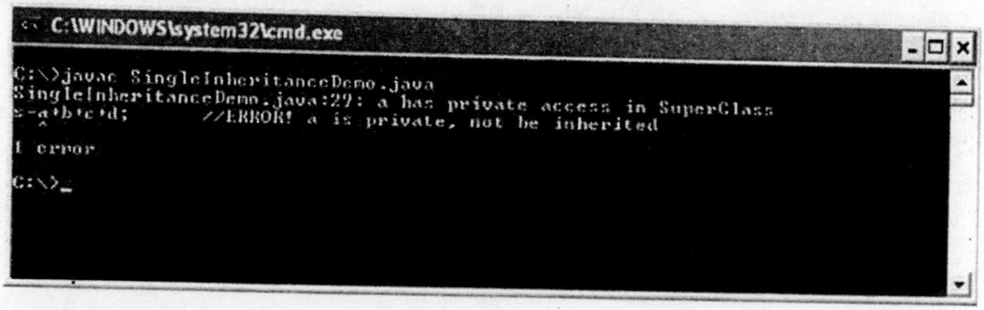

But if you want the value of attribute 'a' in subclass then either implement a method that returns its value or define it as default or public. The modified version of the program is given below with definition of a method that returns the value of attribute 'a' of *SuperClass*.

Example:

```
//Create a superclass
class SuperClass
{
private int a;
protected int b;
public int c;
int d;
public void storeab()
{
a = 41;
b = 41;
c = 41;
d = 41;
}
public void showab()
{
System.out.println("SuperClass contents :: a = "+a+" b = "+b+" c =
    "+c+" d = "+d);
}
public int geta()
{
return a;
}
}
//Create a sub class by extending class SuperClass
class SubClass extends SuperClass
{
int s;
public void sumab()
{
s=geta()+b+c+d; //Get the value of SuperClass private member using method
s = b+c+d;
}
public void shows()
{
System.out.println("SubClass contents:: s = "+s);
}
}
public class SingleInheritanceDemo
{
public static void main(String ar[])
{
```

```
SubClass ob = new SubClass();            //Create object of subclass
ob.storeab();                            //Call super class method
ob.sumab();                              //Call own method
ob.showab();                             //Call super class method
ob.shows();                              //Call own method
} }
```

The protected, public and default field 'b', 'c' and 'd' are inherited in subclass.

2.4. MULTIPLE LEVELS OF INHERITANCE

Java supports multilevel inheritance. In multiple, multilevel class hierarchies contain the layers of inheritance. But at each layer, a class is a subclass of the superclass of another, except the last layer. One pictorial representation of such concept is given below.

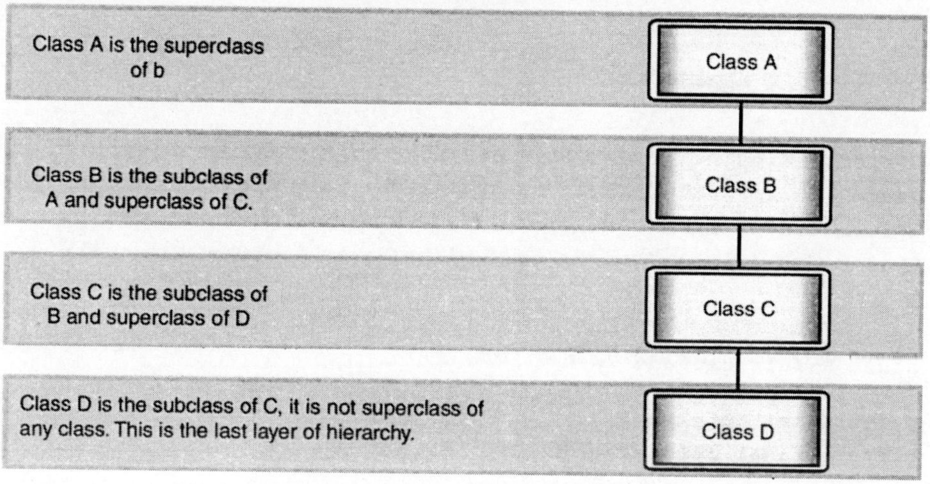

Fig. 2.6. Class hierarchy of 4 classes

Example:

```
class Student
{
private int rollno;
private String name;
private String address;
public void storeDetails(int rno, String sname, String sadd)
{
rollno = rno;
name = sname;
address = sadd;
}
public void showDetails()
{
System.out.println("Roll No :: " + rollno);
System.out.println("Name :: " + name);
System.out.println("Address :: " + address);
}
```

```
}
class Test extends Student
{
protected int marksSubject1;
protected int marksSubject2;
protected int marksSubject3;
protected int marksSubject4;
public void storeMarks(int m1, int m2, int m3, int m4)
{
marksSubject1 = m1;
marksSubject2 = m2;
marksSubject3 = m3;
marksSubject4 = m4;
}
public void showMarks()
{
System.out.println("Marks of Subject1 :: " + marksSubject1);
System.out.println("Marks of Subject2 :: " + marksSubject2);
System.out.println("Marks of Subject3 :: " + marksSubject3);
System.out.println("Marks of Subject4 :: " + marksSubject4);
}
}
class Result extends Test
{
private int totalMarks;
private float percentage;
private char grade;
public void evaluateResult()
{
totalMarks = marksSubject1 + marksSubject2 + marksSubject3 +
    marksSubject4;
percentage = (totalMarks*100.00F/600.00F);
if(percentage >=50.00F)
    grade='D';
else
if(percentage >=55.00F && percentage<=60.00F)
    grade = 'C';
else
if(percentage >=61.00F && percentage<=70.00F)
    grade = 'B';
else
if(percentage >=71.00F && percentage<=75.00F)
    grade = 'A';
else
 if(percentage >=76.00F && percentage<=85.00F)
    grade = 'H';
```

```
    else
        grade = 'S';
    }
    public void showResult()
    {
    showDetails();
    showMarks();
    System.out.println("Total Marks :: " + totalMarks);
    System.out.println("Percentage :: " + percentage);
    System.out.println("Grade :: " + grade);
    }
    }
    public class MultilevelInheritanceDemo1
    {
    public static void main(String ar[])
    {
    Result ob = new Result();
    ob.storeDetails(1001, "Alice","New York");
    ob.storeMarks(135,130,132,138);
    ob.evaluateResult();
    ob.showResult();
    }
}
```

There can be few problems in multilevel inheritance, if the classes are not properly designed by the designer, or proper keywords are not used.

- How the superclass parameterized constructor will be called?
- If superclass and subclass protected or public data members have same name then, superclass data member will be hidden. How to distinguish or to access the superclass data member in the subclass?
- If subclass member methods have the same signature as the superclass member method then how to call the superclass member methods?

The ***super*** keyword is the solution of all the above problems, to refer to its immediate superclass. The *super* has two general forms as given below:

- To call the superclass constructor
- To access the hidden member of the superclass.

Example:

```
class Student
{
private int rollno;
private String name;
private String address;
public Student(int rno, String sname, String sadd)
{
rollno = rno;
name = sname;
```

```
address = sadd;
}
public void show()
{
System.out.println("Roll No :: " + rollno);
System.out.println("Name :: " + name);
System.out.println("Address :: " + address);
}
}
class Test extends Student
{
protected int marksSubject1;
protected int marksSubject2;
protected int marksSubject3;
protected int marksSubject4;
public Test(int rno, String sname, String sadd,int m1, int m2, int m3,
    int m4)
{
super(rno,sname,sadd);                    //Call superclass constructor
marksSubject1 = m1;
marksSubject2 = m2;
marksSubject3 = m3;
marksSubject4 = m4;
}
public void show()
{
super.show();                             //Call superclass method
System.out.println("Marks of Subject1 :: " + marksSubject1);
System.out.println("Marks of Subject2 :: " + marksSubject2);
System.out.println("Marks of Subject3 :: " + marksSubject3);
System.out.println("Marks of Subject4 :: " + marksSubject4);
}
}
class Result extends Test
{
private int totalMarks;
private float percentage;
private char grade;
public Result(int rno, String sname, String sadd,int m1, int m2, int
    m3, int m4)
{
super(rno,sname,sadd,m1,m2,m3,m4); //Call superclass
                                   //constructor
totalMarks = marksSubject1 + marksSubject2 + marksSubject3 +
    marksSubject4;
percentage = (totalMarks*100.00F/600.00F);
```

```
if(percentage >=50.00F)
    grade='D';
else
if(percentage >=55.00F && percentage<=60.00F)
    grade = 'C';
else
if(percentage >=61.00F && percentage<=70.00F)
    grade = 'B';
else
if(percentage >=71.00F && percentage<=75.00F)
    grade = 'A';
else
if(percentage >=76.00F && percentage<=85.00F)
    grade = 'H';
else
    grade = 'S';
}
public void show()
{
super.show();   //Call superclass method
System.out.println("Total Marks :: " + totalMarks);
System.out.println("Percentage :: " + percentage);
System.out.println("Grade :: " + grade);
}
}
public class MultilevelInheritanceDemo2
{
public static void main(String ar[])
{
Result ob = new Result(1001, "Alice","New York",135,130,132,138);
ob.show();
}
}
```

The above code show how to call the superclass constructor and how to call the superclass method if that signature is same as the subclass method signature.

Example:

```
class Abc
{
int i;
public Abc(int i)
{
this.i = i;
}
}
class Xyz extends Abc
```

```
{
int i,j;
public Xyz(int a,int b)
{
super(a);
j=b;
i = ++(super.i);
}
public void showAll()
{
System.out.println("Superclass 'i' :: "+super.i);
System.out.println("Subclass 'i' :: " + i + "   'j' ::  "+ j);
}
}
public class MultilevelInheritanceDemo
{
public static void main(String ar[])
{
Xyz ob = new Xyz(1,2);
ob.showAll();
} }
```

Output:

```
Superclass 'i' :: 2
Subclass 'i' :: 2   'j' ::  2
```

In the above code, in subclass the data member of superclass 'i' hides with subclass data member 'i', so that using *super* keyword it access.

2.5. INTRODUCTION TO POLYMORPHISM

Another important feature of the Java is polymorphism. A famous phrase 'one interface, multiple methods' or 'once instance, multiple forms' is for polymorphism. General classes of actions are defined, and according to the exact nature of situation the specific action is called. A set of related activities are designed in a generic interface, and same interface is used to specify a general class of action. To select the specific action is the responsibility of the compiler's job. Polymorphism allows you to create a sensible, readable, clean, and resilient code.

There are two following ways to implements polymorphism

- Overloading
- Run-time polymorphism
  - Dynamic method dispatching
  - Overridden

Overloading

In a class, if there are two methods having the same name but with different set of parameters, the methods are called overloaded method, and the concept is called overloading. Overloading is one of the mechanisms by which Java implements polymorphism.

In overloaded method, method names are same, but their type of parameters, number of parameters and their order are different. Return types doesn't matter that can be same or different. Whenever any overloaded method is called, then Java uses types and/or number of arguments to decide the exact version of the overloaded method to call, which parameters match with the argument used in the method call.

Example:

```
public class OverloadingDemo
{
public static void main(String ar[])
{
//Call all the version of sum
sum(1,2);
sum(1,2.5F);
sum(6.7F,1.2);
}
//Overload the method of two integer
public static void sum(int a, int b)
{
System.out.println("Sum = " + (a+b));
}
//Overload the method of one integer and float
public static void sum(int a, float b)
{
System.out.println("Sum = " + (a+b));
}
//Overload the method of one float and double
public static void sum(float a, double b)
{
System.out.println("Sum = " + (a+b));
}
}
```

Output:

```
Sum = 3
Sum = 3.5
Sum = 7.899999809265137
```

Yet, whenever an overload method is called, then Java search for a method of same types of parameters, as passed as argument in method call, but always it is not possible to get the exact match. In some cases Java automatically use the type conversion implicitly, that plays a major role in overloading.

Example:

```
public class OverloadingDemo
{
public static void main(String ar[])
{
//Call all the version of sum
```

```
sum(1, 2);
sum(1, 2.5F);
sum(6.7F, 1.2);
sum(6.9F, 2.3F);        //Passing two float value, but there is
    //matched overloaded method
}
//Overload the method of two integer
public static void sum(int a, int b)
{
System.out.println("Sum = " + (a+b));
}
//Overload the method of one integer and float
public static void sum(int a, float b)
{
System.out.println("Sum = " + (a+b));
}
//Overload the method of one float and double
public static void sum(float a, double b)
{
System.out.println("Sum = " + (a+b));
}
//Overload the method of two double
public static void sum(double a, double b)
{
System.out.println("Sum = " + (a+b));
}
}
```

Output:

```
Sum = 3
Sum = 3.5
Sum = 7.899999809265137
Sum = 9.200000047683716
```

In the given code, the method call sum(6.9F, 2.3F) passed two float argument, but there is no method of same type of parameters. Here, Java implicit type conversion plays an important role, it convert the float to double, and sum(double a, double b) method is called.

Overridden

In a class hierarchy, when there is a method in the subclass having the signature same as the signature of a method of its superclass, the subclass method is called override method of the superclass, this concept is called overriding. In overriding the superclass method will be hidden by the subclass overridden method. Overridden method supports the run-time polymorphism.

Example:

```
class Abc {
int i, j;
Abc(int a, int b) {
```

```
i = a;
j = b;
}
void show() {
System.out.println("i :: " + i + " j :: " + j);
}
}
class Xyz extends Abc {
int k;
Xyz(int a, int b) {
super(a, b);
k = a+b;
}
//Overrides show() in Abc
void show() {
System.out.println("k :: " + k);
}
}
public class OverrideDemo {
public static void main(String ar[]) {
Xyz ob = new Xyz(1, 2);
ob.show();    //Call show() in Xyz
}
}
```

Output:

```
k :: 3
```

In the above code, when show() method is invoked by an object of class Xyz, the version of show() defined in Xyz class is called, that is overridden version of the method defined in Abc. If you want to call or access the superclass method then using super keyword you can call it.

Example:

```
class Abc {
int i, j;
Abc(int a, int b) {
i = a;
j = b;
}
void show() {
System.out.println("i :: " + i + " j :: " + j);
}
}
class Xyz extends Abc {
int k;
Xyz(int a, int b) {
super(a, b);
```

```
        k = a+b;
        }
        //Overrides show() in Abc
        void show() {
        super.show();        //Call superclass method
        System.out.println("k :: " + k);
        }
        }
        public class OverrideDemo {
        public static void main(String ar[]) {
        Xyz ob = new Xyz(1, 2);
        ob.show();    //Call show() in Xyz
        }
    }
```

Output:

```
        i  :: 1  j  :: 2
        k  :: 3
```

Why overridden method? Some time, generalize class define a method that will be common to all subclasses, but in subclasses it will have some different implementation with its (subclass) need, then the superclass method has to override. It's another way to implements the "one interface, multiple forms".

Example:

```
        class Employee
        {
           private int Empid;
           private String EmpName;
          public Employee(int eid, String ename)
          {
            Empid = eid;
            EmpName = ename;
          }
          public void display()
          {
            System.out.println("Employee Name is: " + EmpName);
            System.out.println("Employee ID is: " + Empid);
          }
        }
        class EmpAdhoc extends Employee
        {
          private int noofworkingdays;
          private float dailywages;
          private float netsalary;
          public EmpAdhoc(int eid, String ename, int workday, float wages)
          {
            super(eid, ename);
```

```
            noofworkingdays= workday;
            dailywages=wages;
             calSalary();
        }
          public void calSalary()
          {
            netsalary=noofworkingdays * dailywages;
          }

        public void display()          //Override method
        {
          super.display();
          System.out.println("No. of working days :: " +
                                          noofworkingdays);
          System.out.println("Daily wages :: " + dailywages);
          System.out.println("Net Salary :: " + netsalary);
        }
      }
      public class OverrideDemo {
      public static void main(String ar[]) {
      EmpAdhoc ob = new EmpAdhoc(1001,"Alice",25, 600.00F);
      ob.display();
      }
}
```

Output:

```
Employee Name is: Alice
Employee ID is: 1001
No. of working days :: 25
Daily wages :: 600.0
Net Salary :: 15000.0
```

In the above code the display() method is overridden, and in both classes it is showing the corresponding data members. So that, signatures are same, but implementation are different.

Dynamic Method Dispatching

In overriding, there is a major problem of calling the method of the superclass. Because the subclasses object call only their own member method. The only way to call the superclass method by using the super keyword in the subclass member method, or write another method which are responsible to call superclass method using super keyword. So that, to gives power to overriding, the dynamic method dispatching is another and very powerful mechanism to resolves overriding problem at run time. Dynamic method dispatching is important, and implements run-time polymorphism.

Example:

```
class Parent
{
public void display()
{
```

```
System.out.println("I am in class Parent");
}
}

class Child1 extends Parent
{
public void display()              //Override Method
{
System.out.println("I am in class Child1");
}
}

class Child2 extends Parent
{
public void display()              //Override Method
{
System.out.println("I am in class Child2");
}
}
public class DynamicMethodDispatchDemo
{
public static void main(String ar[])
{
Parent p = new Parent();          // Object of type Parent
Child1 c1 = new Child1();         // Object of type Child1
Child2 c2 = new Child2();         // Object of type Child2

Parent ref;                       // Reference variable of type Parent

ref = p;                          //ref stores an Parent object
ref.display();                    //Calls display method of Parent

ref = c1;                         //ref stores a Child1 object
ref.display();                    //Calls display method of Child1

ref = c2;                         //ref store to a Child2 object
ref.display();                    //Calls display method of Child2
}
}
```

Output:

```
I am in class Parent
I am in class Child1
I am in class Child2
```

2.6. ABSTRACT CLASSES

Some time there is a situation in which you feel the need of a superclass that has only declaration of few or all methods, with definition of few or none methods. But it is necessary that it must not be a completely defined or implemented class. The methods are declared only and specified by abstract type modifier called abstract method. It is the responsibility of the concrete subclass to implement all abstract method of superclass. To declare any method as abstract use the following form:

abstract accessmodifer returntype methodName(<parameterlist>);

A class can have one or more abstract methods, that class must also be declared as abstract. To declare a class as abstract, simply add abstract keyword before the class keyword. Abstract class cannot be instantiated, i.e. you cannot create an object with the new operator of abstract class. The construct cannot be abstract, and no method can be abstract static. Any concrete subclass has to implements all the abstract method of its abstract superclass. Abstract classes can be used to create object references.

Example:

```
abstract class Employee
{
   private int Empid;
   private String EmpName;

   public Employee(int eid, String ename)
   {
    Empid = eid;
    EmpName = ename;
   }
  public void display()
  {
    System.out.println("Employee Name is: " + EmpName);
    System.out.println("Employee ID is: " + Empid);
   }
abstract void calcSalary();          //Abstract Method
}
class EmpAdhoc extends Employee
{
  private int noofworkingdays;
  private float dailywages;
  private float netsalary;
  public EmpAdhoc(int eid, String ename, int workday, float wages)
  {
   super(eid, ename);
   noofworkingdays= workday;
   dailywages=wages;
    calcSalary();
}
  public void calcSalary()    //Definition of Abstract Method
  {
    netsalary=noofworkingdays * dailywages;
```

```
    }
  public void display()        //Override method
  {
    super.display();
    System.out.println("No. of working days :: " +
                                  noofworkingdays);
    System.out.println("Daily wages :: " + dailywages);
    System.out.println("Net Salary :: " + netsalary);
   }
  }
 class Salaried extends Employee
 {
    private float hra;
    private float da;
    private float basicsalary;
    private float netsalary;
    Salaried(int eid, String ename, float basicsal)
   {
     super(eid, ename);
     basicsalary=basicsal;
     calcSalary();
   }
   public void display()
   {
     super.display();
     System.out.println("Basic Salary :: " + basicsalary);
     System.out.println("HRA :: " + hra);
     System.out.println("DA :: " + da);
     System.out.println("Net Salary :: " + netsalary);
   }
  public void calcSalary()  //Definition of Abstract Method
   {
     hra=basicsalary* 0.40f;
     da=basicsalary* 0.60f;
     netsalary= basicsalary+hra+da;
    }
  }
  public class AbstractDemo {
  public static void main(String ar[]) {
  EmpAdhoc eaObj = new EmpAdhoc(1001,"Alice",25, 600.00F);
  eaObj.display();
  Salaried salObj = new Salaried(2001,"Bob",60000.00F);
  salObj.display();
  }
}
```

Output:

```
Employee Name is: Alice
Employee ID is: 1001
No. of working days :: 25
Daily wages :: 600.0
Net Salary :: 15000.0
Employee Name is: Bob
Employee ID is: 2001
Basic Salary :: 60000.0
HRA :: 24000.0
DA :: 36000.0
Net Salary :: 120000.0
```

2.7. USE OF 'FINAL'

The *final* keyword has the following uses in the inheritance.

- If you want the method must not be overridden then defined that method as final.

Example:

```
class Abc
{
final public void dispaly()
{
System.out.println("I am in Abc");
}
}
class Xyz extends Abc
{
public void dispaly()
{
System.out.println("I am in Xyz");
}
}
```

At the time of compiler will show the error message given in the image.

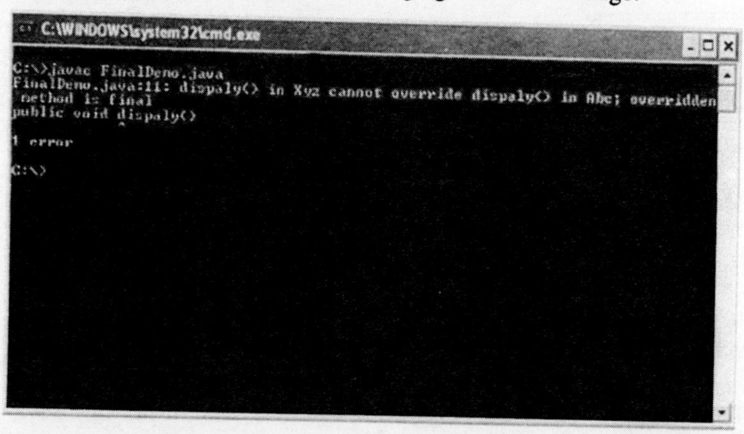

➠ If you want to prevent a class from being inherited then declare that as final.

Example:

```
final class Abc
{
public void dispaly()
{
System.out.println("I am in Abc");
}
}
class Xyz extends Abc
{
public void dispaly()
{
System.out.println("I am in Xyz");
}
}
```

At the time of compiler will show the error message given in the image.

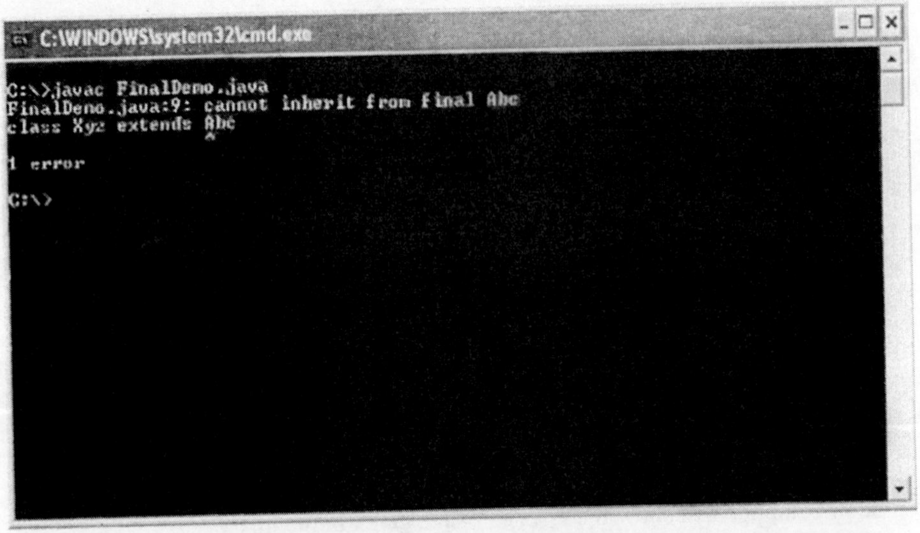

➠ If you want to create a constant then also use final with the variable declaration.

➠ The methods declare as final, compiler is free to make inline calls for those methods. If a small final method is called, the compiler copies the bytecode of the final method as inline code to the calling method. The final method calls are resolved at compile time called, *early binding*.

2.8. INTERFACE

An *interface* is a way of describing what classes should do, without specifying how they should do it. A class can implement more than one *interface*. In Java, an *interface* is not a class but a set of requirements for the class that we want to conform to the interface. All the methods of an *interface* are by default *public*. So, it is not required to use the keyword *public* when declaring a method in an *interface*. Interfaces can also have more than one method. Interfaces can also define constants but do not implement methods. An interface is defined like a class. Its general form is:

```
access_specifier interface InterfaceName
{
  Return-type Method-1(Parameters);
  Return-type Method-2(Parameters);
  Type variable-1=value;
  Type variable-2=value;
   .
   Return-type Method-N(Parameters);
   Type variable-N=value;
  }
```

To make a class implement an *interface*, we must declare that the class intends to implement the given interface and we need to supply definitions for all methods using the *implements* keyword. When an interface has been defined, any of the classes can implement that interface. The general form of the class that includes an implements clause is:

```
class classname extends superclass implements interface
{
 //class body
}
```

If we say that the *Employee* class implements the *Comparable* interface, this can be written as:

```
class Employee implements Comparable
```

Interface is a fully abstract class. Using the keyword interface, you can easily define the interface. Interface classes specify what a class must do, instead of how to do it. Interface body contains the variables, but they are not counted as instance variables, and methods without any body, i.e. only the declaration of the methods are there.

> **DO YOU KNOW**
>
> Interfaces are used to implement *multiple inheritance* in Java.

Interface has to be implemented using *implements* keyword. If the interface is implemented in a concrete class then, in concrete class all the abstract method of the interface must be implemented. But if it is implemented in abstract class then only required methods have to be defined, others can be left, but in concrete class this abstract class will be extended, rest all the methods have to be implemented.

Interface works as a contract among the team members of a project. Some time there is a need to define fixed layout or model that has to be followed by everybody, yet each will have its own implementation, own code without bothering about the others code. They follow the same prototype. For example, a display () method can be a part of any class, but its definition will vary.

A class can extends one class and implements one or more interfaces simultaneously, the way how multiple inheritances can be possible in Java. An interface can extends any number of interfaces but one interface cannot implements another interface, because if any interface is implemented then its methods must be defined, and interface never has the definition of any method. Interface cannot be instantiated, but it can work as super class of any other class, so that, its reference variable can be created to store or refer to its sub-ordinate classes objects.

Implementation of Interfaces

The interface contains only method declarations. Interface does not allow the implementation of methods, so that method declaration is followed by a semicolon, but no braces. All methods declared in the interface are by default implicitly public, so no need to use public as prefix in declaration. Interface contains the data members as constants. These constant values are implicitly public, static, and final, these modifiers can be omitted.

Interface declaration can be headed by interface modifiers, those are as follows:

- public
- abstract
- strict floating point

Syntax to declare an interface:

```
modifier interface_name {
return_type methodName1(parameter-list);
return_type methodName2(parameter-list);
type final-varName1 = value;
type final-varName2 = value;
// ...
}
```

Example:

```
interface Abc
{
void displayMsg();
}
public class DemoInterface implements Abc
{
public void displayMsg()
{
System.out.println("This is implemented method of Abc interface");
}
}
```

An interface can extends any number of interfaces. A demonstration is given below.

```
interface A
{
//...
}
interface B extends A
{
//...
}
interface C
{
//...
}
interface D extends B, C
{
//...
}
```

```
interface E
{
//...
}
class Abc
{
//...
}
class Xyz extends Abc implements D, E
{
//...
}
```

REVIEW QUESTIONS

Q. 1. What is an Interface? Explain with a suitable example, how multiple inheritance is implemented? What are abstract classes? **[MDU, May 2007, Dec. 2007]**

Ans. An abstract class can have instance methods that implement a default behavior. An Interface can only declare constants and instance methods, but cannot implement default behavior and all methods are implicitly abstract. An interface has all public members and no implementation. An abstract class is a class which may have the usual flavors of class members but has some abstract methods.

Q. 2. If a method is declared protected, where can the method be accessed?

Ans. A protected method can only be accessed by classes or interfaces of the same package or by subclasses of the class in which it is declared.

Q. 3. If you can add methods and fields to a subclass, why cann't you remove them?

Ans. If the methods and fields from the subclass have to be removed, then the subclass could no longer be used as an instance of its superclass. The subclass might have used any relation at all to the superclass if all the methods and fields of the superclass were removed.

Q. 4. What are interfaces? How do they support multiple inheritance?

Ans. Interfaces are the classes that are declared with the keyword interface and contain only abstract methods and static variables. The abstract methods and the static variables that are defined in an interface are by default public. Interface is a way of achieving multiple inheritance in Java.

❑❑❑

3
CHAPTER

Packages and Exception Handling

A class is a single entity that contains related members (data and methods) of same type of objects. If we write all the related classes, abstract classes individually, that will be in scattered format. Later if we want to use only those classes as a bundle of related classes, we can't use them easily. As a solution of this problem, Java gives a concept called *package,* that facilitate to bundle all the related classes and abstract classes in a single unit, package. A condition that is abnormal and arises at run time in the code sequence is called exception. A run time error is called exception in java. There are few languages, that do not support exception handling; then to handle them programmer has to write the code. For each exception and error, code writing is a tedious job. Java's exception is the solution of this problem, because to some extent Java provides in-build exception handling for certain type of errors, that often happen or occur. The chapter is dedicated to the topic exception handling, and good exploration of knowledge about this topic is included in this chapter.

INSIDE THIS CHAPTER

3.1. UNDERSTANDING PACKAGES

In Java, package is an encapsulation mechanism by which classes, abstract classes, interfaces, enums, and subpackages are grouped. Package is counted as a single unit of software. It can be distributed independently and combined with other packages of another application to create that application. As stated earlier, all related members are combined into a single package. For further categorization or classification within the package, subpackages are created.

Packages are specific reusable independent software module. There are following reasons to bundle classes and interface in a package.

- To easily determine the type of related components (classes and interfaces). By the classification or categorization, any programmer can easily identify the classes and interfaces on the basis of common functionality.

- The programmer can easily find them, by knowing their position with their set of task. The common types of classes and interfaces are placed in the same package.
- The type name (class or interface) must not conflict with other package type name, because at the same time if both packages are in use, compiler must not confuse.
- There must not be any restriction to access any type from another type within the same package. But assign the restricted access from types outside the package.

3.2. DEFINING A PACKAGE

Now as we know, a package is a collection of classes and interfaces. A package does not contain the source code of any type of Java file. Each Java source code file is a collection of one or more classes, or one or more interface with their members. After compilation, a separate class file is created for individual class or interface, either these are part of one Java file or stored in individual file. That class file (byte code) is placed in a particular related package using a package declaration.

Declaration Syntax:

```
package <fully qualified package name>;
```

The package declaration statement must be the first statement of the source file, and at least one package declaration must be in a source file. The package name is stored in the byte code (class file). The dot (.) notation is used to distinctively spot (identify) the package name in the package hierarchy. One naming convention for a package is that, package name is written in all lowercase. It will escape the confliction with classes and interfaces name.

Example:

```
package mypkg;
```

Here, *mypkg* is the package name. Moreover a file belonging to same package can include the same package name.

You can also create a hierarchy of packages, i.e. package within a package called subpackage for further classification and categorization of classes and interfaces.

```
package mypkg.mypkg2.mypkg3;
```

Here, mypkg2 is a subpackage in package mypkg, mypkg3 is a subpackage in subpackage mypkg2.

If there is no package declaration in a source file, then the compiled file is stored in the default package (unnamed package), and that is the current working directory.

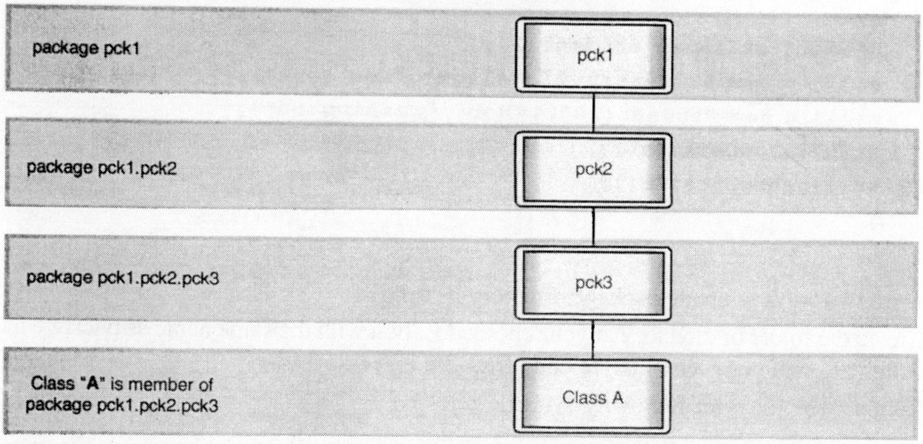

Fig. 3.1. Package Hierarchy

In the above package hierarchy class A is identified by the name pck1.pck2.pck3.A .This is called the *fully qualified type name* of the class (type). A fully qualified type name (name includes class or interface name) is comprises of *fully qualified package name*. According to above hierarchy pck1.pck2.pck3 is the fully qualified package name.

3.3. PACKAGING UP CLASSES

Now, you must have a model or a blue print of package in your mind. Let's take one example how to create package with keeping in mind all the concepts that we have discussed in last sections.

Example:

```
package pck1;
class Student
{
private int rollno;
private String name;
private String address;
public Student(int rno, String sname, String sadd)
{
rollno = rno;
name = sname;
address = sadd;
}
public void showDetails()
{
System.out.println("Roll No :: " + rollno);
System.out.println("Name :: " + name);
System.out.println("Address :: " + address);
}
}
public class DemoPackage
{
public static void main(String ar[])
{
Student st[]=new Student[2];
st[0] = new Student(1001,"Alice","New York");
st[1] = new Student(1002,"Bob","Washington");
st[0].showDetails();
st[1].showDetails();
}
}
```

There are two ways to create package directory as follows:

1. Create the folder or directory at your choice location with the same name as package name. After compilation of copy *.class* **(byte code file)** file into this folder.
2. Compile the file with following syntax.

```
javac -d <target location of package> sourceFile.java
```

The above syntax will create the package given in the *sourceFile* at the *<target location of pacakge>* if it is not yet created. If package already exist then only the ***.class* (byte code file)** will be stored to the package given in *sourceFile*.

Steps to compile the given example code:

- Compile the code with the command on the command prompt.

```
javac -d . DemoPackage.java
```

- The command will create the package at the current location with the name pck1, and contains the file DemoPackage.class and Student.class
- To run write the command given below

```
java pck1.DemoPackage
```

Note: The DemoPackate.class is now stored in pck1 package. So that we've to use *fully qualified type name* to run or access it.

Output:

```
Roll No :: 1001
Name :: Alice
Address :: New York
Roll No :: 1002
Name :: Bob
Address :: Washington
```

Note: If you put multiple types in a single source file, only one can be public, and it must have the same name as the source file. For example, you can define public class Circle in the file Circle.java, define public interface Draggable in the file Draggable.java, define public enum Day in the file Day.java, and so forth.

You can include non-public types in the same file as a public type (this is strongly discouraged, unless the non-public types are small and closely related to the public type), but only the public type will be accessible from outside of the package. All the top-level, non-public types will be *package private*.

3.4. UNDERSTANDING CLASSPATH

The packages are referred by directories. When the command is given to execute or run Java application, its first work is to search the corresponding class. But, the question is that "where will it search the class file?'. By default the Java run-time system search the class file at the current location. But if we talk about packages, it must not be stored at the same location. For those we've to provide the path, where packages are stored, and at the time of run, Java run-time system will search the class files, the path is called the CLASSPATH.

To show the CLASSPATH value the command given below at command prompt

```
set CLASSPATH
```

CLASSPATH can be set in two ways.

1. By giving the command at command prompt.

```
set CLASSPATH=<package path>
```

Suppose for the example give in section 3.3, use the following command for compilation.

The pck1 will be created in d:\, so that CLASSPATH will be set upto the parent of package **pck1**, *i.e.* d:\

2. Set the CLASSPATH using environment variable in windows following steps have to be taken: First open the control panel, and select the System, highlighted with value shown in name of the icon

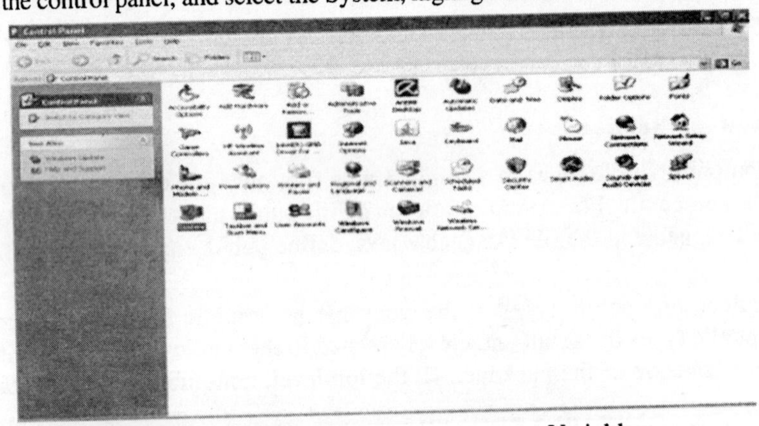

➠ Click on Advanced tab, and then click on Environment Variable.

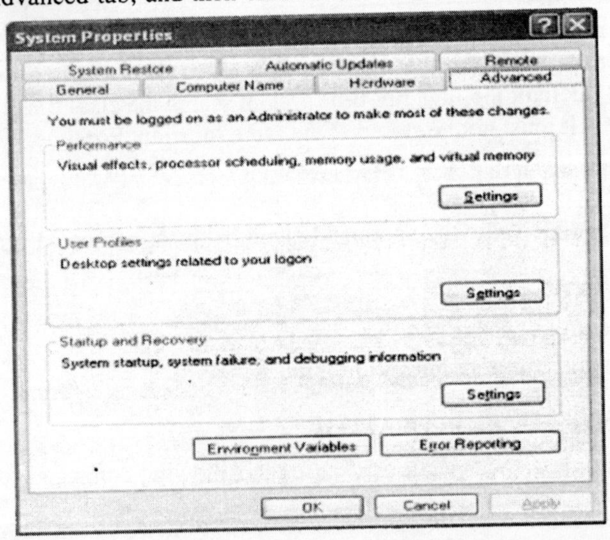

➠ Click on New Button in Use variables for Owner area.

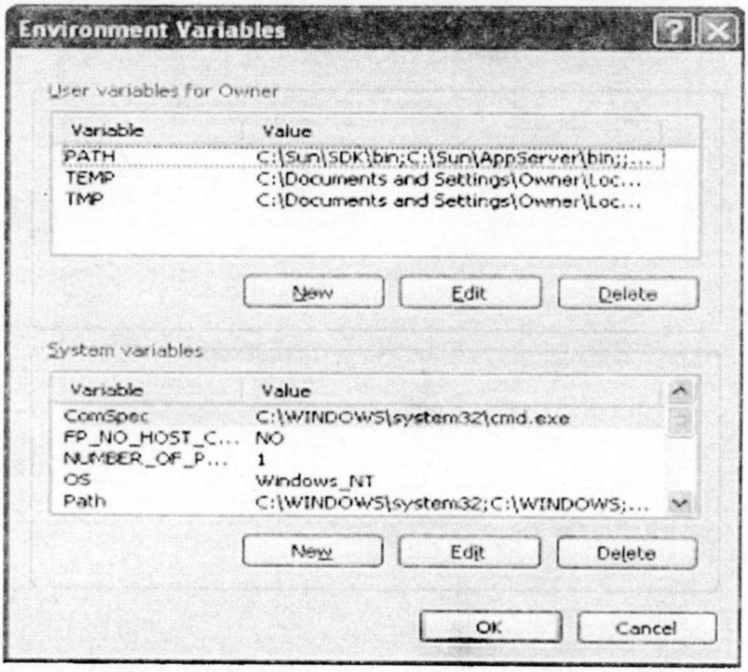

➠ Give the environment variable name CLASSPATH and it's value d:\ , because the pck1 is stored in d:\ .

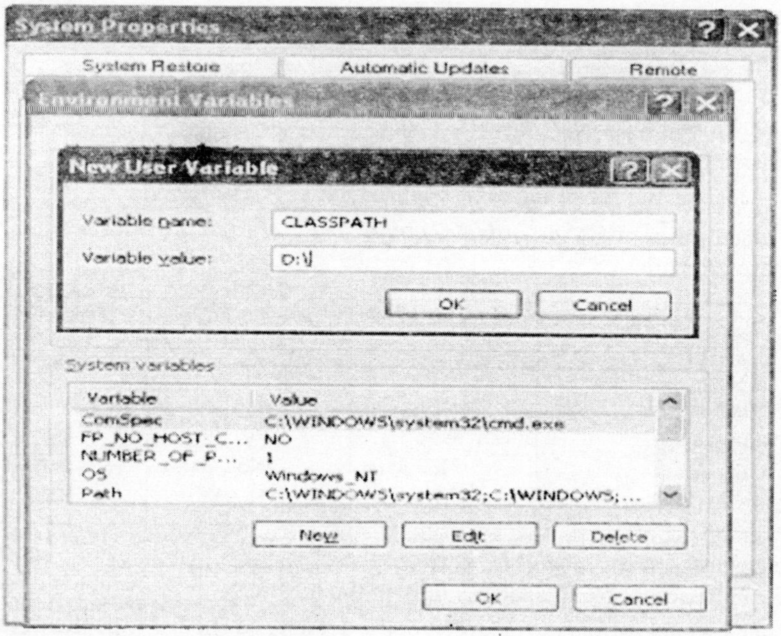

➠ Now the CLASSPATH environment variable has been set. Close all the windows related to this.

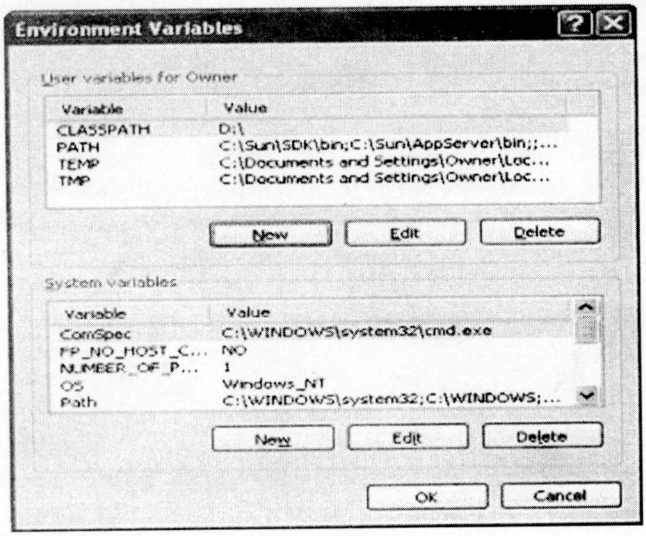

3.5. ACCESS PROTECTION IN PACKAGES

In the previous chapters, we've studied about the access control mechanism. Access modifiers define the scope of the class and its members (data and methods). For example, private members are accessible within the same class members (methods). Java provides many levels of security that provides the visibility of members (variables and methods) within the classes, subclasses, and packages.

Packages are meant for encapsulating, it works as a containers for classes and other subpackages. Class acts as containers for data and methods. There are four categories, provided by Java regarding the visibility of the class members between classes and packages:

- Subclasses in the same package
- Non-subclasses in the same package
- Subclasses in different packages
- Classes that are neither in the same package nor subclasses

The three main access modifiers ***private, public*** and ***protected*** provides a range of ways to access required by these categories.

	Private	No Modifier	Protected	Public
Same Class	Yes	Yes	Yes	Yes
Same package subclass	No	Yes	Yes	Yes
Same package non-subclass	No	Yes	Yes	Yes
Different package subclass	No	No	Yes	Yes
Different package non-subclass	No	No	No	Yes

Simply remember, private cannot be seen outside of its class, public can be access from anywhere, and protected can be accessible in subclass only in the hierarchy.

A class can have only two access modifier, one is default and another is public. If the class has default access then it can only be accessed within the same package by any other code. But if the class has public access then it can be access from any where by any other code.

Example:

```
//PCKG1_ClassOne.java
package pckg1;
public class  PCKG1_ClassOne{
int a = 1;
private int pri_a = 2;
protected int pro_a = 3;
public int pub_a = 4;
public PCKG1_ClassOne() {
System.out.println("base class constructor called");
System.out.println("a = " + a);
System.out.println("pri_a = " + pri_a);
System.out.println("pro_a = " + pro_a);
System.out.println("pub_a = " + pub_a);
}
}
```

The above file PCKG1_ClassOne belongs to package pckg1, and contains data members with all access modifiers.

//PCKG1_ClassTwo.java

```
package pckg1;
class PCKG1_ClassTwo extends PCKG1_ClassOne {
PCKG1_ClassTwo() {
System.out.println("derived class constructor called");
System.out.println("a = " + a);
// accessible in same class only
// System.out.println("pri_a = " + pri_a);
System.out.println("pro_a = " + pro_a);
System.out.println("pub_a = " + pub_a);
}
}
```

The above file PCKG1_ClassTwo belongs to package pckg1, and extends PCKG1_ClassOne, which belongs to the same package.

//PCKG1_ClassInSamePackage

```
package pckg1;
class PCKG1_ClassInSamePackage {
PCKG1_ClassInSamePackage() {
PCKG1_ClassOne co = new PCKG1_ClassOne();
System.out.println("same package class constructor called");
System.out.println("a = " + co.a);
// accessible in same class only
// System.out.println("pri_a = " + co.pri_a);
System.out.println("pro_a = " + co.pro_a);
System.out.println("pub_a = " + co.pub_a);
}
}
```

The above file PCKG1_ClassInSamePackage belongs to package pckg1, and having an instance of PCKG1_ClassOne.

//DemoPackage1.java:

```
package pckg1;
//Demo package pckg1
// Instantiati the various classes in pckg1.
public class DemoPackage1 {
public static void main(String ar[]) {
PCKG1_ClassOne ob1 = new PCKG1_ClassOne();
PCKG1_ClassTwo ob2 = new PCKG1_ClassTwo();
PCKG1_ClassInSamePackage ob3 = new PCKG1_ClassInSamePackage();
}
}
```

The above file DemoPackage1 belongs to package pckg1, and having an instance of all classes in pckg1.

//PCKG2_ClassOne.java

```
package pckg2;
class PCKG2_ClassOne extends pckg1.PCKG1_ClassOne {
PCKG2_ClassOne() {
System.out.println("derived class of other package constructor
                                                    called");

// accessible in same class or same package only
// System.out.println("a = " + a);
// accessible in same class only
// System.out.println("pri_a = " + pri_a);
System.out.println("pro_a = " + pro_a);
System.out.println("pub_a = " + pub_a);
}
}
```

The above file PCKG2_ClassOne belongs to package pckg2. extends PCKG1_ClassOne, which belongs to pckg1, and it is trying to access data members of the class PCKG1_ClassOne.

//PCKG2_ClassInOtherPackage

```
package pckg2;
class PCKG2_ClassInOtherPackage {
PCKG2_ClassInOtherPackage() {
pckg1.PCKG1_ClassOne co = new pckg1.PCKG1_ClassOne();
System.out.println("other package constructor");
// accessible in same class or same package only
// System.out.println("a = " + co.a);
// accessible in same class only
// System.out.println("pri_a = " + co.pri_a);
// accessible in same class, subclass of same or other package
// System.out.println("pro_a = " + co.pro_a);
System.out.println("pub_a = " + co.pub_a);
}
}
```

The above file PCKG2_ClassInOtherPackage belongs to package pckg2, and having an instance of PCKG1_ClassOne of package pckg1, trying to access its some data members.

//DemoPackage2.java

```
// Demo package pckg2.
// Instantiation of the various classes in pckg2.
package pckg2;
public class DemoPackage2 {
public static void main(String ar[]) {
PCKG2_ClassOne ob1 = new PCKG2_ClassOne();
PCKG2_ClassInOtherPackage ob2 = new PCKG2_ClassInOtherPackage();
}
}
```

The above file DemoPackage2 belongs to package pckg2, and having an instance of all classes of pckg2.

3.6. STANDARD PACKAGES

Java provides a number of in-built packages, contains a number of classes for a specific jobs. Here we'll give a brief introduction about java.lang package and java.util package.

java.lang

It is automatically imported into all programs. The classes and interfaces belongs to this package, are mostly in use and are essential to most of the java programs. This package is a set of fundamental classes and interface, and are widely in use.

Few java.lang classes are as following:

Boolean	Long	StackTraceElement
Byte	Math	StrickMath
Character	Number	String
Class	Object	StringBuffer
ClassLoder	System	Thread
Float	Runtime	ThreadGroup
Integer	Short	Throwable

Few java.lang interfaces are as follows:

- Cloneable
- Comparable
- Runnable
- CharSequence

java.util

This package is also very useful and most demanded. It's most powerful features is the collection. A collection is a group of object of either type. It provides the classes that are used to store the object of either type. In addition to collection, it contains other wide range of classes and interface that supports others functionalities.

Few java.util classes are listed here.

Abstractcolleciton	AbstractList	AbstractMap
ArrayList	AbstractSet	Arrays

BitSet
Dictionary
HashMap
LinkedHshSet
Properties
StringTokenizer
Currency
EventObject
Hashtable
LinkedList
Random
Timezone
Date
EventListenerProxy
LinkedHashMap
Locale
Stack
Vector

The java.util defines the following interfaces.

Collection
Comparator
Enumeration
EventListener
Iterator
List
ListIterator
Map
Map.Entry
Observer
RandomAccess
Set
SortedMap
SortedSet

3.7. ADDING CLASSES FROM A PACKAGE TO YOUR PROGRAM

The packages are used for categorization of the same type of classes and interface in a single unit. There is no core or in-built classes that belong to unnamed default package. To use any classes or interface in other class, we need to use it with their fully qualified type name. But some time, we've the need to use all or not all the classes or interface of a package then it's a tedious job to use in such a way discussed. Java supports *imports* statement to bring entire package, or certain classes into visibility. It provides flexibility to the programmer to save a lot of time just by importing the classes in his/her program, instead of rewriting them.

In a Java source file, **import** statements occur immediately following the **package** statement (if it exists) and must be before of any class definitions. This is the general form is as follows:

```
import pkg1(.pkg2).[classname|*];
```

For example

```
import java.util.Vector;
import java.io.*;
import java.awt.*;
import java.awt.event.*;
```

The star (*) increases the compilation time, if the package size is very large. A good programming methodology says that give the fully qualified type name explicitly, i.e. import only those classes or interface that you want to use instead of importing whole packages. Yet the start will not increase the overhead on run-time performance of code or size of classes.

Example:

//Student.java

```
package student;
class Student
{
private int rollno;
private String name;
private String address;
public Student(int rno, String sname, String sadd)
{
```

```
rollno = rno;
name = sname;
address = sadd;
}
public void show()
{
System.out.println("Roll No :: " + rollno);
System.out.println("Name :: " + name);
System.out.println("Address :: " + address);
}
}
//Test.java
package student;
class Test extends Student
{
protected int marksSubject1;
protected int marksSubject2;
protected int marksSubject3;
protected int marksSubject4;
public Test(int rno, String sname, String sadd,int m1, int m2, int
m3, int m4)
{
super(rno,sname,sadd);          //Call superclass constructor
marksSubject1 = m1;
marksSubject2 = m2;
marksSubject3 = m3;
marksSubject4 = m4;
}
public void show()
{
super.show();                   //Call superclass method
System.out.println("Marks of Subject1 :: " + marksSubject1);
System.out.println("Marks of Subject2 :: " + marksSubject2);
System.out.println("Marks of Subject3 :: " + marksSubject3);
System.out.println("Marks of Subject4 :: " + marksSubject4);
}
}
```

//Result.java

```
package student;
public class Result extends Test
{
private int totalMarks;
private float percentage;
private char grade;
public Result(int rno, String sname, String sadd,int m1, int m2,
int m3, int m4)
```

```
{
super(rno,sname,sadd,m1,m2,m3,m4); //Call superclass
                                                              //
constructor
totalMarks = marksSubject1 + marksSubject2 + marksSubject3 +
marksSubject4;
percentage = (totalMarks*100.00F/600.00F);
if(percentage >=50.00F)
   grade='D';
else
if(percentage >=55.00F && percentage<=60.00F)
   grade = 'C';
else
if(percentage >=61.00F && percentage<=70.00F)
   grade = 'B';
else
if(percentage >=71.00F && percentage<=75.00F)
   grade = 'A';
else
 if(percentage >=76.00F && percentage<=85.00F)
   grade = 'H';
else
   grade = 'S';
}
public void show()
{
super.show();       //Call superclass method
System.out.println("Total Marks :: " + totalMarks);
System.out.println("Percentage :: " + percentage);
System.out.println("Grade :: " + grade);
}
}
//ImportPackageDemo.java
import student.Result;
public class ImportPackageDemo
{
public static void main(String ar[])
{
Result ob = new Result(1001, "Alice","New York",135,130,132,138);
ob.show();
}
}
```

In the source file ImportPackageDemo, *import student.Result* is the first statement, it will import the class Result from the student package.

3.8. THE IDEA BEHIND EXCEPTION

Any complete software is not always fully bug and error free. Some time, these errors are generated logically, so by any side effect of any other module of the application. These errors cannot be trapped at the testing time, because these are generated at run time. Some errors are logical and not shown at compile time of the code. Some errors are left at the time of testing.

Exception provides a mechanism to detect the origination of errors, cause of error and forward the flow in other direction without halting the system. The control flow will be redirected to handle it instead of shutting down the system.

Whenever any error is generated the corresponding exception will be fired. An exception is thrown when an unexpected abnormal error is occurred. If an exception is uncaught then the current thread of execution will be terminated, but before termination, ThreadGroup is given an opportunity to handle the exception. If there is no solution against the fired exception, then only information and details about the exception can be shown.

Whenever any exception is thrown then runtime environment create an instance of the corresponding exception then send to the corresponding catch block.

3.9. TYPES OF EXCEPTION

All exceptions are subclasses of the in-built class *Throwable*. So that Throwable is the top most classes in the exception class hierarchy. There are two subclasses of Throwable, one is *Exception* class and second is *Error* class.

The Exception class is used for the exceptional conditions that are handled by the programmer and they should catch it. RuntimeException is the important subclass of the Exception class. These types of exception fired are automatically defined for the programs.

The Error is another branch of Throwable; these are the exceptions that are not expected to be caught under normal conditions of your programs. These are the errors occurred at run time and handled by the run time environment. Stack overflow is an example.

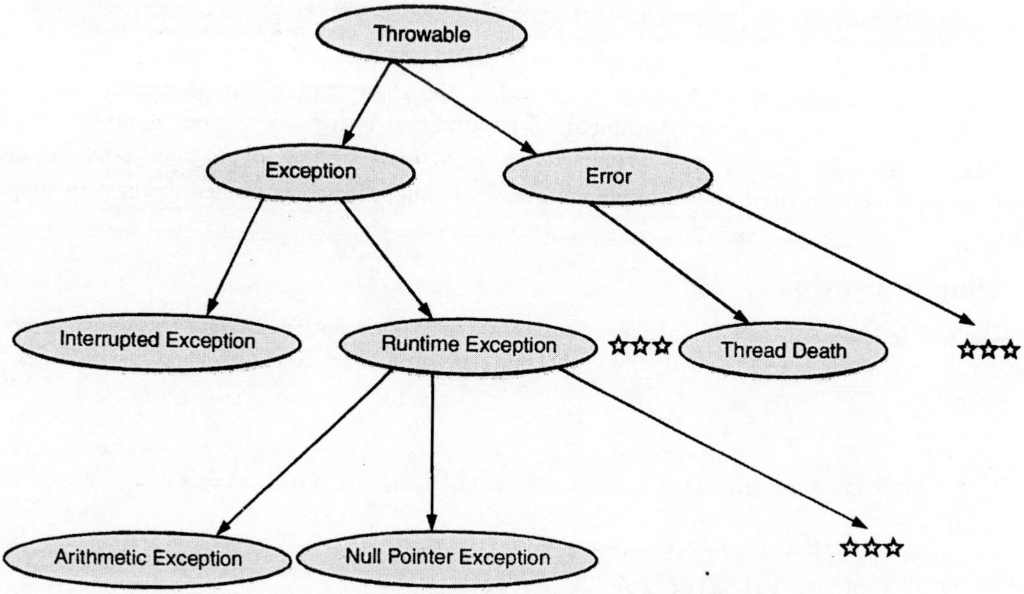

Fig. 3.2. Hierarchy of Throwable class

3.10. DEALING WITH EXCEPTION

Before taking the tour on how to handle the exceptions, take a look on a program without using the exception handling.

Code: ***Code Without Exception***

```
class DemoProg
{
public static void main(String ar[])
{
int a,b,c=0;
a=9;
b=0;
c = a / b;
System.out.println("The value of c :: " + c);
}
}
```

In the above given code there is not any syntax error. But a logical error is there, that the value of b is 0 (zero), and no number can be divided by zero. This error can not be checked at the time of compilation.

But as we run it, an exception will be fired, and the system generated message is given in window given below.

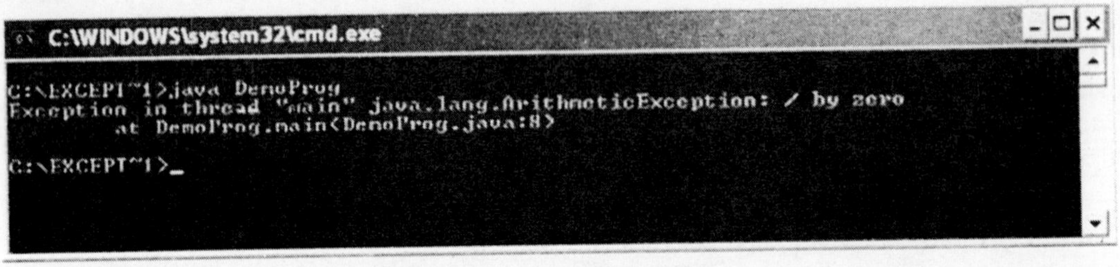

The message is showing the message generated by the JVM, stating that An exception is fired in "main" method of type "java.lang.ArithmeticException" that is for any number is divided by zero, showing in message as " / by zero" and the line number "8" in method "main" of program "DemoProg.java".

So as the exception fired, the run time environment caught it and generated the system supported message.

Handling Exception

Now we study how to handle the exception. To handle the exception we have to sandwich the code between the try block braces, and should catch the exception. The exception handling block is given below for it:

```
try {
// the of statements or code have to monitor for errors.
}
catch (ExceptionType1 exObj) {
// exception handling for ExceptionType1
// the set of action have to be taken when exception has been caught
}
```

```
catch (ExceptionType2 exObj) {
// exception handling for ExceptionType2
// the set of action have to be taken when exception has been caught
}
// ...
finally {
// the set of action have to be executed before end of try block
   either exception fired or not
}
```

Whatever the code is written in the try block under observation and monitor, as any exception will be fired, will be thrown, that will be handled by the corresponding related catch block. The catch block will receive the corresponding Exception object. It's the block of code that will execute for the fired exception or the type of exception object that has been received by it. So, the programmer must know how much exception can be thrown and what type of, accordingly the number of corresponding catch statements have to be added with the try block.

Now we see the same example with exception handling.

Example: Code With Exception Handling Show Own Message

```
class DemoProg
{
public static void main(String ar[])
{
int a,b,c=0;
a=9;
b=0;
try
{
c = a / b;
}catch(ArithmeticException ae)
{
System.out.println("Divided by zero.");
}
System.out.println("The value of c :: " + c);
}
}
```

Now, in the immediately above given code the expression"c=a/b" is in the try block, and ArithemeticException is caught. Now see what will be the output of the code if we run it.

A user given message Divided by zero." is showing, that was given in catch block. As the exception was caught the instruction "System.out.println("Divided by zero.")" executed in the corresponding catch block.

But the major difference in the output is that in the last code output the value of c did not print, as the exception was caught by runtime environment, the program was terminated, but here the point to be noted is that exception was fired, and handled here, so that after executing the catch block of the corresponding statement the rest of the code is also executed, it print the value of c.

Displaying Exception Description

Lets have a look on any example that will show the complete detail of the fired exception by printing the value of fired exception object in the catch block.

Example: Code With Exception Handling Show Object Message

```
class DemoProg
{
public static void main(String ar[])
{
int a,b,c=0;
a=9;
b=0;
try
{
c = a / b;
}catch(ArithmeticException ae)
{
System.out.println("System generated message :: \n " + ae);
}
System.out.println("The value of c :: " + c);
}
}
```

If we run the above code the following message will be shown:

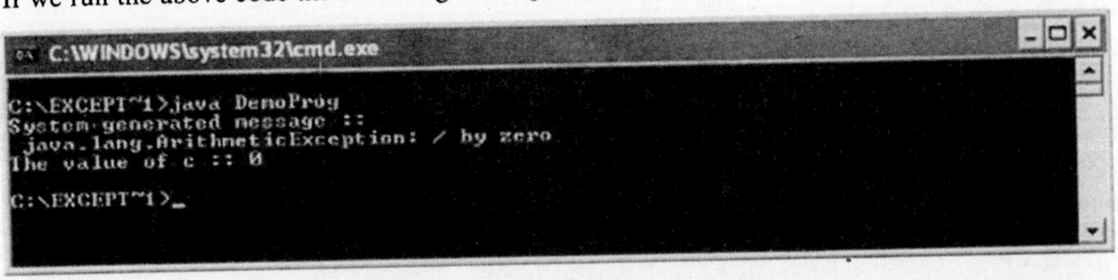

Here the run time system generates only the Exception class name with its attached message. It will not show in which method of which class at which line number it was fired.

Example: Code With Exception Handling Show Complete Description Of Exception

```
class DemoProg
{
public static void main(String ar[])
```

```
{
int a,b,c=0;
a=9;
b=0;
try
{
c = a / b;
}catch(ArithmeticException ae)
{
ae.printStackTrace();
}
System.out.println("The value of c :: " + c);
}
}
```

There is a method of Throwable class printStackTrace() that gives the complete stack trace, i.e. complete detail about the fired exception, the related message, exception class name, the method order in which it was called and handled, the class and the method name in which it fired, and with the line number.

If we run the immediate above example then the output will given in the above window.

Use of Exception Object

The Exception class is superclass of all exception category classes. So, if exception is unknown, then in catch block the Exception class object can be used to receive either type of thrown exception object.

Example:

```
class DemoProg
{
public static void main(String ar[])
{
int a,b,c=0;
a=9;
b=0;
try
{
c = a / b;
}catch(Exception e)
{
e.printStackTrace();
}
System.out.println("The value of c :: " + c);
}
}
```

When you compile the immediate above code same error message will be generated as the last code example.

Multiple Catch

There may be possibility that more than one exception will be fired in single try block of code. The programmer must know that which type of exceptions can be thrown, and for each and every type of exception, there must be a catch block.

Example: Multiple catch with example

```
class DemoProg
{
public static void main(String ar[])
{
int a,b,c=0;
a=9;
b=ar.length-1;
try
{
System.out.println("Passed parameters\n");
System.out.println(ar[0]);
c = a / b;
}catch(ArrayIndexOutOfBoundsException ae)
{
System.out.println("Array index outof bounds Exception");
}
catch(ArithmeticException ae)
{
System.out.println("Divided by zero");
}
System.out.println("The value of c :: " + c);
}
}
```

In the above code two exception can be thrown, one is ArrayIndexOutOfBoundsExcption when there is no argument is passed and we try to access the ar[0] value. And another is ArithmeticException when the value of b is zero, with the same condition when no argument is passed. So for each type of exception there are catch block, and each is handling one type of exception.

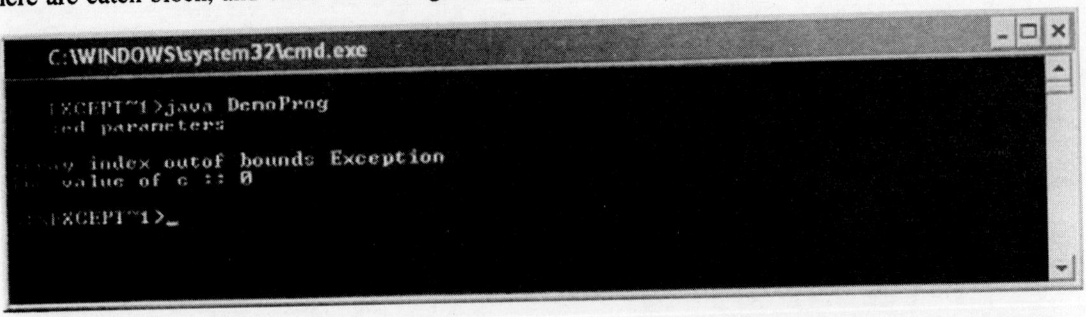

If no argument is passed to the code then ArrayIndexOutOfBoundsException will be thrown and the correspond catch block is executed.

If argument is passed then the value of b is zero (0), the ArithmeticException will be thrown, and corresponding catch block is executed.

Nested Try Block

The try statement can be nested, i.e. a try inside the block of another try. The try are stored in a stack as the order they are entering in its own block. At any depth, any inner try block doesn't has the matched catch handler of thrown exception, then that catch handler will search in the parent try-catch block. This process of searching of appropriate catch handler continues until one of the matched catch handler is found. If no catch statements found, then the Java run time environment will handle the exception.

Example : Each try has it's own catch handler

```
class DemoProg
{
public static void main(String ar[])
{
int a,b,c=0;
a=9;
b=ar.length-1;
try
{
System.out.println("Passed parameters\n");
System.out.println(ar[0]);
try
{
c = a / b;
}catch(ArithmeticException ae)
{
System.out.println("Divided by zero");
}
System.out.println("The value of c :: " + c);
}catch(ArrayIndexOutOfBoundsException ae)
{
System.out.println("Array index outof bounds Exception");
}
}
}
```

Example: Here catch handler will be search in parent try block

```
class DemoProg
{
public static void main(String ar[])
{
int a,b,c=0, x=0;
a=9;
b=ar.length-1;
try
{
c = a / b;
System.out.println("value of c :: " + c);
try
{
System.out.println("Passed parameters\n");
System.out.println(ar[0]);
x = a / (x-x); //Arithmetic exception will be thrown, but not in inner
    try
}
catch(ArrayIndexOutOfBoundsException ae)
{
System.out.println("Array index outof bounds Exception");
}
}catch(ArithmeticException ae)
{
System.out.println("Divided by zero");
}
}
}
```

Output:

```
C:\WINDOWS\system32\cmd.exe
C:\EXCEPT~1>javac DemoProg.java

C:\EXCEPT~1>java DemoProg
value of c :: -9
Passed parameters

Array index outof bounds Exception

C:\EXCEPT~1>java DemoProg abc
Divided by zero.

C:\EXCEPT~1>
```

Use of throw

As any exception is occurred, Java run time systems creates an instance of the corresponding exception and throw implicitly. However, sometime there is a need of throw an exception explicitly, then throw statement is used.

The general statement of **throw** is shown here:

```
throw ThrowableInstance;
```

The *ThrowableInstance* must be an object of either type. It can be of Throwable or any subclass of Throwable. Any non-subclass of Throwable cannot be used. There are two ways to obtain the Throwable object, one is as a parameter into a catch clause, or by creating one instance by new operator.

The subsequent statements after will never execute after the throw statement if the exception is thrown. The searching of the corresponding catch handler will remain same as discussed in nest try block.

Example:

```
class DemoProg
{
public static void main(String ar[])
{
int a=0;
a=ar.length;
try
{
if (a<2 || a>2)
throw new RuntimeException("Given parameters are less then required");
System.out.println("Sum of given number :: " +
   (Integer.parseInt(ar[0])+Integer.parseInt(ar[1])));}catch(Exception e)
{
System.out.println(e);
}
}
}
```

If we run this code without passing any argument or more then two argument from the command line, then the output is

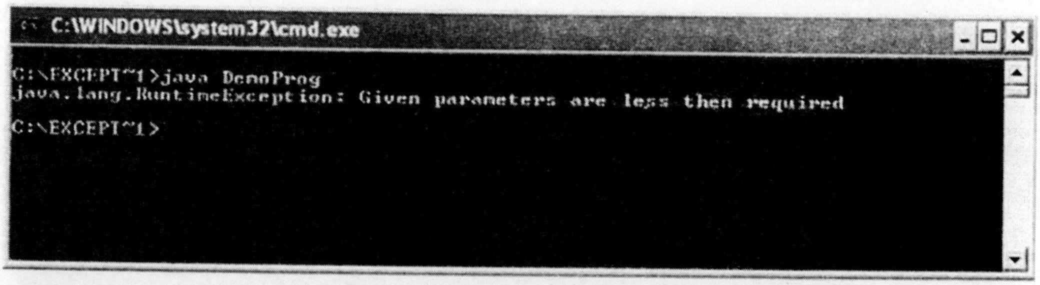

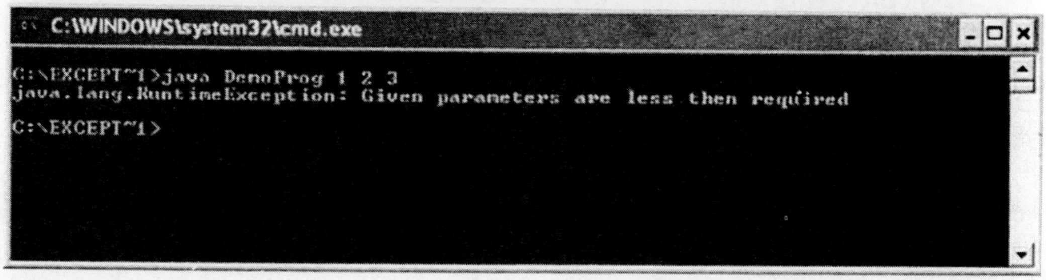

If we pass exact two argument as numbers then the output is

Use of throws

Some time it happens that the method can be a cause of any type of exception but it doesn't handle it. It must specify that the caller method must handle that exception. Do it by including the throws clause in the method declaration or definition. A throws clause contains the list of exceptions that may be thrown. All exceptions that method can throw must be declared in throws list. This is necessary for all exceptions, except those of type Error or RuntimeException, or any of their subclasses. Unchecked exception must be in throws list, if it is thrown by the method. If they are not included then the compile time error will be generated.

This is the general syntax form of a method declaration that includes a **throws** clause:

```
type method_name(parameter_list) throws exception_list
{
// body of method
}
```

Here, *exception_list* is a comma-separated list of all the exceptions that a can be thrown by method.

Example:

```
class DemoProg
{
public static void main(String ar[])
{
int a=0;
a=ar.length;
try
{
check(a);
System.out.println("Sum of given number ::
    "+(Integer.parseInt(ar[0])+Integer.parseInt(ar[1])));
}catch(Exception e)
{
System.out.println(e);
}
}
public static void check(int x) throws Exception
{
if(x<2 || x>2)
```

```
throw new RuntimeException("Given parameters are less then required");
}
}
```

Output:

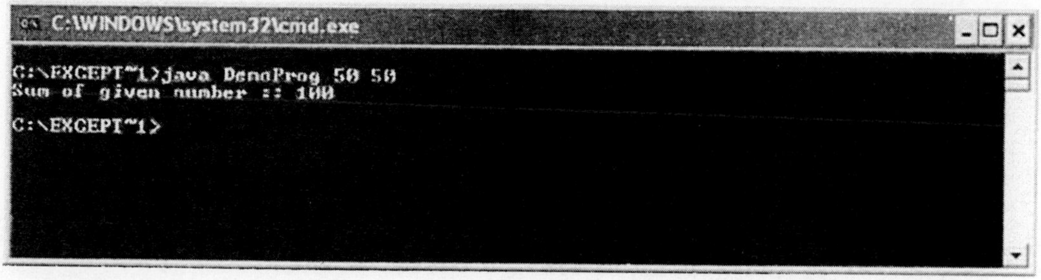

Use of finally

When exceptions are thrown, then execution of the code is not linear, there is a interruption. The subsequent statements will never execute from where the exception will be fired in the block. But if we want that, a set of statement must executes either exception throw or not, then the set of statement have to be written or encapsulate in the finally block.

For example:

```
{
//1 open a file
//2 read file
//3 write file
//4 write file
//5 read file
//6 write file
//7 close file
}
```

In the above set of statement, the statement is responsible to close the file. But if at the 6th statement if any kind of exception throw related to file, I/O, or network (any), then the 7th statement will never executes. The file remains open. To solve this problem, finally block was introduced. The 7th statement must be written in finally block, and the it will execute always, either exception fire or not.

Example:

```
class DemoProgFinally
{
static void method1() {
try {
System.out.println("inside method1");
throw new RuntimeException("From method1");
} finally {
System.out.println("method1 finally executed");
}
}
```

```
static void method2() {
try {
System.out.println("inside method2");
return;
} finally {
System.out.println("method2 finally executed");
}
}
static void method3() {
try {
System.out.println("inside method3");
} finally {
System.out.println("method3 finally executed");
}
}
public static void main(String args[]) {
try {
method1();
} catch (Exception e) {
System.out.println("Exception caught in main");
}
method2();
method3();
}
}
```

Output:

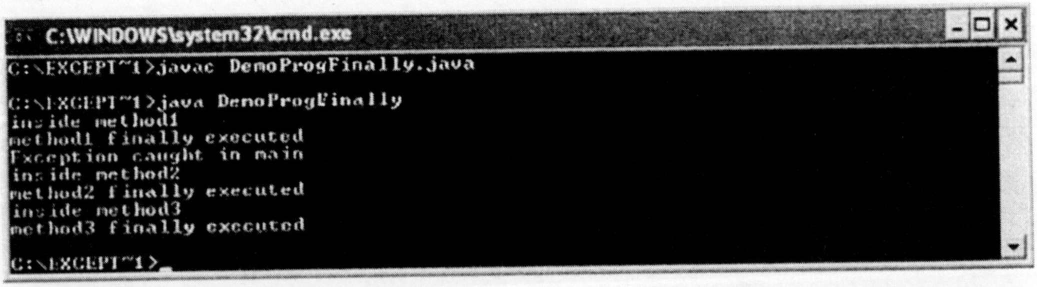

3.11. EXCEPTION OBJECTS

Exception class is the superclass of all subclasses of Exception category. As we know, the superclass reference variable can have the object of its subclasses. By this concept, if the exception to be thrown is unknown or we want to write on one catch handler for all type of exception, then all throws exception can be received in Exception object. Examples are given below.

Example:

```
class DemoProg
{
public static void main(String ar[])
```

```
{
int a,b,c=0;
a=9;
b=ar.length-1;
try
{
Class.forName("sun.jdbc.odbc.JdbcOdbcDriver");
System.out.println("Passed parameters\n");
System.out.println(ar[0]);
c = a / b;
}catch(ArrayIndexOutOfBoundsException ae)
{
System.out.println("Array index outof bounds Exception");
}catch(ArithmeticException ae)
{
System.out.println("Divided by zero");
}catch(Exception e)
{
System.out.println(e);
}
System.out.println("The value of c :: " + c);
}
}
```

The above code can fire ClassNotFoundException if "sun.jdbc.odbc.JdbcOdbcDriver" class not found, and there is not corresponding match catch handler. It will be handle or caught by the catch handler that receiving in throws exception in Exception object, that is the last catch handler.

Output:

Example: One catch handler for all exceptions

```
class DemoProg
{
public static void main(String ar[])
{
int a,b,c=0;
a=9;
b=ar.length-1;
try
```

```
{
Class.forName("sun.jdbc.odbc.JdbcOdbcDriver");
System.out.println("Passed parameters\n");
System.out.println(ar[0]);
c = a / b;
}catch(Exception e)
{
System.out.println(e);
}
System.out.println("The value of c :: " + c);
}
}
```

But most important thing is that, the catch handler that is using the Exception class object for all type of exceptions, must be in the last of series of multiple catch handler, otherwise an error message will be generated by the compiler at the compilation time of the program. Related example is given below.

Example:

```
class DemoProg
{
public static void main(String ar[])
{
int a,b,c=0;
a=9;
b=ar.length-1;
try
{
Class.forName("sun.jdbc.odbc.JdbcOdbcDriver");
System.out.println("Passed parameters\n");
System.out.println(ar[0]);
c = a / b;
}catch(Exception e)
{
System.out.println(e);
}catch(ArrayIndexOutOfBoundsException ae)
{
System.out.println("Array index outof bounds Exception");
}catch(ArithmeticException ae)
{
System.out.println("Divided by zero");
}
System.out.println("The value of c :: " + c);
}
}
```

Output:

```
C:\WINDOWS\system32\cmd.exe
C:\EXCEPT~1>javac DemoProg.java
DemoProg.java:17: exception java:lang.ArrayIndexOutOfBoundsException has already
 been caught
}catch(ArrayIndexOutOfBoundsException ac)
 ^
DemoProg.java:20: exception java:lang.ArithmeticException has already been caugh
t
}catch(ArithmeticException ac)
 ^
2 errors

C:\EXCEPT~1>
```

3.12. USER DEFINED EXCEPTION

Java's in-built exceptions can handle most common errors. But they are not able to handle the application specific situations. These situations are not matched with predefined conditions by Java. These are user defined condition for her/his application or module. Java provides the easiest way to define its own exception just by extends Exception class, and system treat your class as exception class. Exception class does not define any method of its own.

Example:

```
class UserException extends Exception
{
String msg;
public UserException(String msg)
{
this.msg = msg;
}
public String toString()
{
return "Error occured : " + msg;
}
}
public class DemoException
{
public static void main(String ar[])
{
int a = ar.length;
try{
if(a == 0)
throw new UserException("No argument received. Please");
else if (a >2 )
throw new UserException("Extra arguments are passed.");
else if(a == 1)
throw new UserException("Few argument missing.");
```

```
else if (a==2)
System.out.println("Sum of given value is :: " +
                    (Integer.parseInt(ar[0])+Integer.parseInt(ar[1])));
}catch(UserException ue)
{
System.out.println(ue);
}
}
}
```

Output:

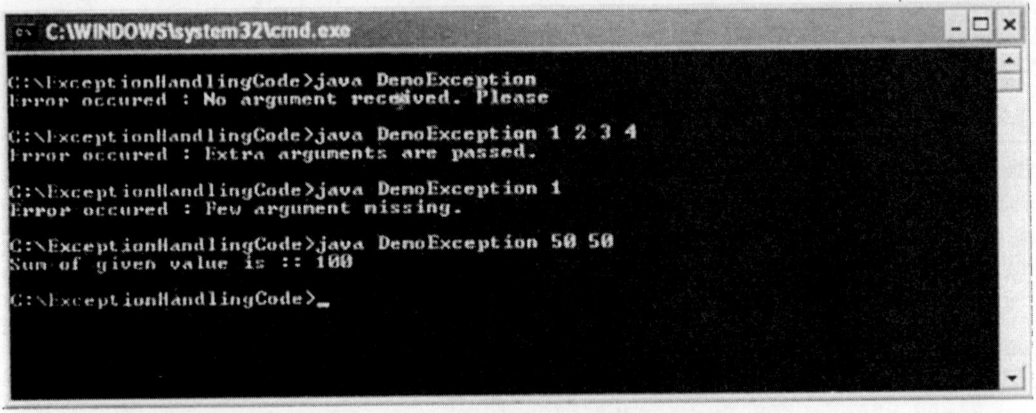

The exception will be thrown on the following three conditions:

- When there is no argument
- When number of argument are more then two
- When there is only one argument

As the program received two arguments then it will print the sum of passed value.

3.13. CHECKED AND UNCHECKED EXCEPTIONS

The exception must have to check or catch and not handled by system itself are called *Checked* exceptions. Those exceptions are handled by system itself, if not handled by the programmer are called *Unchecked* exception.

List of Unchecked Exception Defined in Exception meaning

Exception	Meaning
Class Not Found Exception	Throws when class not found.
Illegal Access Exception	Throws when an access to a resource (class) is denied.
Instantiation Exception	Throws when an attempt to create an object of an abstract class or interface.
Interrupted Exception	Throws when one thread has been interrupted by another thread.
No Such Field Exception	Throws when a requested field of any resource (class, interface) does not exist.
NoSuchMethodException	Throws when a requested method of class does not exist.

List of Unchecked Exception Defined in RuntimeException meaning

Exception	Meaning
NullPointerException	Throws when invalid use of a null reference is used
NumberFormatException	Throws when invalid conversion of a string to a numeric format.
SecurityException	Throws when there is an attempt to violate security.
StringIndexOutOfBounds	Throws when an attempt to access index outside the bounds of a string.
UnsupportedOperationException	Throws when an unsupported operation was encountered on any type of data member.

REVIEW QUESTIONS

Q. 1. Write a program to create a user defined Exception called "No match Exception" that throws an arbitrary message when a string is not equal to "INDIA". **[MDU, Dec. 2007]**

Ans.

```
import java.io.*;
class MyException extends Exception
{
  String msg;
  MyException(String msg)
  {
    this.msg=msg;
  }
  public String toString()
  {
    return msg;
  }
}
public class Exception_Prog
{
  public static void main(String args[]) throws IOException
  {
    String str;
    BufferedReader br = new BufferedReader(new InputStreamReader
   (System.in));
    System.out.println("Enter the string");
   try
   {
     str=br.readLine();
     if(!str.equals("INDIA"))
     throw new MyException("No Match Exception");
     System.out.println("Entered data is :" +str);
   }
  catch(MyException me)
   {
```

```
            System.out.println(me);
        }
    }
}
```

Q. 2. What exceptional errors during the execution of a program? Discuss the usage of "finally" block? **[MDU, May 2008]**

Ans. An exceptions are conditions that occur because of bad input etc. e.g. FileNotFoundException will be thrown if the specified file does not exist. Or a NullPointerException will take place if you try using a null reference. In most of the cases it is possible to recover from an exception (probably by giving user a feedback for entering proper values etc.).

Q. 3. Explain Inheritance & Exception handling in JAVA with examples. **[MDU, May 2012]**

Ans. Refer Art. 2.1 and Art. 3.7.

Q. 4. What are Checked and UnChecked Exception?

Ans. A checked exception is some subclass of Exception (or Exception itself), excluding class RuntimeException and its subclasses. Making an exception checked forces client programmers to deal with the possibility that the exception will be thrown. *e.g.,* IOException thrown by java.io.FileInputStream's read() method. Unchecked exceptions are RuntimeException and any of its subclasses. Class Error and its subclasses also are unchecked. With an unchecked exception, however, the compiler doesn't force client programmers either to catch the exception or declare it in a throws clause. In fact, client programmers may not even know that the exception could be thrown. eg, StringIndexOutOfBoundsException thrown by String's charAt() method· Checked exceptions must be caught at compile time. Runtime exceptions do not need to be. Errors often cannot be.

Q. 5. What is a user-defined exception?

Ans. There are many exceptions defined by Java which are used to track the run-time exceptions and act accordingly. Users can also define exceptions that can be thrown in the same way as Java exceptions.

Q. 6. When are exception objects created?

Ans. The exception objects are created when an unexpected condition occurs. When we use throw to send a new exception, or when Java library uses throw to send a new exception, or when Java runtime signals a Runtime Exception; then the exception objects are created.

❑❑❑

4
CHAPTER

Multithreading in Java

A program under execution is called a process. Process is complete, acquired set of resources to run until it completes, and has its own memory area. A system that has merely single execution core, then only one process can execute at a time. Yet now a days, more than one execution core are in the system. But in case of single core only one process can run at a time and it acquires all the required resources.

INSIDE THIS CHAPTER

Now a day's every user wants that he/she can run more than one application on his/her system. That can be possible with the help of implementing the concept of *multitasking*. Multitasking can be of two types: *process-based* and *thread-based.* This can be done in two conditions: one is that a computer has more than one core processing units (called multiprocessor), and second is timesharing, assign a fixed time slice to each process.

The second approach of multitasking is thread-based. But before we go in-depth of this approach, first understand what the *thread* is. A thread is the smallest unit of the single program, which is dedicated to a specific job. Thread is a lightweight process. Few resources are required to create a thread rather than a process. Each thread has a separate path of execution. Multitasking with thread-based model provides an environment in which single program (application) can carry out more than one tasks simultaneously. For example word processor; simultaneously it performs the formatting on the text as you type, it checks the spelling and grammar as you type, as you entered a character from keyboard the spell check thread pause and allows the thread to read the character and print on screen, as the reading completed of a word spell check thread is notified by the reading thread that it has completed its job and move to waiting state, now spell check will be resumed. Now a days almost all operating system supports multithreading.

Multitasking is implemented by multithreading, or we can say the multithreading is an extended version of multitasking using threads. Thus, a specialized form of multitasking is called *multithreading*. Multithreading requires less overhead rather than multitasking process. It reduces the overheads of inter-process communication because threads are lightweight process. Context switching between the thread is also less costly. Because threads share the same address space of the process it belongs.

Multithreading facilitates you to write an efficient program that makes you the optimum utilization of resource including CPU.

4.1. THE JAVA THREAD MODEL

The Java language and its run-time system was designed keeping in mind about multithreading. The run-time system depend upon multithreading. Java provides asynchronous thread environment, this helps to increase the utilization of CPU.

Multithreading is best in all cases in contrast with single-thread model. Single-thread system uses an approach of event loop with polling. According to this approach a single thread in the system runs in an infinite loop. Polling the mechanism, that selects a single event from the event queue to choose what to do next. As the event is selected, then event loop forwards the control to the corresponding required event handler. Nothing else can be happened, until the event handler returns. Because of this CPU time is wasted. Here, only one part of the complete program is dominating the whole system, and preventing the system to execute or start any other process. In single-thread model one thread blocks all other threads until its execution completes. On other waiting or idle thread can start and acquire the resource which is not in use by the current thread. This causes the wastage of resources.

Java's multithreading provides benefit in this area by eliminating the loop and polling mechanism, one thread can be paused without stopping the other parts of the program. If any thread is paused or blocked, still other threads continue to run.

As the process has several states, similarly a thread exists in several states. A thread can be in the following states:

- *Ready to run (New):* First time as soon as it gets CPU time.
- *Running:* Under execution.
- *Suspended:* Temporarily not active or under execution.
- *Blocked:* Waiting for resources.
- *Resumed:* Suspended thread resumed, and start from where it left off.
- *Terminated:* Halts the execution immediately and never resumes.

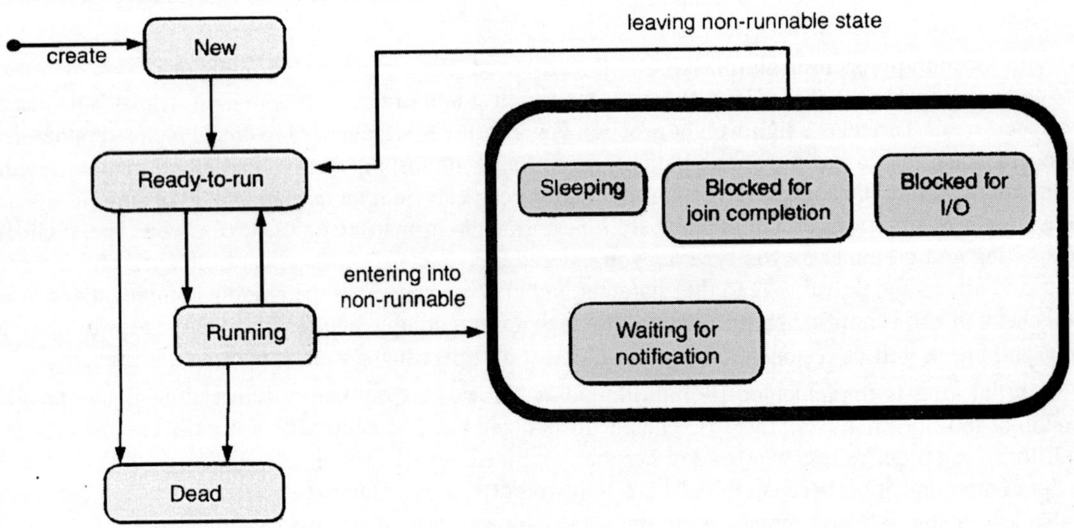

Fig. 4.1

Java thread model can be defined in the following three sections:

Thread Priorities

Each thread has its own priority in Java. Thread priority is an absolute integer value. Thread priority decides only when a thread switches from one running thread to next, called *context switching*. Priority does increase the running time of the thread or gives faster execution.

Synchronization

Java supports an asynchronous multithreading, any number of thread can run simultaneously without disturbing other to access individual resources at different instant of time or shareable resources. But some time it may be possible that shareable resources are used by at least two threads or more than two threads, one has to write at the same time, or one has to write and other thread is in the middle of reading it. For such type of situations and circumstances Java implements synchronization model called *monitor*. The monitor was first defined by C.A.R. Hoare. You can consider the monitor as a box, in which only one thread can reside. As a thread enter in monitor, all other threads have to wait until that thread exits from the monitor. In such a way, a monitor protects the shareable resources used by it being manipulated by other waiting threads at the same instant of time. Java provides a simple methodology to implement synchronization.

Messaging

A program is a collection of more than one thread. Threads can communicate with each other. Java supports messaging between the threads with lost-cost. It provides methods to all objects for inter-thread communication. As a thread exits from synchronization state, it notifies all the waiting threads.

4.2. UNDERSTANDING THREADS

In Java multithreading is built upon two important entities given by the language: one is *Thread* class with its method, and second is *Runnable* interface. The *Runnable* interface defines only single method *run*(), that contain the code executed by the thread in the classes it will be implemented. A *Runnable* object is passed to the Thread constructor. Another entity is Thread class, which is the subclass of the *Runnable* interface. It implements *Runnable*, though its *run*() method do nothing. As such you can not refer the insubstantial state of any running thread, you have to deal with the proxy of it. *Thread* encapsulates the thread of execution, so that you must have an instance of *Thread*. For this purpose either extends *Thread* class or implements *Runnable* interface. The Thread class defines several method, few of them are discussed below:

Method Name	Functionality
getName	Returns the name of thread.
getPriority	Returns a thread's priority.
isAlive	Use to determine if a thread is still running.
join	Force to parent thread to wait for a child thread to terminate.
run	An entry point for the thread, and contain the code that executed when thread run.
sleep	Suspend a thread for a giver certain period of time.
start	Start a thread by calling its run method.

4.3. THE MAIN THREAD

As any Java program comes under execution, first thread starts immediately, called the *main thread*. There are two importance's of main thread as follows:

- It is the parent of all the threads of the program and all other "child" threads will be spawned form it.
- It must be the last thread to finish the execution of the program.

As your program start, the main thread is created automatically, under the control of Thread object. But, if you want the reference of the main thread then there is a static method *curentThread()* of Thread class, which returns the reference of thread in which it is called. Its general form is shown here:

```
static Thread currentThread( )
```

Example:

```
class MainThreadDemo {
    public static void main (String ar[] ) {
        Thread t = Thread.currentThread();
        System.out.println("Current thread: " + t);
        t.setName("Main Thread Demo");
        System.out.println("New name of main Thread: " + t);
try
{
for(int n=1;n<=10;i++)
{
System.out.print(i + '\t');
Thread.sleep(500);
}
}catch(InterruptedException ie)
{
System.out.println(ie);
}
    }
}
```

Output:

```
Current thread: Thread[main,5,main]
New name of main Thread: Thread[Main Thread Demo,5,main]
1  2  3  4      5      6      7      8      9      10
```

Output is showing three line statement. In first statement within square brackets "[]" first value is the name of thread "main", second is priority "5" and the last is the name of threadgroup "main", the current (main) thread belongs to. Similarly in the second printed statement first value is name of the thread after modification "Main Thread Demo", second is priority "5" and the last is thread group name "main". The default created group by java is "main". In the last it will print 1 to 10 and each digit after 0.5 second.

In the above example *setName()* is used to change the name of the thread. You can get the name of the thread by calling *getName()*. Both of these methods are of *Thread* class. There prototype is given below:

```
final void setName(String threadName)
final String getName( )
```

To give a pause to the thread *sleep()* method is used in the given example. The argument passed to *sleep()* method are in milliseconds. There are two versions of *sleep()* methods of *Thread* class as following:

```
static void sleep(long time_in_milliseconds) throws InterruptedException
static void sleep(long time_in_milliseconds, int additional_nanoseconds)
            throws InterruptedException
```

4.4. CREATING A THREAD

In Java, a user defined thread can be created to implement the Runnable interface. The Runnable interface has only one method run() that you must have to implement in your class, that you want to create as thread. The run() is the method in which that code is written, which is counted as thread task.

The content of the *run()* method will be counted as thread portion, but the method written outside the *run()* counts as part of the main thread. Both the new thread and main thread can run concurrently. The thread terminates with the termination of run method. Control is returned to the caller method of the run method, or can say from where it starts. The *run()* method should not be called directly, the *start()* method of thread class is called, and that is responsible to call *run()* for the corresponding thread.

Another way to create a thread is to inherit the Thread class. Thread class has a number of methods. Few of them are static and few are non-static. As per the need you can override non-static methods. But *run()* method must override, because it defines new thread functionality.

4.5. EXTENDING THREAD

A new thread is created by creating a new class that extends Thread, and then create an instance of that class. The extending class must override the method *run()*, which is the entry point of the new thread (user defined). To begin the execution of the new thread, it must call *start()* method. Inside the *run()* method, you'll write the code, which is considered as the functionality of the thread. The *run()* method can call other method of the same class or other class using their objects. It is an entry point of the thread. As the *run()* method returns (exit or terminate) the thread ends.

Example:

```
class NewThreadDemo extends Thread
{

public NewThreadDemo()
{
super("User Define Thread");
System.out.println("New userdefined child thread created " + this);
start(); // Start the thread
}

public void run()
{
try {
for(int i = 1; i <=10; i++) {
System.out.println("UserDefined Child Thread: " + i);
Thread.sleep(500);
}
} catch (InterruptedException e) {
System.out.println("UseDefine Child interrupted.");
}
System.out.println("Exiting UserDeinfed child thread.");
}
}

class ThreadDemo
{
```

```
public static void main(String args[])
{
new NewThreadDemo(); // create a new thread
try {
for(int i = 1; i <10; i++)
{
System.out.println("Main Thread: " + i);
Thread.sleep(500);
}
} catch (InterruptedException e) {
System.out.println("Main thread interrupted.");
}
System.out.println("Main thread exiting.");
}
}
```

Output:

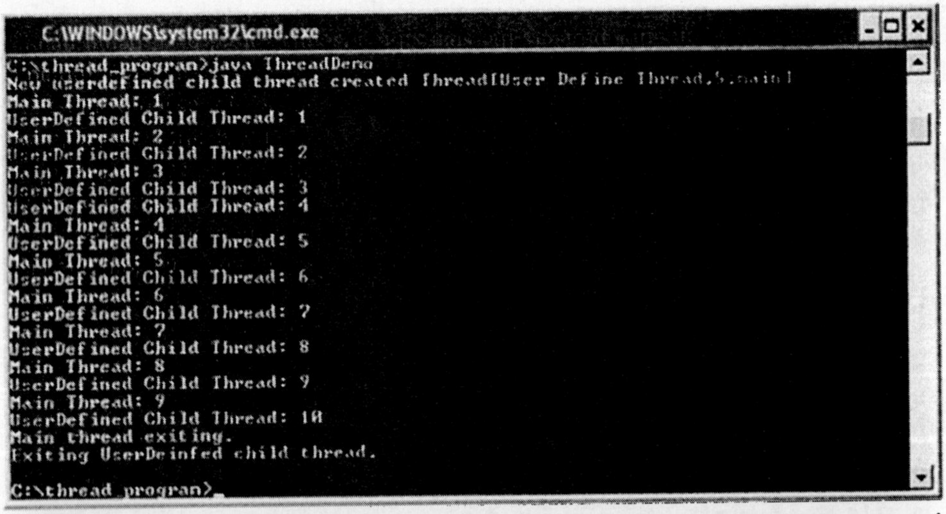

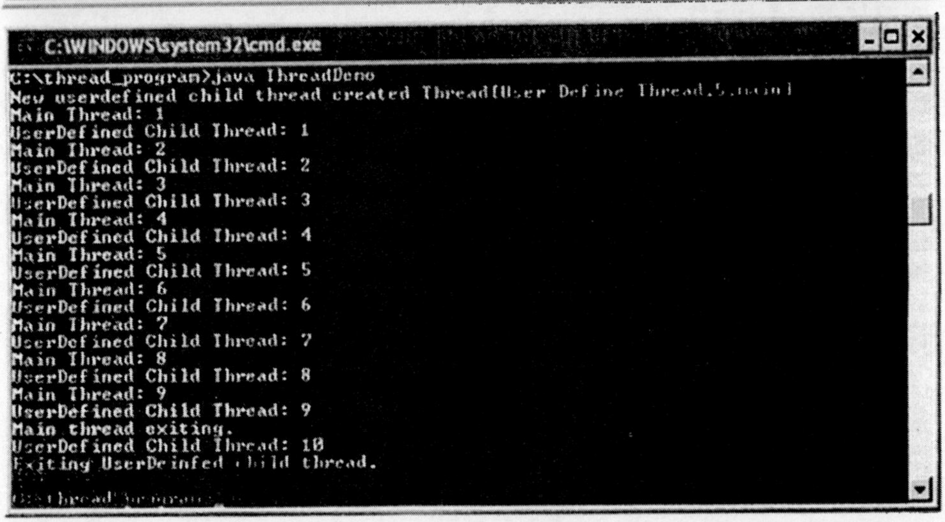

There are two output of the same thread code. Because it is not always sure that a thread executes in the same manner. The output can be same or vary every time. The *super()* inside the NewThreadDemo constructor invokes the following form of Thread constructor:

```
public Thread(String thread_Name)
```

4.6. IMPLEMENTING RUNNABLE

By implementing the Runnable interface the thread creation is the easiest way. You can construct a thread on any type of object that implements Runnable. To implement Runnable interface, a new class need only to implement run() method. When you implement the Runnable interface then create an instance of Thread from within that class. The Thread class has several constructor.

```
public Thread()
public Thread(String thread_Name)
public Thread(Runnable threadObj)
public Thread(Runnable threadObj, String thread_Name)
public Thread(ThreadGroup groupObj, Runnable threadObj)
public Thread(ThreadGrcup groupObj, Runnable threadObj, String name)
```

After the new thread is created, the *start( )* method has to be used to start running, otherwise it will not be started. In essence, *start( )* executes a call to *run( )*.

Example:

```
class NewThreadDemo implements Runnable
{
Thread t;
NewThreadDemo()
{
t=new Thread(this, "User Define Thread");    //A thread object is
created
System.out.println("New userdefined child thread created " + this);
t.start(); // Start the thread
}

public void run()
{
try
{
for(int i = 1; i <=10; i++)
{
System.out.println("UserDefined Child Thread: " + i);
Thread.sleep(500);
}
}
catch (InterruptedException e)
{
System.out.println("UseDefine Child interrupted.");
}
```

```
System.out.println("Exiting UserDeinfed child thread.");
}
}

class ThreadDemo
{
public static void main(String args[])
{
new NewThreadDemo(); // create a new thread
try
{
for(int i = 1; i <10; i++)
{
System.out.println("Main Thread: " + i);
Thread.sleep(500);
}
}
 catch (InterruptedException e)
 {
 System.out.println("Main thread interrupted.");
 }
 System.out.println("Main thread exiting.");
 }
 }
```

Output:

The program was run three times, and all time different output was generated.

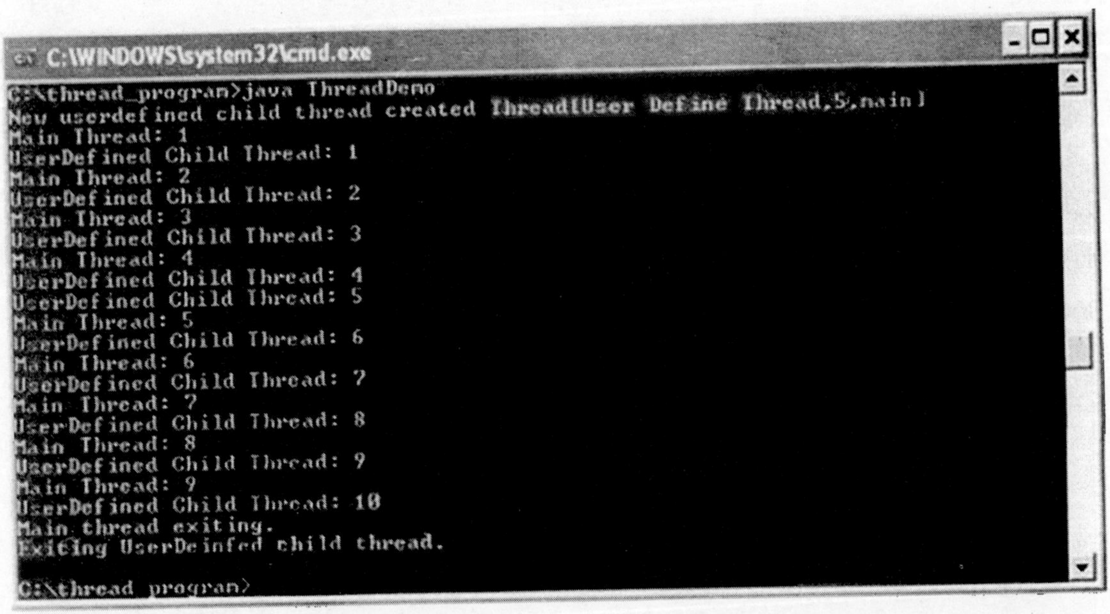

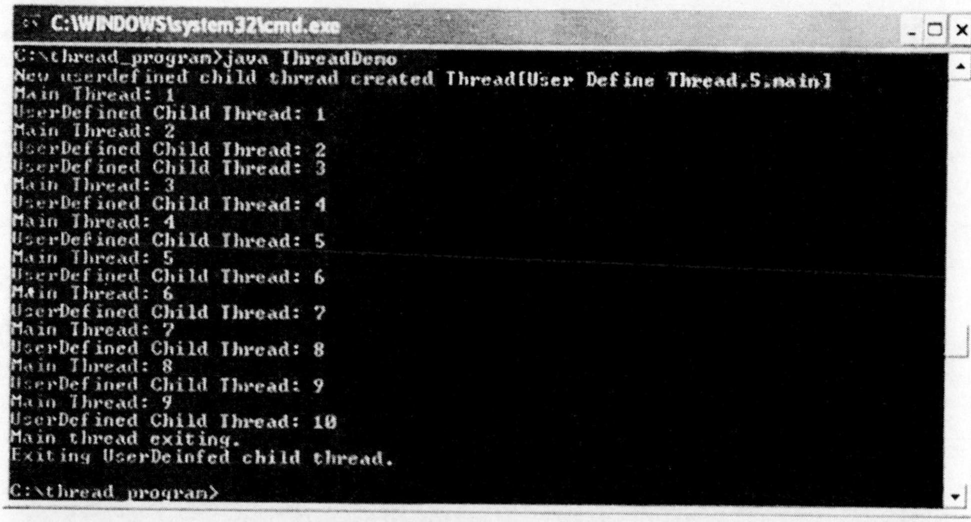

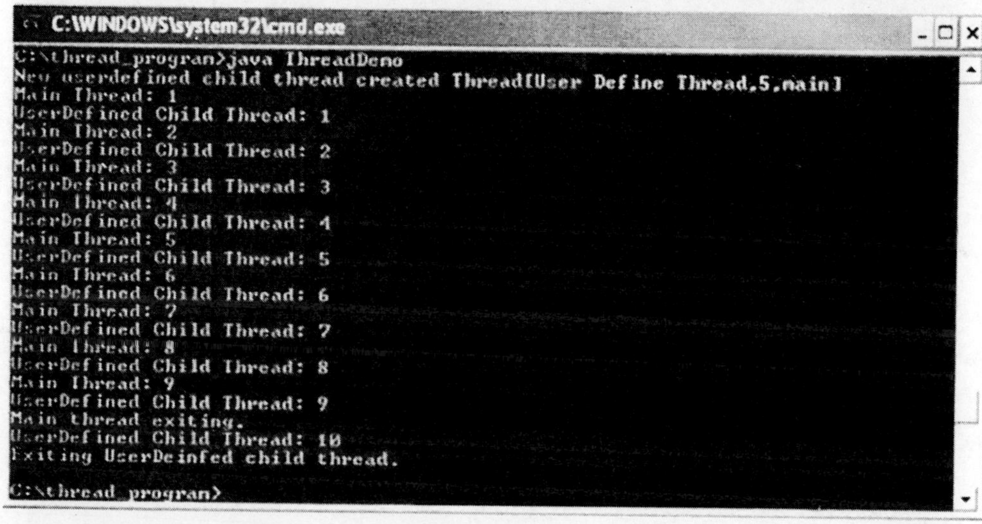

4.7. MULTITHREADING

Till the last section we've studied about single thread that is created using Runnable interface or Thread class. You can spawn as many threads as you want.

Example:

```
class NewThreadDemo implements Runnable
{
Thread t;

NewThreadDemo(String threadName)
{
t=new Thread(this, threadName); //A thread object is created
System.out.println("New userdefined child thread created " +
threadName);
```

```
t.start(); // Start the thread
}

public void run()
{
try
{
for(int i = 1; i <=10; i++)
{
System.out.println("Child Thread: [" +Thread.currentThread().getName()
+ "] " + i);
Thread.sleep(500);
}
}

catch (InterruptedException e)
{
System.out.println("UseDefine Child interrupted.");
}
System.out.println("Exiting UserDeinfed child thread.");
}
}

class MultiThreadDemo
{
public static void main(String args[])
{
new NewThreadDemo("Child_One"); // create a new thread
new NewThreadDemo("Child_Two"); // create a new thread
new NewThreadDemo("Child_Three"); // create a new thread
new NewThreadDemo("Child_Four"); // create a new thread

try
{
for(int i = 1; i <10; i++)
{
System.out.println("Main Thread: " + i);
Thread.sleep(500);
}
}

catch (InterruptedException e)
{
System.out.println("Main thread interrupted.");
}
System.out.println("Main thread exiting.");
}
}
```

Output:

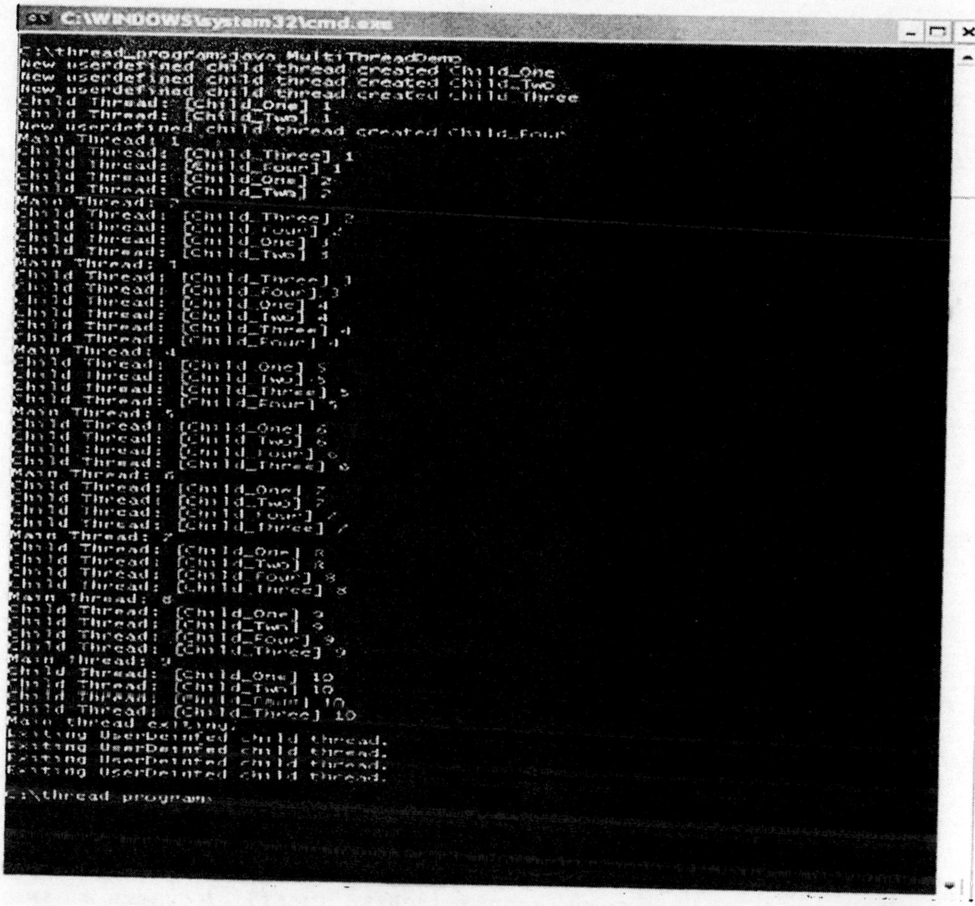

Four child threads are created in the program, they run in the order in which those get the CPU time by run-time system.

About join() and isAlive()

In the immediately above example there is a problem that the main thread is exited before its child thread. Conceptually and logically the parent thread must die or exit after all its child thread exits. May be the parent thread works has been completed, but still it has to wait for its child thread to complete their task and exit. This can be possible, by join() method. All child threads have to join, to prevent the main or parent thread exit before them. Now main or parent thread must wait, until all child thread exits. The isAlive() method is used to determine that whether a given thread is finished or running. If isAlive() return true means the thread is running, else return false means the thread is finish.

Example:

```
class NewThreadDemo implements Runnable
{
Thread t;
NewThreadDemo(String threadName)
{
```

```
t=new Thread(this, threadName);  //A thread object is created
System.out.println("New userdefined child thread created " + threadName);
t.start(); // Start the thread
}

public void run()
{
try
{
for(int i = 1; i <=10; i++)
{
System.out.println("Child Thread: [" +Thread.currentThread().getName()
                                                            + "] " + i);

Thread.sleep(200);
}
}
 catch (InterruptedException e)
{
System.out.println("UseDefine Child interrupted.");
}
System.out.println("Exiting UserDeinfed child thread.");
}
}

class MultiThreadDemo
{
public static void main(String args[])
{
NewThreadDemo ch1=new NewThreadDemo("Child_One"); // create a new thread
NewThreadDemo ch2=new NewThreadDemo("Child_Two"); // create a new thread
NewThreadDemo ch3=new NewThreadDemo("Child_Three"); // create a new thread
NewThreadDemo ch4=new NewThreadDemo("Child_Four"); // create a new thread


System.out.println("Child Thread One is alive: " + ch1.t.isAlive());
System.out.println("Child Thread Two is alive: "+ ch2.t.isAlive());
System.out.println("Child Thread Three is alive: "+ ch3.t.isAlive());
System.out.println("Child Thread Four is alive: "+ch4.t.isAlive());

try
{
System.out.println("Waiting for child threads to finish.");
ch1.t.join();
ch2.t.join();
ch3.t.join();
ch4.t.join();
}

 catch (InterruptedException e)
```

```
{
System.out.println("Main thread Interrupted");
}
System.out.println("Child Thread One is alive: " + ch1.t.isAlive());
System.out.println("Child Thread Two is alive: "+ ch2.t.isAlive());
System.out.println("Child Thread Three is alive: "+ ch3.t.isAlive());
System.out.println("Child Thread Four is alive: "+ch4.t.isAlive());
System.out.println("Main thread exiting.");
}
}
```

In the code all child thread are join(). If see the output, the main thread exited in the mid of the running code. It wait for all child to be exit. At last it exits from the code.

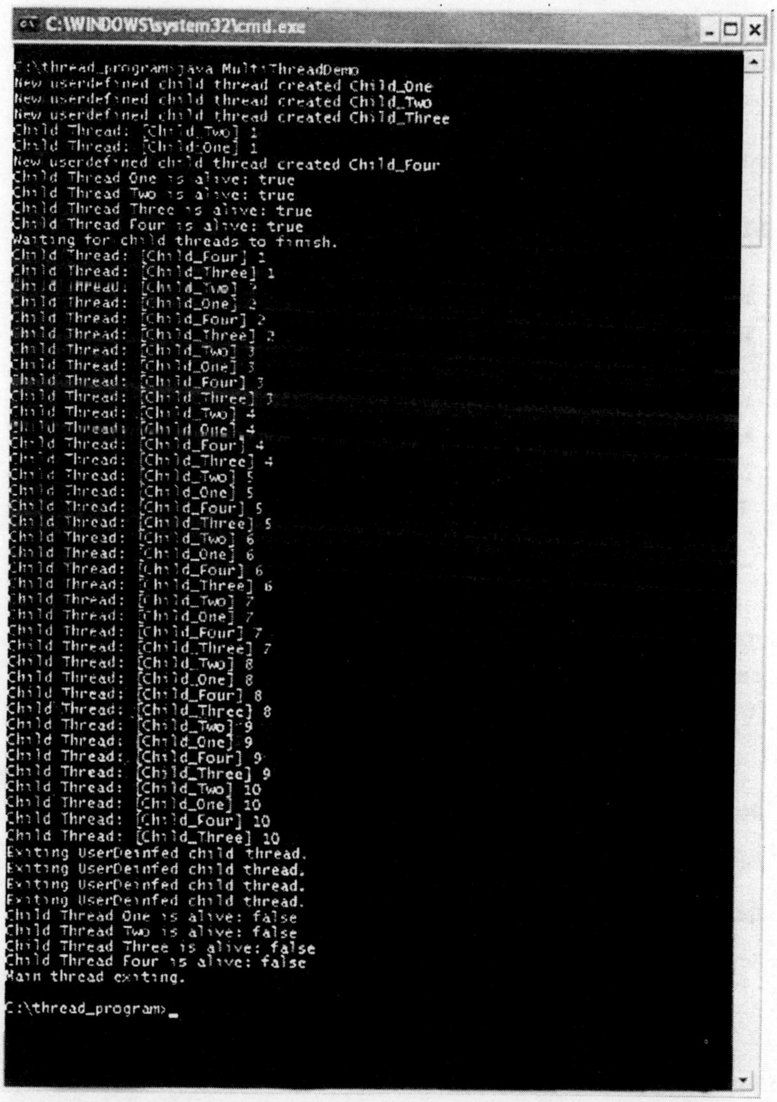

4.8. THREAD PRIORITIES

Priority allow the scheduler to take the decision when the thread should be allowed to run, and in which order. The higher priority threads get more CPU time then lower priority threads. A lower priority thread can be preempted by higher priority thread. The threads having equal priority get equal CPU time.

To set a thread's priority, use the *setPriority( )* method, which belongs to *Thread* class. Prototypes is as

```
final void setPriority(int priorityValue)
```

Here, *priorityValue* give the new priority for the calling thread. The value of *priorityValue* must be within the range MIN_PRIORITY and MAX_PRIORITY. Values of these constants are 1 and 10, respectively. A thread default priority is NORM_PRIORITY, which is currently 5. These priorities constants are defined as *final* variables within *Thread.*

You can get current priority value of the thread by calling the *getPriority*( ) method of *Thread*, shown here.

```
final int getPriority( )
```

Example:

```
public class Main
{
    public void setPrioritiesOnThreads()
    {
        Thread thread1 = new Thread(new TestThread(1));
        Thread thread2 = new Thread(new TestThread(2));

        thread1.start();
        thread2.start();

        try
        {
            thread1.join();
            thread2.join();
        }
        catch (InterruptedException ex)
        {
            ex.printStackTrace();
        }
        System.out.println("Done.");
    }

    public static void main(String[] args)
    {
        new Main().setPrioritiesOnThreads();
    }

    class TestThread implements Runnable
    {
```

```
        int id;
      public TestThread(int id)
    {
         this.id = id;
      }

       public void run()
     {
         for (int i = 1; i <= 10; i++)
         {
                System.out.println("Thread" + id + ": " + i);
             }
         }
    }
}
```

Output:

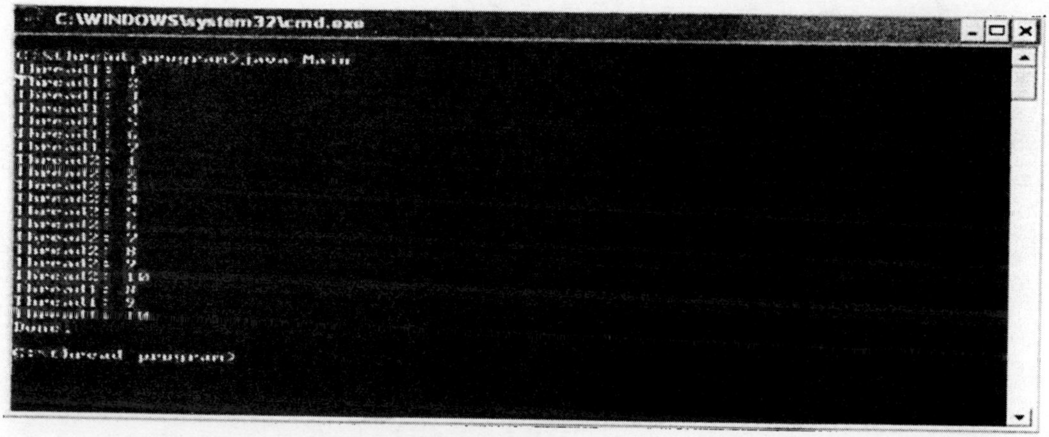

The output is without priority assigning to the thread. Both thread are executing not in exact order, but executing concurrently.

If we assign the priority to the thread then the code is given below.

Example:

```
public class Main
{
    public void setPrioritiesOnThreads()
    {
        Thread thread1 = new Thread(new TestThread(1));
        Thread thread2 = new Thread(new TestThread(2));

        thread1.setPriority(Thread.MAX_PRIORITY);
        thread2.setPriority(Thread.MIN_PRIORITY);

        thread1.start();
        thread2.start();
```

```
        try
        {
            thread1.join();
             thread2.join();
        }
    catch (InterruptedException ex)
    {
            ex.printStackTrace();
     }

       System.out.println("Done.");
      }
    public static void main(String[] args)
    {
       new Main().setPrioritiesOnThreads();
    }

class TestThread implements Runnable
{
     int id;
     public TestThread(int id)
    {
       this.id = id;
        }

         public void run()
      {
        for (int i = 1; i <= 10; i++)
          {
                System.out.println("Thread" + id + ": " + i);
             }
        }
    }
}
```

Output:

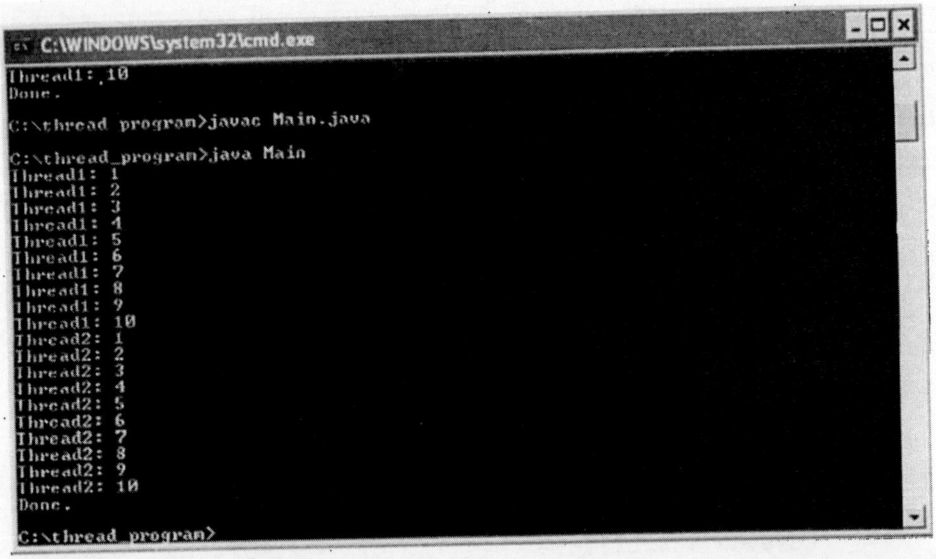

Now the thread1 has the highest priority so it executes first.

4.9. SYNCHRONIZATION

When two or more threads wants to access a shared resource, then it must ensure that the resource will be used by only one thread at an instant of time. The mechanism of this process is called *synchronization*. Synchronization is the concept of the monitor or semaphore. Monitor works as mutex and restrict to one thread to own a monitor at a given time. As the thread acquires the lock, all the threads that want to acquire the monitor will be suspended. As the first thread exits from the monitor, one thread will acquire monitor from the waiting list of threads.

Synchronization of code can be in two ways, but both use **synchronized** keyword.

- Using synchronized method
- Using synchronized statement

Example:

```
public class SynchronizedMethodDemo extends Object
{
private static int c = 1;

private static void print(String msg)
{
String threadName = Thread.currentThread().getName();
System.out.println(threadName + ": " + msg);
}

public static synchronized int getCount()
{
int i = c;
c++;
return i;
}

public static void main(String[] args)
{
try
{
Runnable runnable = new Runnable() { public void run() {
    System.out.println("count=" + getCount());
}
};
Thread thread1 = new Thread(runnable, "Thread1");
thread1.start();
Thread.sleep(500);
Thread thread2 = new Thread(runnable, "Thread2");
thread2.start();
Thread.sleep(500);
Thread thread3 = new Thread(runnable, "Thread3");
thread3.start();
```

```
Thread.sleep(500);
Thread thread4 = new Thread(runnable, "Thread4");
thread4.start();
}
 catch(Exception x )
{
}
}
}
```

Output:

4.10. INTER-THREAD COMMUNICATION

If we talk about the mechanism of synchronization, then as one thread exits from the monitor then it must inform the waiting threads that it has left the monitor, now suspended thread can proceed to acquire the lock on the resources or entered in the monitor. If that is not possible then the waiting thread will always be in the waiting list. So, to solve this problem threads must communicate with each other.

Java provides a set of methods by which they can communicate to each other.

Object class has some final methods for such purposes. All these methods can be called only from the synchronized context.

- wait( ) tells the calling thread to leave the monitor and go to sleep until some other thread enters the same monitor and calls notify( ).
- notify( ) give a wake up signal or call to the first thread that called wait( ) on the same object.
- notifyAll( ) give a wake up signal or call to all the threads that called wait( ) on the same object. The highest priority thread will run first.

These methods are declared within **Object**, as shown here:

```
final void wait( ) throws InterruptedException
final void notify( )
final void notifyAll( )
```

Example:

```
/** Example from Sun Thread tutorial */
public class CubbyHole
{
    private int contents;
    private boolean available = false;
```

```
        public synchronized int get()
        {
            while (available == false)
            {
                try
                {
                    wait();
                }
                catch (InterruptedException e) { }
            }
            available = false;
            notifyAll();
            return contents;
        }

        public synchronized void put(int value)
        {
            while (available == true)
            {
                try
                {
                    wait();
                }
                catch (InterruptedException e) { }
            }
            contents = value;
            available = true;
            notifyAll();
        }
    }
```

REVIEW QUESTIONS

Q. 1. What is multithreading? Explain the life cycle of a thread.

[MDU, May 2011, Dec. 2007, Dec. 2009]

Ans. A thread is a stream of code execution. This means that a single program can perform more than one task at the same time. Thread-based multitasking deals with the programming details. A thread is actually the coding construct that does not affect the architecture of the application program. A thread can be loosely defined as a separate stream of execution that takes place simultaneously.

Q. 2. What is a daemon thread and which method is used to create the daemon thread?

Ans. Daemon thread is a low priority thread which runs intermittently in the back ground doing the garbage collection operation for the java runtime system. setDaemon method is used to create a daemon thread.

Q. 3. What is the difference between process and Threads?

Ans. Process is a program in execution. The execution of peocesses must be sequential. On the other hand thread is a light-weighted process which is executed in parallel with other threads. The thread shares some of the resources with other threads but certain things are unique to as particular thread like process id and program counter.

Q. 4. What are three ways in which a thread can enter the waiting state?

Ans. A thread can enter the waiting state by invoking its sleep() method, by blocking on I/O, by unsuccessfully attempting to acquire an object's lock, or by invoking an object's wait() method. It can also enter the waiting state by invoking its (deprecated) suspend() method.

Q. 5. How does the run() method in runnable work?

Ans. It may help to think of the method link the main method in standard single threaded applications. The run method is a standard entry point to run or execute a class. The method is normally only executed in the context of an independent Thread, but is a normal method in all other respects.

Q. 6. What's the difference between a thread created as a subclass of java.lang.Thread and one created as an implementation of Runnable?

Ans. As threads, both behave the same. However a thread that implements runnable can also extend another class.

❑❑❑

CHAPTER 5

Input/Output in Java

A computer user need a basic application to write data to the standard output device or any other source like file, network connection and read from the standard input or any other source like file, network connection. For providing read and write facilities in the Java application, Java introduced the I/O Streams. I/O will be used by the programmer to write an application to and read from a source, and represents to input or output resource. A resource (for read or write) can be of any kind, e.g disk files, devices, other programs; and network channel etc.

INSIDE THIS CHAPTER

In Java, we deal with different kinds of data, e.g. bytes, primitive data types, object etc. So Java streams support most of these types of data to read and write.

Programmer does not bothered about how they are working internally, only he/she has to use the corresponding stream and call the methods. To read the data from any resource *input stream* is used, and to write the data on any resource *output stream* is used.

5.1. I/O BASIC

I/O is the mechanism provided by the Java language, which facilitates the programmer to write an application to give I/O support to files and networks as well as devices too. I/O system is unified and consistent. In fact, in last few chapters you've usedSystem.out.*print*() and System.out.*println*() methods to give output on screen, but there are a lot of I/O methods that have not been used extensively. If you understand the fundamental and ground structure of I/O then to be a master in I/O is easy.

Streams

Java provides I/O operations through streams in I/O dependent applications. A *stream* is an abstraction of operations on resources to store (write) or fetch (read) information. Java stream is linked to a physical device by the I/O system. All streams work with the same approach, even if they are connected with different devices. Thus, I/O class methods can be applied to any kind of device. Thus, we can say that input stream is an abstraction for input from a keyboard, a disk file, and a network socket. Similarly, an output stream is an abstraction to refer a disk file, console, and a network socket. Streams provide an abstraction at the level at which programmer does not need to differentiate every part of code, that the statement is performing input/output operation on keyboard or a network connection.

5.2. BYTE AND CHARACTER STREAMS

Java provides two types of stream: byte and character. Here we introduce both, *Byte Stream* and *Character Stream*. Byte Streams provides a convenient way to read and write data in the form of bytes. Character Streams provides a method to read and write data in the form of character, and the character set if Unicode, so it can be internationalized. In few cases, character streams gives better performance and use the byte streams.

At the lowest level of communication with devices is still a byte-oriented communication, so only byte streams are used there.

The Byte Stream Classes

The byte stream class hierarchy is headed by two abstract classes: *InputStream* and *OutputStream.* These abstract classes have a number of concrete classes, able to work or perform I/O operations on various devices, such as network connection, a disk file, or even memory buffers. InputStream and OutputStream two most important methods are *read()* and *write()* respectively, that are used to read and write data in the form of bytes. These methods belong to their corresponding classes and are abstract in nature, so that, these methods are overridden by the subclass of the corresponding byte stream super class.

The Character Stream Classes

The byte stream class hierarchy is headed by two abstract classes: *Reader* and *Writer*, and handle Unicode character streams. Each abstract class has a number of concrete classes. Character stream classes have two most important and used methods *read()* and write(), that deal with character data. These methods are overridden by the derived character stream classes.

5.3. I/O CLASSES

The classes and interface in java.io are divided into following five groups:

- The general classes for different types of bytes and character stream and converting between them.
- A range of classes that define various types of stream – filtered streams, buffered streams piped streams.
- To read and write primitive values and string, data stream classes are in use.
- A set of classes and interfaces for interacting with files in a system.
- Classes and interfaces for object serialization.

Java implements I/O stream with class hierarchies defined in the java.io package.

InputStream

It is an abstract class, and declares methods to read bytes from the source. InputStream is the superclass of most of the byte input stream classes. All methods throw IOException on any error.

```
public abstrace int read() throws IOException
```

Reads a single byte of data from source and returns integer in the range 0 to 255 equivalent to read byte.

```
public int raed(byte ar[], int startIndex, int length) throws IOException
```

Reads bytes and store into a part of a byte array, and it allows to read *length* bytes from the srouce, and start writing from *startIndex* position.

`public int raed(byte ar[]) throws IOException`

Reads all bytes from the source and store from 0.

`public long skip(long length) throws IOException`

Skips length number of bytes of source from the current position of the cursor location, and returns the actual number of skipped bytes. If returns negative then no byte is skipped.

`public int available() throws IOException`

Returns number of bytes that can be read or skipped.

`public void close() throws IOException`

Close the input stream.

OutputStream

It is an abstract class, provides abstraction for writing bytes to a destination srouce.

`public abstract void write(int a) throws IOException`

Writes a as a byte. The byte is passes as an int, only the lowest 8 bits of the integer are written. This method blocks until the byte is written.

`public void write(byte ar[], int startIndex, int legnth) throws IOException`

Writes a part of a byte array from byte array *ar[]*, and *length* number of bytes starting at the *startIndex*.

`public void write(byte ar[]) throws IOException`

Write all bytes from the source byte array ar[] to output stream, from 0the positon.

`public void flush() throws IOException`

Flushes the stream.

`public void close() throws IOException`

Closes the output stream.

Reader

It is an abstract class provides a character stream, provides facilities to read character from the source.

`public int read() throws IOException`

Method reads a single character and returns it as an integer in the range 0 to 65536. It returns -1, if no character is available because the end of the stream has been reached.

`public abstract int read(char ar[], int startIndex, int length)`

Method reads *length* number of characters from the source and store into target character array ar[], at *startIndex.*

`public int read(char []ar) throws IOException`

Method read all data and store in character array ar[].

`public long skip(long count) throws IOException`

Method skips *count* number of characters of input or until the end of the stream is sense.

```
public boolean ready() throws IOException
```

Method returns *true* if source is ready to read, i.e. there is at least one character available to read.

```
public abstract void close() throws IOException
```

Method is responsible to close the stream.

Writer

It is also an abstract class provides to write data to the source.

```
public void write(int a) throws IOException
```

Method writes a as a character to the target source.

```
public abstract void write(char []ar, int startIndex, int length)
          throws IOException
```

Method write *length* number of characters array *ar[]* to the target source at *startIndex*.

```
public void write(char ar[]) throws IOException
```

Method write the complete character array to target source.

```
public void write(String strObj, int startIndex, int length) throws IOException
```

Method write *length* number of characters from string *strObj* to target source at *startIndex*.

```
public void write(String strObj) throws IOException
```

Method write all string content from *strObj* to target source.

```
public abstract void flush() throws IOException
```

Method flushes the stream.

```
public abstract void close() throws IOException
```

Method closes the stream.

InputStreamReader and OutputStreamWriter

InputStreamReader and OutputStreamWriter are used to translate or convert between Unicode and byte stream using specified encoding or local encoding of the system.

An *InputStreamReader* object reads data form a byte stream source and produces the corresponding Unicode characters.

```
public InputStreamReader(InputStream inObj)
```

Creates an object of *InputStreamReader* to read from the given *InputStream* (byte stream) object *inObj* by applying default encoding.

```
public InputStreamReader(InputStream inObj, String encoding)
                                                        throws
          UnsupportedEncodingException
```

Creates an object of *InputStreamReader* to read from the given *InputStream* object *inObj* by applying given encoding scheme.

An *OutputStreamWriter* object reads data from character stream source and produces the corresponding byte character.

```
public OutputStreamWriter(OutputStream outObj)
```

Creates an object of *OutputStreamWriter* to write to the given *OutputStream* object *outObj* by applying default encoding.

```
public OutputStreamWriter(OutputStream outObj, String encoding)
                                                        throws
        UnsupportedEncodingException
```

Creates an object of *OutputStreamWriter* to write data to the given *OutputStream* object *outObj* by applying given encoding.

BufferedInputStream and BufferedOutputStream

The Buffered byte stream classes buffer their data into itself, which avoid directly reading and writing on the next linked stream. These classes are often used in combination with File stream. Because as it holds the file, the number of disk access reduces, there is only a need to interact with buffer object.

BufferedInputStream object encapsulated with InputStream object and used to read data from the linked source. *BufferedInputStream* also supports *skip()* and *reset()* methods. The *BufferedInputStream.markSupported()* method returns true, if it supports the mark.

It has two constructor:

```
BufferedInputStream(InputStream inputStreamObj)
```

Simply creates an object with a default buffer size.

```
BufferedInputStream(InputStream inputStreamObj, int bufferSize)
```

Creates an object of buffered stream with a given buffer size.

BufferedOutputStream is similarly to any *OutputStream* with an exception of added *flush()* method. It doesn't provide any additional method unlike buffered input.

It has two constructor:

```
BufferedOutputStream(OutputStream ouputStreamObj)
```

Creates an object of buffered stream of 512 bytes.

```
BufferedOutputStream(OutputStream ouputStreamObj, int bufferSize)
```

Creates an object of buffered stream of given *bufferSize*

BufferedReader and BufferedWriter

The concept is same as given for buffered byte stream, the only different is these classes works on character stream, i.e. read the character and write the character from and to the target source respectively.

BufferedReader increase the performance of reading and writing data of buffer. It has two constructors:

```
BufferedReader(Reader inputStreamObj)

BufferedReader(Reader inputStreamObj, int bufferSize)
```

BufferedWriter adds a *flush()* method that can be guarantee that data buffer are physically written to the definite output stream. It has two constructors:

```
BufferedWriter(Writer outputStreamObj)
BufferedWriter(Writer outputStreamObj, int bufferSize)
```

FileInputStream

FileInputStream is an subclass of *InputStream* that is use to read data from the target source in form of bytes. Following are the often used constructor of the class:

```
FileInputStream(String filePath)
FileInputStream(File fileObj)
```

FileOutputStream

FileOutputStream is an subclass of *OutputStream* that is use to write data to the target source in form of bytes. Following are the often used constructor of the class:

```
FileOutputStream(String filePath)
FileOutputStream(File fileObj)
FileOutputStream(String filePath, boolean append)
FileOutputStream(File fileObj, boolean append)
```

FileReader

FileReader is a subclass of *Reader* class that is used to read data from the file in character format. Its few constructors are as following:

```
FileReader(String filePath)
FileReader(File fileObj)
```

FileWriter

FileWriter is a subclass of Writer, and use to write to a file. Its commonly used constructors are shown here:

```
FileWriter(String filePath)
FileWriter(String filePath, boolean append)
FileWriter(File fileObj)
FileWriter(File fileObj, boolean append)
```

5.4. READING CONSOLE INPUT

Java provides two streams to read the data from the console, either by byte stream classes or character stream classes. But now a day's, byte stream is not recommended because it has to use some deprecated methods, although if data has to be read from low level then byte streams are preferable. To read data from the console using character-oriented stream is the easiest way that makes your program internationalized.

In Java, *System.in* is default bind with standard input device keyboard. Reading can be done by reading from *System.in*. To acquire a character stream that links with standard input, you have to encapsulate *System.in* in *BufferedReader* object, to generate or create a character stream to read the data from the console. *BuffereredReader* supports a buffered input stream. Its most commonly used constructor is given below:

```
BufferedReader(Reader inputReaderObject)
```

Here, *inputReaderObject* is the stream that is connected to the instance of *BufferedReader* that is being created.

In the classes hierarchy of character stream *Reader* is an abstract class and superclass of those character stream classes for reading the data. *InputStreamReader*s is its concrete subclass, which converts the bytes to characters. To get an object of *InputStreamReader* that is connected with *System.in*, you've to use the following construct of *InputStreamreader*:

```
InputStreamReader(InputStream inputStreamObject)
```

Because *System.in* is an object of type *InputStream*, it can be used for *inputStreamObject*. By putting all these together, the following line of code creates a *BufferedReader* that is linked to the standard input device: keyboard

```
BufferedReader br = new BufferedReader(new InputStreamReader(System.in));
```

Here, **br** is an object of character-based stream that is linked to the console through **System.in.**

Reading Characters

To read a character form the Buffered Reader object, use the following version of read() method.

```
int read( ) throws IOException
```

The method read() will return a integer value corresponding the character from the input stream and if end of the stream come across then it will return -1.

Example: To read characters.

```
import java.io.*;
class CharacterReadDemo
{
   public static void main(String args[]) throws IOException
   {
      char ch;
       BufferedReader brObj = new BufferedReader(new
        InputStreamReader(System.in));
       System.out.println("Enter characters, 'Q' to quit.");
       do
       {
          ch = (char) brObj.read();
          System.out.println(ch);
        } while(ch != 'Q');
      }
   }
```

Output:

Reading String

To read a string form the Buffered Reader object, use the following version of readLine() method.

```
String readLine( ) throws IOException
```

Example:

```
import java.io.*;
class StringReadDemo
{
    public static void main(String args[]) throws IOException
    {
        BufferedReader brObj = new BufferedReader(new
        InputStreamReader(System.in));
        String strObj;
        System.out.println("Enter lines of string : ['exit to quit']
        ");
        do
        {
            strObj = brObj.readLine();
            System.out.println(strObj);
        } while(!strObj.equals("exit"));
    }
}
```

Output:

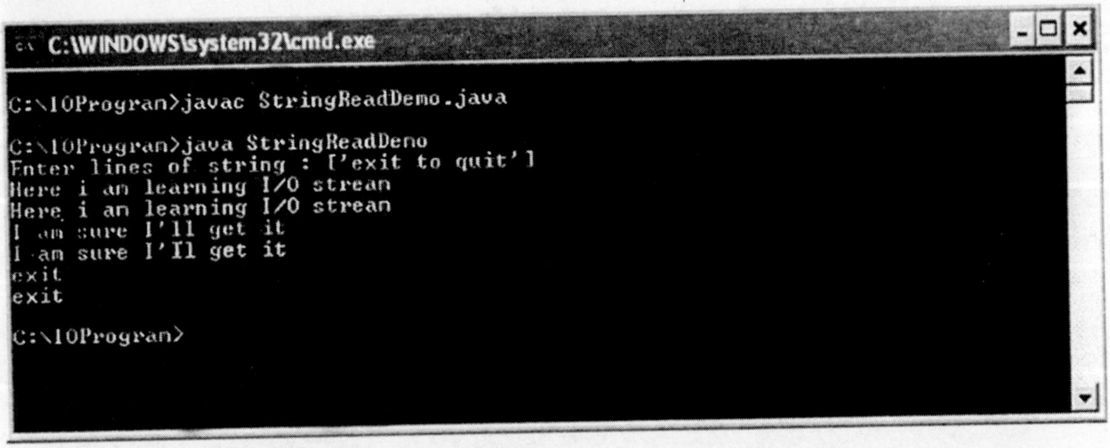

In it enter a line of text and press enter key, it will be print, the process will repeat until you entered exit.

5.5. WRITING TO CONSOLE OUTPUT

In last chapters System.out.println() and System.out.print() are only used method for output on console. But there is a class PrintStream a drived class of OutputStream having a implemented method write(), it can be used to write to the console. The simplest form of write() is :

```
void write(int byteValue)
```

The method writes to the stream in byte format by *byteValue*. The *byteValue* is declared as an integer but only low-order eight bits are written.

Example:

```
class DataWriteDemo
{
public static void main(String args[])
{
int x;
x = 'A';
System.out.write(x);
System.out.write('\t');
x='P';
System.out.write(x);
System.out.write('\n');
}
}
```

Output:

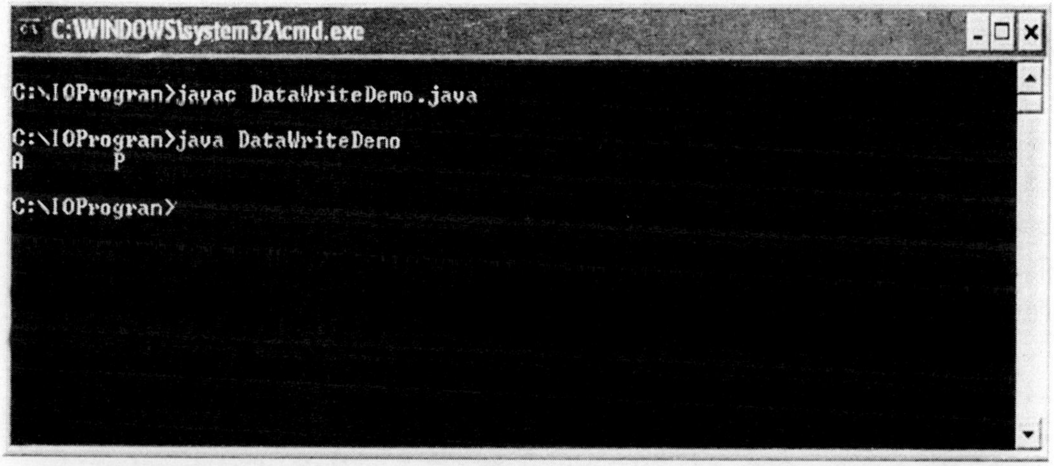

5.6. READING AND WRITING ON FILES

Java makes available a set of classes and methods that allows you to read from and write to files. Java supports both streams and methods to read from and write to file respectively. API's are already discussed earlier in the section above, so that here we'll have a number of examples related to read and write file.

Example: Read file using FileInputStream

```
import java.io.*;
class DemoShowFile
{
public static void main(String ar[]) throws IOException
{
int ch;
```

```
FileInputStream finObj;
try
{
finObj = new FileInputStream("DemoShowFile.java");
}
catch(FileNotFoundException e)
{
System.out.println("File Not Found");
return;
}
do
{
ch = finObj.read();
if(ch != -1)
System.out.print((char) ch);
} while(ch != -1);
finObj.close();
}
}
```

Output

Print contents of own source file, read character by character and write character by character.

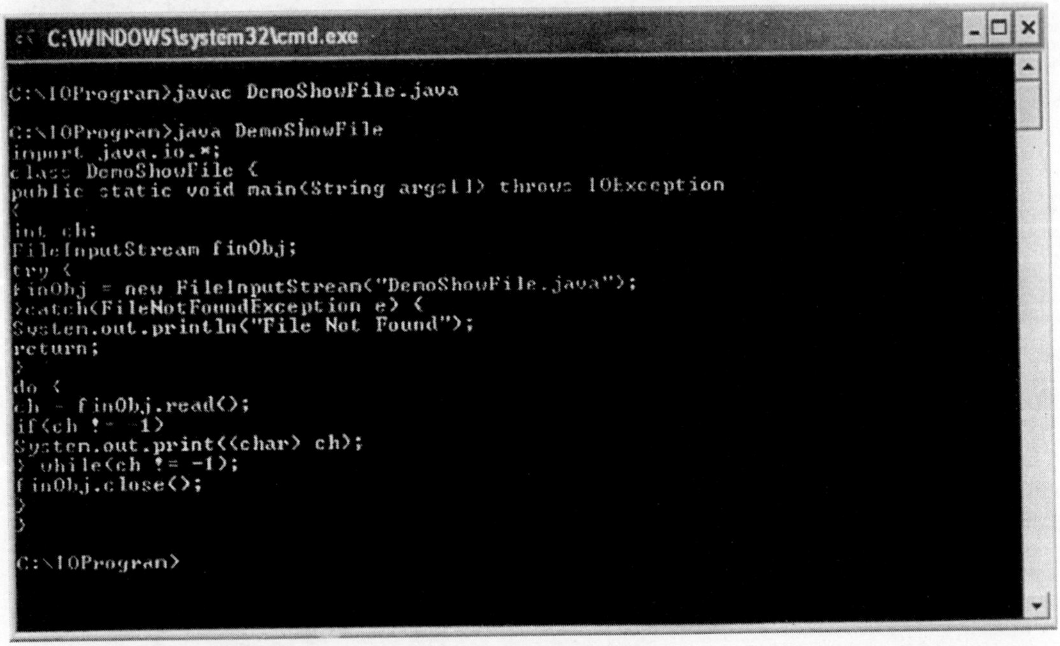

Example: Read file using FileReader

```
import java.io.*;
class DemoFileRead
{
```

```
public static void main(String ar[]) throws IOException
{
FileReader frObj = new FileReader("DemoFileRead.java");
BufferedReader brObj = new BufferedReader(frObj);
String str;
while((str  = brObj.readLine())!=null)
System.out.println(str);
frObj.close();
}
}
```

Output

Read the file content line by line, and print in the same manner.

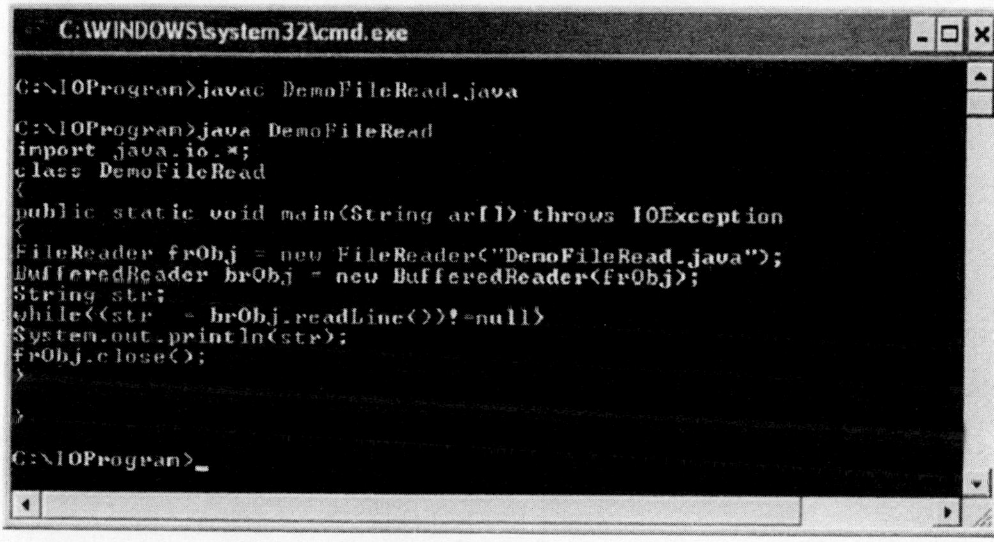

Example: Writing data using FileOutputStream

```
import java.io.*;
class DemoCopyFile
{
public static void main(String ar[]) throws IOException
{
int i;
FileInputStream finObj;
FileOutputStream foutObj;
try
{
 //Open source file
try
{
finObj = new FileInputStream(ar[0]);
}
```

```
catch(FileNotFoundException e)
{
System.out.println("File " + ar[0] + " not Found");
return;
}

//Open target file
try
{
foutObj = new FileOutputStream(ar[1]);
}
 catch(FileNotFoundException e)
{
System.out.println("File " +ar[1]+" not found");
return;
}
}
catch(ArrayIndexOutOfBoundsException e)
{
System.out.println("Argument missing");
return;
}

//Code to a copy File

try
{
do
{
i = finObj.read();
if(i != -1)
 foutObj.write(i);
}while(i != -1);
}
catch(IOException e)
{
System.out.println("Some file error occurd");
}
finObj.close();
foutObj.close();
}
}
```

Output:

Copy content of a.txt to b.txt

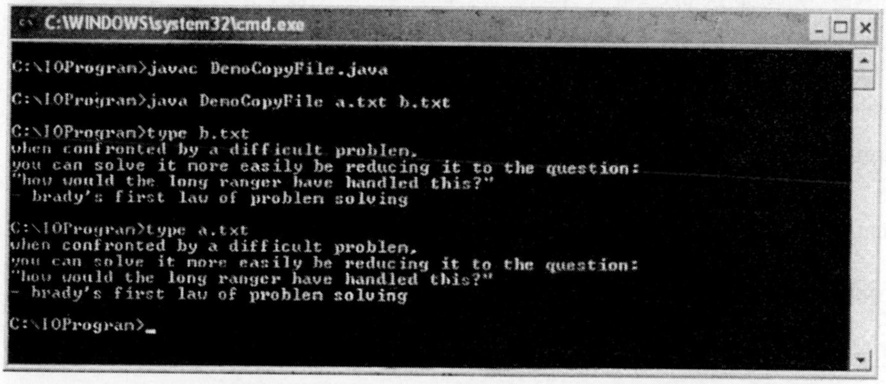

It is clear form the output window given above that content of a.txt has been copied to b.txt.

Example: Writing data using FileWriter

```
import java.io.*;
public class DemoWriteFile
{
    public static void main(String[] args) throws IOException
    {
        File inputFileObj = new File("a.txt");
        File outputFileObj = new File("b.txt");
        FileReader finObj = new FileReader(inputFileObj);
        FileWriter foutObj = new FileWriter(outputFileObj);
        int c;
        while ((c = finObj.read()) != -1)
        foutObj.write(c);
        finObj.close();
        foutObj.close();
    }
}
```

Output:

Copy content of a.txt to b.txt

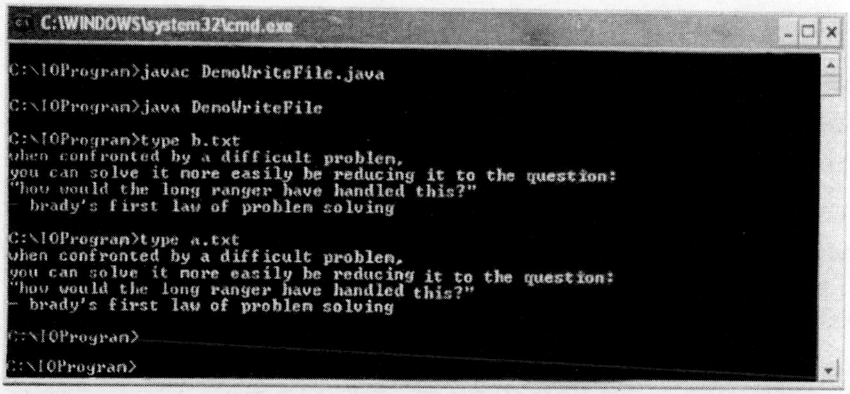

5.7. RANDOM ACCESS FILES

The above concept is used to append the data in a file, e.g. write at end of the file. But RandomAccessFile provides you to read or write data from or to in a file at any location. It supports you to move around the file, and read from it and write to it, where ever you like.

It has these two constructors:

```
RandomAccessFile(File fileObj, String accessType) throws
        FileNotFoundException
```

here, *fileObj* specifies the name of the file to open as a File object.

```
RandomAccessFile(String filename, String accessType) throws
        FileNotFoundException
```

here, the name of the file is passed in filename as string.

Example:

```
RandomAccessFile fileObj = new RandomAccessFile("a.txt", "rw);
```

In both constructors, *accessType* determines what type of file access is permitted. There are following cases, if accessType is:

- "r", then the file can be read, not written
- "rw", then file is opened in read-write mode.
- "rws", then file is opened in read-write mode, both operations can be performed, and every changes to file will update to underlying file written on physical device.

To search or seek a position to perform either operation seek() method is used and it's prototype is given below:

```
void seek(long newPosition) throws IOException
```

Here, *newPosition* specifies the new position, in bytes, of the file pointer from where next read or write operation will occur at the new file position.

The current position of the file pointer can be obtained by calling the *getFilePointer*() method.

Example:

```
RandomAccessFile fileObj = new RandomAccessFile("a.txt", "rw);
fileObj.seek(40);
long pointer = fileObj.getFilePointer();
fileObj.close();
```

RandomAccessFile includes some additional methods. One of them is setLength( ). It has this signature:

```
void setLength(long length) throws IOException
```

This method is used to fix the length of the file.

Reading from a RandomAccessFile is done using one read() methods.

Example:

```
RandomAccessFile fileObj = new RandomAccessFile("a.txt", "rw);
int a = fileObj.read();
fileObj.close();
```

Writing to a RandomAccessFile can be done using one it its many write() methods.

Example:

```
RandomAccessFile file = new RandomAccessFile("c:\\data\\file.txt", "rw);
file.write("Hello World".getBytes());
file.close();
```

Example: Write to a file

```
import java.io.FileNotFoundException;
import java.io.IOException;
import java.io.RandomAccessFile;

public class DemoRandomAccessFile
{
   public static void main(String ar[])
   {
        try
        {
            RandomAccessFile raObject = new RandomAccessFile("a.txt",
        "rw");

           String data[] = new String[5];
           data[0] = "Learning ";
           data[1] = "Random";
           data[2] = "Access";
           data[3] = "File";
           data [4] = " class.";

           for (int i = 0; i < data.length; i++)
           {
              raObject.writeUTF(data[i]);
           }
           raObject.seek(raObject.length());
           raObject.writeUTF("Now I now how to work with
        RandomAccessFile");
           }
        catch (FileNotFoundException e)
         {
            e.printStackTrace();
        }
         catch (IOException e)
         {
            e.printStackTrace();
        }
   }
}
```

Output:

```
C:\WINDOWS\system32\cmd.exe

C:\IOProgram>javac DemoRandomAccessFile.java

C:\IOProgram>java DenoRandomAccessFile

C:\IOProgram>type a.txt
        Learning  ♦Random ♦Access ♦File  class.ou can solve it more easily be re
ducing it to the question:
"how would the long ranger have handled this?"
 brady's first law of problem solving
 ♦Now I now how to work with RandomAccessFile
C:\IOProgram>
```

5.8. STORING AND RETRIEVING OBJECTS FROM FILE

If you want to persist an object of any class on secondary storage then an object has to be written in file and later on can be retrieved. Java supports this feature and provides the classes and interfaces related to it.

ObjectOutput

It is an interface, and it extends the *DataOutput* interface and also supports object serialization. The most important method to write an object is writeObject().

ObjectOutputStream

It is a class and extends the OutputStream class and implements the ObjectOutput interface. It is used to write objects to a stream. A constructor of this class is:

```
ObjectOutputStream(OutputStream outStreamObj) throws IOException
```

The argument *outStreamObj* is an output stream object to which serialized objects will be written. It facilitates the writing to persistent fields.

ObjectInput

It is an interface, and it extends the *DataInput* interface and defines the methods and also supports object serialization. An special method *readObject()* method. It supports deserialization of an object.

ObjectInputStream

It is a class and extends the InputStream class and implements ObjectInput interface. It is responsible for reading objects from a stream. A constructor of this class is:

```
ObjectInputStream(InputStream inStream) throws IOException,
        StreamCorruptedException
```

Example: To write object in file.

```
import java.io.Serializable;
import java.io.FileNotFoundException;
import java.io.FileOutputStream;
import java.io.IOException;
```

```
import java.io.ObjectOutputStream;
class Employee implements Serializable
{
    private String firstName;
    private String lastName;
    private String designation;
    private String department;
    private int basic;
    private int age;
    public Employee()
    {
    }
    public void setFirstName(String firstName)
    {
        this.firstName = firstName;
    }
    public String getFirstName()
    {
        return firstName;
    }
    public void setLastName(String lastName)
    {
        this.lastName = lastName;
    }
    public String getLastName()
    {
        return lastName;
    }
    public void setDepartment(String department)
    {
        this.department = department;
    }
    public String getDepartment()
    {
        return department;
    }
    public void setDesignation(String designation)
    {
        this.designation = designation;
    }
    public String getDesignation()
    {
        return designation;
    }
    public void setBasic(int basic)
    {
```

```
        this.basic = basic;
    }
    public int getBasic()
    {
        return basic;
    }
    public void setAge(int age)
    {
        this.age = age;
    }
    public int getAge()
    {
        return age;
    }
    public String toString()
    {
        StringBuffer buffer = new StringBuffer();
        buffer.append(firstName);
        buffer.append("\n");
        buffer.append(lastName);
        buffer.append("\n");
        buffer.append(department);
        buffer.append("\n");
        buffer.append(designation);
        buffer.append("\n");
        buffer.append(age);
        buffer.append("\n");
        return buffer.toString();
    }
}
public class ObjectWriteDemo
{
    public void writeEmployee(String filename)
    {
        ObjectOutputStream outputStream = null;
        try
        {
            outputStream = new ObjectOutputStream(new
        FileOutputStream(filename));
            Employee empObj = new Employee();
            empObj.setFirstName("Alice");
            empObj.setLastName("Awad");
            empObj.setDepartment("Sales");
            empObj.setDesignation("Sales Head");
            empObj.setBasic(464556);
            empObj.setAge(49);
```

```
            outputStream.writeObject(empObj);
            empObj = new Employee();
            empObj.setFirstName("Kanak");
            empObj.setLastName("Joe");
            empObj.setDepartment("Accounts");
            empObj.setDesignation("Accountant");
            empObj.setBasic(345667);
            empObj.setAge(46);
            outputStream.writeObject(empObj);
        }
        catch (FileNotFoundException ex)
        {
            ex.printStackTrace();
        }
        catch (IOException ex)
        {
            ex.printStackTrace();
        }
        finally
        {
            try
            {
                if (outputStream != null)
                {
                    outputStream.flush();
                    outputStream.close();
                }
            }
            catch (IOException ex)
            {
                ex.printStackTrace();
            }
        }
    }
    public static void main(String[] args)
    {
        new ObjectWriteDemo().writeEmployee("EmpFile.txt");
    }
}
```

Example: To read object from file.

```
import java.io.EOFException;
import java.io.FileInputStream;
import java.io.FileNotFoundException;
import java.io.IOException;
```

```
import java.io.ObjectInputStream;
public class ObjectReadDemo
{
   public void readEmployee(String filename)
   {
       ObjectInputStream inputStream = null;
         try
        {
            inputStream = new ObjectInputStream(new
        FileInputStream(filename));
             Object obj = null;
             while ((obj = inputStream.readObject()) != null)
             {
                if (obj instanceof Employee)
                {
                   System.out.println(((Employee)obj).toString());
                 }
              }
           }
          catch (EOFException ex)
          {
             System.out.println("End of file.");
           }
           catch (ClassNotFoundException ex)
           {
             ex.printStackTrace();
           }
           catch (FileNotFoundException ex)
            {
             ex.printStackTrace();
             }
            catch (IOException ex)
            {
             ex.printStackTrace();
              }
            finally
            {
              try
              {
                  if (inputStream != null)
                  {
                      inputStream.close();
                  }
              }
            catch (IOException ex)
            {
```

```
                    ex.printStackTrace();
               }
          }
     }
     public static void main(String[] args)
     {
          new ObjectReadDemo().readEmployee("EmpFile.txt");
     }
}
```

Output:

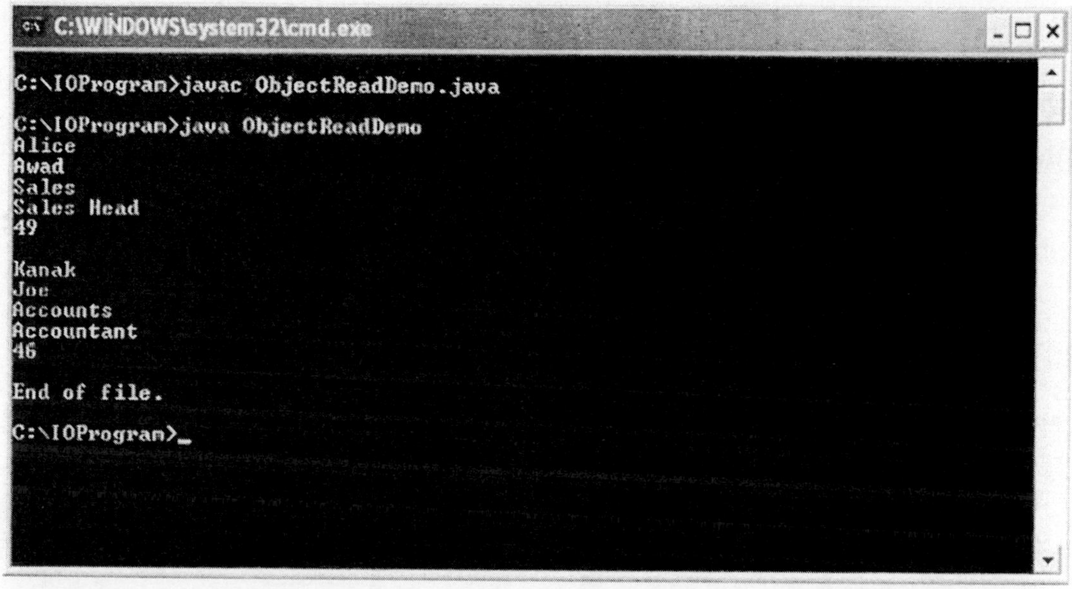

5.9. STREAM BENEFITS

Java provides stream interfaces that facilitate the programmer to write the complex and heavy code to deal with I/O very easily, by given abstraction. Java programs are written using high-level InputStream, OutputStream, Reader and Writer classes. Serialization plays an important role in I/O operations on an object to persist it.

5.10. COLLECTIONS

A collection is a single object representing a group of objects. The objects in the collection are called elements. They are contained in the java.util package. Collections are used to maintain references to the object of type *Object*. Collections were designed to provide high-performance to the processes. The collection framework defines several interfaces as:

- **Collection** - Enables you to work with groups of objects; it is at the top of the collections hierarchy
- **List** - Extends **Collection** to handle sequences (lists of objects)
- **Set** - Extends **Collection** to handle sets, which must contain unique elements

- **SortedSet** - Extends **Set** to handle sorted sets

Collections also use the **Comparator, Iterator, ListIterator** and **RandomAccess. Comparator** defines how two objects are compared. **Iterator** and **ListIterator** enumerate the objects within a collection. By implementing **RandomAccess**, a list indicates that it supports efficient, random access to its elements.

REVIEW QUESTIONS

Q. 1. Why are streams used in java?

Ans. In java streams are used to process input and output data as a sequence of bytes, which does not assume a specific character content or encoding. Byte content can be read from network source, files and other source, and bytes copies between multiple input and output.

Streams types can be sub-classed to add filtering, to mark a point in the stream and re-read those bytes, or to skip a number of bytes. Streams also enable serlializable objects to stored and re-constructed using ObjectInputStream and ObjectOutputStream types.

Q. 2. What's the difference between a file and a filename? Which is more closely related to the java.io.File class?

Ans. A file is the actual data located at a particular location on a disk. A filename is used to refer to the file. The name of the file can be changed without changing the data. The java.io.File class is closer to a filename than a file.

❑❑❑

6

CHAPTER

AWT and Applets in Java

INSIDE THIS CHAPTER

An applet is a Java program that can be used to create dynamic web pages on the browser with the help of HTML tags. All applets are subclasses of **Applet** class. Thus, all applets must import **java.applet**. Applets must also import **java.awt**. Applets differ from Java applications in the following aspects:

- An applet is a program that has to be run from a web browser while a Java application is a stand-alone program that can run by itself.
- We can use all the features of a web browser that is being used for creating an applet.
- When a user loads a page containing a Java applet in its browser, the browser downloads the applet on the local machine and executes the bytecode on the local system.
- Applets cannot communicate with other servers on the network.
- Applets are restricted from using the libraries from other languages such as C++.
- Applets execute under strict security limitations that disallow certain operations whereas Applications have no inherent security restrictions.
- Applets don't have the main() method as in applications. Instead they operate on an entirely different mechanism where they are initialized by init(),started by start(),stopped by stop() or destroyed by destroy().

The class hierarchy for Applet class is as follows:

```
java.lang.Object
      ↓
java.awt.Component
      ↓
java.awt.Container
      ↓
java.awt.Panel
      ↓
java.applet.Applet
```

Fig. 6.1. Applet class hierarchy

We will discuss about applet life cycle and using HTML tags with Applet in the coming sections.

6.1. AWT

Abstract Window Toolkit was the driving force behind java's popularity. We can create and display buttons, labels, menus, text fields, combo boxes, and other user-interface controls. Most popular AWT classes are:

- **Applet** : Creates an Applet
- **Button** : Creates a Button
- **Canvas** : Creates a Canvas for Drawing
- **Checkbox** : creates a Checkbox
- **Choice** : Creates a Choice Control
- **Label** : Creates a Label
- **Menu** : Creates a Menu
- **ComboBox** : Creates a combo box
- **List** : Creates a list control
- **Frame** :Creates a frame for windowed applications
- **Dialog** : Creates a Dialog Box
- **Panel**: Creates a panel that can contain other controls
- **PopupMenu**: Creates a Pop-up Menu
- **RadioButton** : Creates a radio Button
- **Scrollbar** : Creayes a Scrollbar
- **ScrollPane** : Creates a Scrollable surface
- **TextArea**: Creates a 2-D Text control
- **TextPane** : Creates a Text surface
- **TextField** : Creates a 1-D Text field
- **Window** : Creates a Freestanding window

The features of Abstract Window Toolkit will be discussed in detail in coming chapters.

6.2. APPLET BASICS

An *applet* is a class file that is specially written to display graphics in a web browser. Applets are embedded in web pages using the HTML tag <Applet>. Java applets are downloaded automatically and run by the web browser. Any web-browser that supports Java, such as *Applet Viewer*, can be used to transport *applets* to the Internet. An applet can perform many functions such as graphics, play sound, animation, arithmetic operations, play games etc.

Applets differ from application programs in the sense that they do not use *main()* program for initiating the execution of code. They do not run independently but from inside a web page that makes an efficient use of HTML tags. To build an applet code, we need to include certain packages like *java.awt* and *java.applet* that contains the *Graphics* class that provides methods where the output of all *applet* operations are performed. A simple program using *applet* looks like:

```
import java.awt.*;
import java.applet.*;
/* <applet code="NewApplet" width=500 height=200>
    </applet>*/
public class NewApplet extends Applet
```

```
{
  public void paint(Graphics g)
  {
    g.drawString("New Applet Program", 20, 100);
  }
}
```

The statement below draws the string *New Applet Program* at the pixel position (20, 100).

```
g.drawString("New Applet Program", 20, 100);
```

To execute Java Applet, *appletviewer* is used. We pass the source filename as an argument to *appletviewer* because it will read the <applet> tag given in comment part of the source file. From there, it will find which class has to execute from the attribute *code*, what will be the width and the height of applet window, read from width and height attribute of <applet> tag respectively. The screenshot below explains the compilation and execution of a Java Applet.

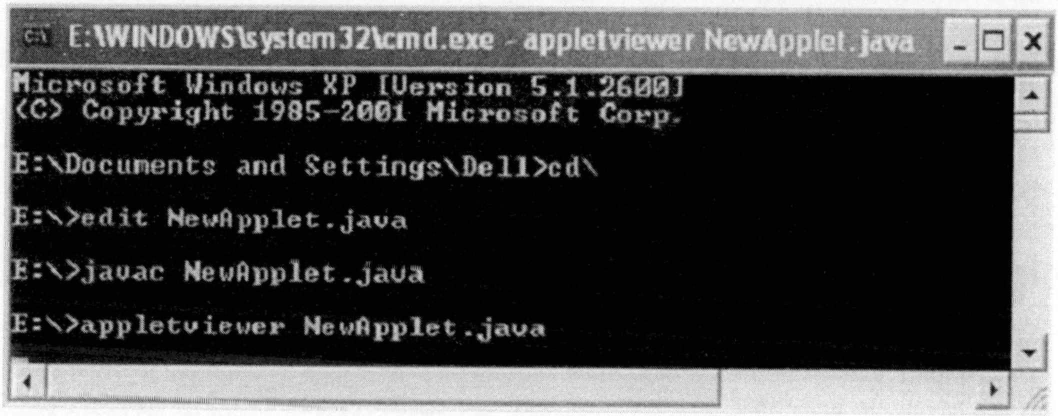

The screenshot below shows the starting of an Applet.

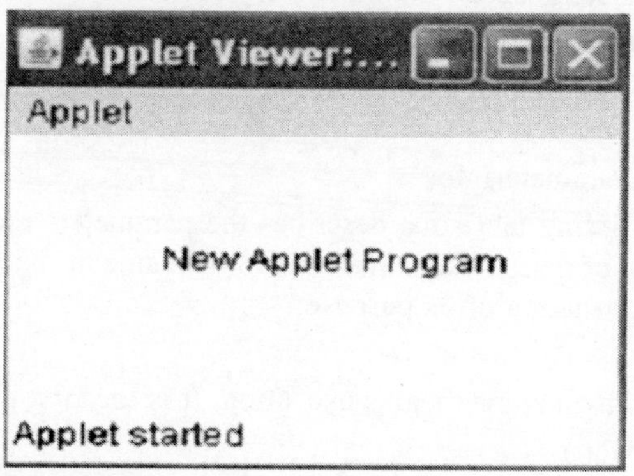

The **Applet** class defines the methods described below. **Applet** provides all necessary support for applet execution, such as starting and stopping. It also provides methods that load and display images, and methods that load and play audio clips. **Applet** extends the AWT class **Panel**. In turn, **Panel** extends **Container**, which extends **Component**. These classes provide support for Java's window-based, graphical interface. Thus, **Applet** provides all of the necessary support for window-based activities.

Method: void destroy( )

Description: Called by the browser just before an applet is terminated. Your applet will override this method if it needs to perform any cleanup prior to its destruction.

Method: AccessibleContext getAccessibleContext( )

Description: Returns the accessibilty context for the invoking object.

Method: AppletContext getAppletContext( )

Description: Returns the context associated with the applet.

Method: String getAppletInfo( )

Description: Returns a string that describes the applet.

Method: AudioClip getAudioClip(URL *url*)

Description: Returns an **AudioClip** object that encapsulates the audio clip found at the location specified by *url*.

Method: AudioClip getAudioClip(URL *url*, String *clipName*)

Description: Returns an **AudioClip** object that encapsulates the audio clip found at the location specified by *url* and having the name specified by *clipName*.

Method: URL getCodeBase( )

Description: Returns the URL associated with the invoking applet.

Method: URL getDocumentBase( )

Description: Returns the URL of the HTML document that invokes the applet.

Method: Image getImage(URL *url*)

Description: Returns an **Image** object that encapsulates the image found at the location specified by *url*.

Method: Image getImage(URL *url*, String *imageName*)

Description: Returns an **Image** object that encapsulates the image found at the location specified by *url* and having the name specified by *imageName*.

Method: Locale getLocale( )

Description: Returns a **Locale** object that is used by various locale-sensitive classes and methods.

Method: String getParameter(String *paramName*)

Description: Returns the parameter associated with *paramName*. **null** is returned if the specified parameter is not found.

Method: String[ ] [ ] getParameterInfo( )

Description: Returns a **String** table that describes the parameters recognized by the applet. Each entry in the table must consist of three strings that contain the name of the parameter, a description of its type and/or range, and an explanation of its purpose.

Method: void init( )

Description: Called when an applet begins execution. It is the first method called for any applet

Method: boolean isActive( )

Description: Returns **true** if the applet has been started. It returns **false** if the applet has been stopped.

Method: static final AudioClip newAudioClip(URL *url*)

Description: Returns an **AudioClip** object that encapsulates the audio clip found at the location specified by *url*. This method is similar to **getAudioClip( )** except that it is static and can be executed without the need for an **Applet** object

Method: void play(URL *url*)

Description: If an audio clip is found at the location specified by *url*, the clip is played.

Method: void play(URL *url*, String *clipName*)

Description: If an audio clip is found at the location specified by *url* with the name specified by *clipName*, the clip is played

Method: void resize(Dimension *dim*)

Description: Resizes the applet according to the dimensions specified by *dim*. **Dimension** is a class stored inside **java.awt**. It contains two integer fields: **width** and **height**.

Method: void resize(int *width*, int *height*)

Description: Resizes the applet according to the dimensions specified by *width* and *height*.

Method: final void setStub(AppletStub *stubObj*)

Description: Makes *stubObj* the stub for the applet. This method is used by the run-time system and is not usually called by your applet. A *stub* is a small piece of code that provides the linkage between your applet and the browser.

Method: void showStatus(String *str*)

Description: Displays *str* in the status window of the browser or applet viewer. If the browser does not support a status window, then no action takes place.

Method: void start( )

Description: Called by the browser when an applet should start (or resume) execution. It is automatically called after **init( )** when an applet first begins.

Method: void stop( )

Description: Called by the browser to suspend execution of the applet. Once stopped, an applet is restarted when the browser calls **start( )**.

6.3. RESTRICTIONS TO AN APPLET

The following are the restrictions that are placed on Java applets by different web browsers such as Internet Explorer and Netscape Navigator:

- Applets are not allowed to read from or write to files on the local system.
- Applets are not allowed to execute any applications on the local computer.
- Applets are not allowed to create DLLs on the local machine.
- Applets are allowed only to communicate with the server on which they are the resident or with other servers only via Internet or the client's local computer. This means that an applet cannot make network connections.
- Applets cannot load libraries.

6.4. APPLET ARCHITECTURE

An applet is a window-based program. Its architecture is different from the simple console-based programs. There are a few key concepts that the user must understand. First, applets are event driven. An applet resembles a set of interrupt service routines. An applet waits until an event occurs. The AWT notifies the applet about an event by calling an event handler that has been provided by the applet. Once this is done, the applet must take an appropriate action and then quickly return control to the AWT. This is a very important point. The applet should not enter a "mode" of operation in which it maintains control for an extended period. Rather, it must perform some specific actions in response to the events and then return control to the AWT run-time system. In the cases where the applet needs to perform a repetitive task on its own, we must start an additional thread of execution. Secondly, the user initiates interaction with an applet. The user interacts with the applet as and when required. These interactions are sent to the applet as events to which the applet must respond. Applets contain various controls, such as push buttons and check boxes. When the user interacts with one of these controls, an event is generated. While the architecture of an applet is not as easy to understand as that of a console-based program, Java's AWT makes it as simple as possible.

6.5. APPLET LIFE CYCLE

Java *applet* inherits features from the class *Applet*. Thus, whenever an *applet* is created, it undergoes a series of changes from initialization to destruction. Various stages of an *applet* life cycle are depicted in the figure below:

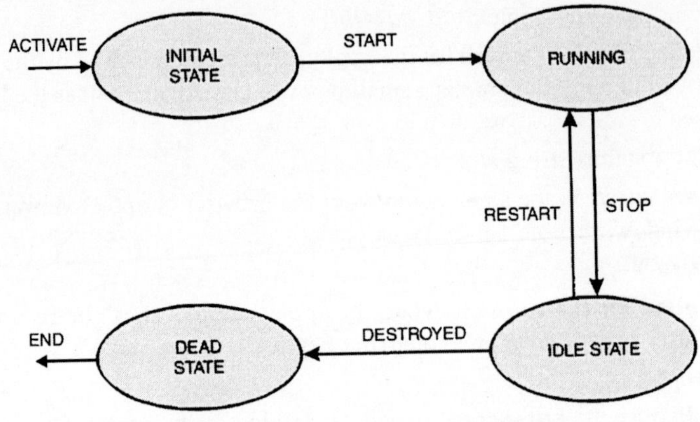

Fig. 6.2. Life Cycle of an Applet

Initial State

When a new *applet* is born or created, it is activated by calling *init()* method. At this stage, new objects to the *applet* are created, initial values are set, images are loaded and the colors of the images are set. An *applet* is initialized only once in its lifetime. It's general form is:

```
public void init( )
{
//Action to be performed
}
```

Running State

An *applet* achieves the running state when the system calls the *start()* method. This occurs as soon as the *applet* is initialized. An *applet* may also start when it is in idle state. At that time, the *start()* method is overridden. It's general form is:

```
public void start( )
{
//Action to be performed
}
```

Idle State

An *applet* comes in idle state when its execution has been stopped either *implicitly or explicitly*. An *applet* is *implicitly* stopped when we leave the page containing the currently running applet. An *applet* is *explicitly* stopped when we call *stop()* method to stop its execution. It's general form is:

```
public void stop( )
{
//Action to be performed
}
```

Dead State

An *applet* is in dead state when it has been removed from the memory. This can be done by using *destroy()* method. It's general form is:

```
public void destroy( )
{
//Action to be performed
}
```

Apart from the above stages, Java applet also possess *paint( )* method. This method helps in drawing, writing and creating colored backgrounds of the applet. It takes an argument of the graphics class. To use the graphics, it imports the package java.awt.Graphics

6.6. SIMPLE APPLET DISPLAY METHODS

Applets are displayed in a window and they use the AWT to perform input and output functions. To output a string to an applet, use **drawString( )**, which is a member of the **Graphics** class. Typically, it is called from within either **update( )** or **paint( )**. It has the following general form:

```
void drawString(String message, int x, int y)
```

Here, *message* is the string to be output beginning at *x, y*. In a Java window, the upper-left corner is location 0, 0. The **drawString( )** method will not recognize newline characters. If we want to start a line of text on another line, we must do it manually, specifying the X,Y location where we want the line to begin.

To set the background color of an applet's window, we use **setBackground( )**. To set the foreground color, we use **setForeground( )**. These methods are defined by **Component**, and have the general forms:

```
void setBackground(Color newColor)

void setForeground(Color newColor)
```

Here, *newColor* specifies the new color. The class **Color** defines the following values that can be used to specify colors:

```
Color.black Color.magenta Color.blue  Color.orange  Color.cyan
Color.pink Color.darkGray    Color.red    Color.gray    Color.white
Color.green    Color.yellow   Color.lightGray
```

For example, this sets the background color to blue and the text color to yellow:

```
setBackground(Color.blue);

setForeground(Color.yellow);
```

We can change these colors as and when required during the execution of the applet. The default foreground color is black. The default background color is light gray. We can obtain the current settings for the background and foreground colors by calling **getBackground( )** and **getForeground( )**, respectively. They are also defined by **Component** and bear the general form:

```
Color getBackground( )

Color getForeground( )
```

The example below shows a simple applet to change the color of the foreground and background.

```
import java.awt.*;
import java.applet.*;
/* <applet code="Applet_Prog" width=500 height=550>
</applet>*/
public class Applet_Prog extends Applet
{
 public void paint(Graphics g)
 {
  setBackground(Color.BLUE);
  setForeground(Color.RED);
  g.drawString("Using Colors in An Applet" , 100,250);
 }
}
```

The output of the Applet would look like:

6.7. REQUESTING REPAINTING

One of the important architectural constraints that have been imposed on an applet is that it must quickly return control to the AWT run-time system. It cannot create a loop inside **paint( )**. This would prevent control from passing back to the AWT. Whenever your applet needs to update the information displayed in its window, it simply calls **repaint( )**. The **repaint( )** method is defined by the AWT that causes AWT run-time system to execute a call to your applet's **update( )** method, which in turn calls **paint( )**. The AWT will then execute a call to **paint( )** that will display the stored information. The **repaint( )** method has four forms. The simplest version of **repaint( )** is:

```
void repaint( )
```

This causes the entire window to be repainted. Other versions that will cause repaint are:

```
void repaint(int left, int top, int width, int height)
```

If your system is slow or busy, **update( )** might not be called immediately. If multiple calls have been made to AWT within a short period of time, then **update( ) is not** called very frequently. This can be a problem in many situations in which a consistent update time is necessary. One solution to this problem is to use the following forms of **repaint( )**:

```
void repaint(long maxDelay)

void repaint(long maxDelay, int x, int y, int width, int height)
```

Here, maxDelay specifies the maximum number of milliseconds that can elapse before **update( )** is called. If the time elapses before **update( )** can be called, it isn't called. It is possible for a method other than **paint( )** or **update( )** to output to an applet's window. To do so, it must obtain a graphics context by calling **getGraphics( )** and then use this context to output to the window. For most applications, it is better and easier to route window output through **paint( )** and to call **repaint( )** when the contents of the window change.

6.8. USING THE STATUS WINDOW

An applet can output a message to the status window of the browser or applet viewer on which it is running. For this, it makes a call to **showStatus( )** with the string that we want to be displayed. The status window is a place where the user can give feedback about what is occurring in the applet, suggest options, or report some types of errors. The status window also makes an excellent debugging aid, because it gives an easy way to output information about the applet. The applet below shows the use of **showStatus( )**:

```
import java.awt.*;
import java.applet.*;
/*
<applet code="StatusWindow" width=300 height=50>
</applet>
*/
public class StatusWindow extends Applet
{
  public void init()
 {
   setBackground(Color.pink);
}

public void paint(Graphics g)
{
   g.drawString("You are in main applet window.", 10, 20);
   showStatus("This is the status window.");
 }
}
```

The output of the above Applet code is as follows:

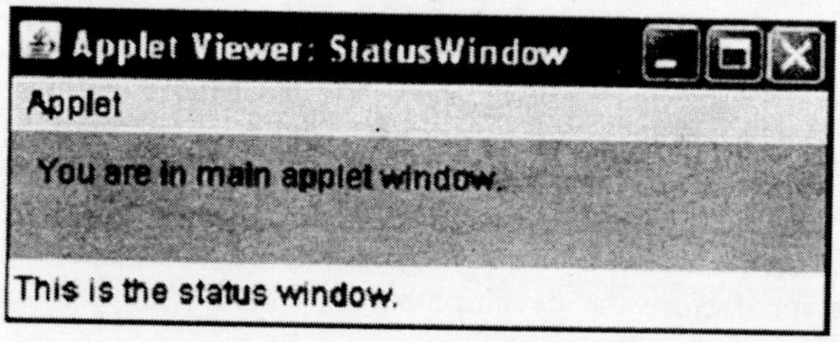

6.9. HTML

HTML is the Hyper Text Markup Language that is used to create web pages. HTML can link words, phrases, block of text or images on to the web pages. HTML uses various tags to mark-up elements. The tags are surrounded by angular brackets, i.e. < and >. The tags are normally in pairs where we have start tag and an end tag For example,

```
<Body>\\ set of statements </Body>
```

To create a normal HTML file, we need to type the following code in the notepad.

```
<HTML>
<head>
<title> HTML Introduction </title>
</head>
<body>
Welcome to HTML Homepage
</body>
</HTML>
```

Save this notepad file as test.html and open your Internet browser. Select "Open" in the file menu. A dialog box will open. Select "Browse" to browse the file test.htm, select the file and click 'OK'. The browser will display the page as follows:

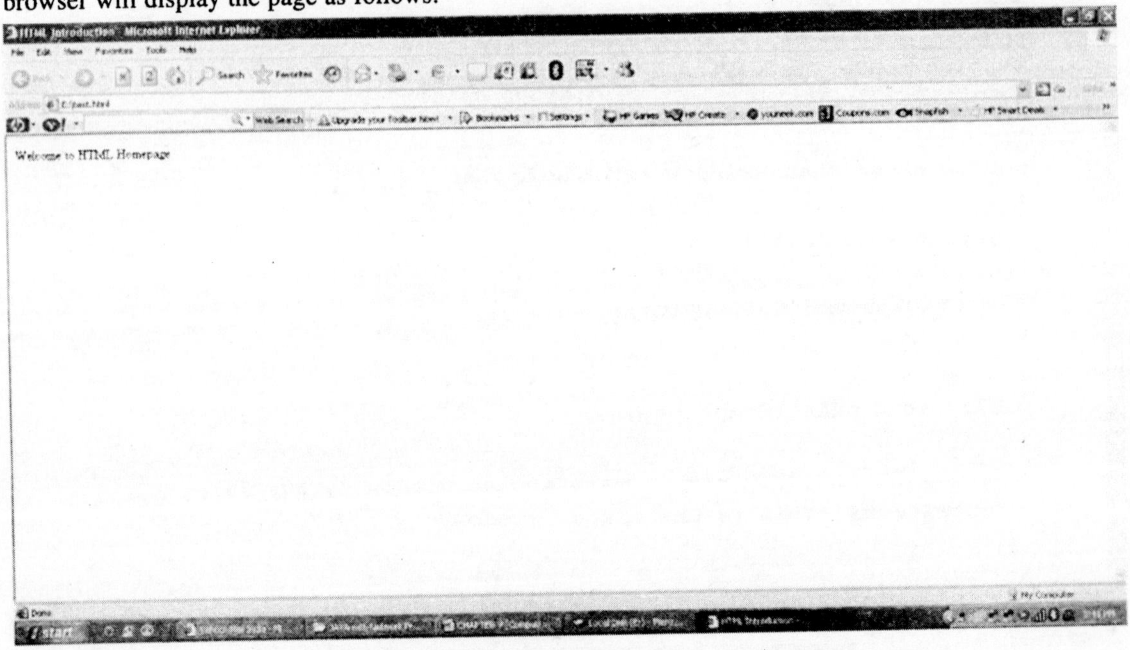

A web page designed using HTML is a static web page. This means that the information present in a web page remains constant. The dynamic behavior of the web pages makes it more attractive. The dynamic behavior can be in the form of scrolling text, messages displayed on the status bar and changing images. To apply such a behavior with HTML, we make use of Java applets.

An applet is a small application that can be embedded in a HTML page providing it the ability to show the dynamic behavior. Before the development of World Wide Web, applets were the built-in programs for writing and drawing that came with windows. Java applets are capable of performing interactive animations, calculations and other tasks without sending the request back to the server.

6.9.1. The HTML APPLET Tag

An applet viewer executes each APPLET tag that it finds in a separate window, while web browsers like Netscape Navigator, Internet Explorer allow many applets on a single page. The syntax for the standard APPLET tag is shown here. The items in the brackets are optional.

```
< APPLET
[CODEBASE = codebaseURL]
CODE = appletFile
[ALT = alternateText]
[NAME = appletInstanceName]
WIDTH = pixels HEIGHT = pixels
[ALIGN = alignment]
[VSPACE = pixels] [HSPACE = pixels]
>
[< PARAM NAME = AttributeName VALUE = AttributeValue>]
[< PARAM NAME = AttributeName2 VALUE = AttributeValue>]
. . .
[HTML Displayed in the absence of Java]
</APPLET>
```

Each component can be described as follows:

CODEBASE: CODEBASE is an optional attribute that specifies the base URL of the applet code, which is the directory that will be searched for the applet's executable class file. The HTML document's URL directory is used as the CODEBASE if this attribute is not specified. The CODEBASE does not have to be on the host from which the HTML document was read.

CODE: CODE is a required attribute that gives the name of the file containing your applet's compiled **.class** file. This file is relative to the code base URL of the applet, which is the directory that the HTML file was in or the directory indicated by CODEBASE if set.

ALT: The ALT tag is an optional attribute used to specify a short text message that should be displayed if the browser understands the APPLET tag but can't currently run Java applets. This is distinct from the alternate HTML you provide for browsers that don't support applets.

NAME: NAME is an optional attribute used to specify a name for the applet instance. Applets must be named in order for other applets on the same page to find them by name and communicate with them. To obtain an applet by name, use **getApplet( )**, which is defined by the **AppletContext** interface.

WIDTH AND HEIGHT: WIDTH and HEIGHT are required attributes that give the size (in pixels) of the applet display area.

ALIGN: ALIGN is an optional attribute that specifies the alignment of the applet. This attribute is treated the same as the HTML IMG tag with these possible values: LEFT, RIGHT, TOP, BOTTOM, MIDDLE, BASELINE, TEXTTOP, ABSMIDDLE, and ABSBOTTOM.

VSPACE AND HSPACE: These attributes are optional. VSPACE specifies the space, in pixels, above and below the applet. HSPACE specifies the space, in pixels, on each side of the applet. They're treated the same as the IMG tag's VSPACE and HSPACE attributes.

PARAM NAME AND VALUE: The PARAM tag allows you to specify applet specific arguments in an HTML page. Applets access their attributes with the **getParameter( )** method.

HANDLING OLDER BROWSERS: Some very old web browsers can't execute applets and don't recognize the APPLET tag. Although these browsers are now nearly extinct, you may need to deal with them occasionally. The best way to design your HTML page to deal with such browsers is to include HTML text and markup within your <applet></applet> tags. If the applet tags are not recognized by your browser, you will see the alternate markup. If Java is available, it will consume the entire markup between the <applet></applet> tags and disregard the alternate markup.

The HTML code to start an applet called **TestApplet** in Java and to display a message in older browsers is as follows:

```
<applet code="TestApplet" width=200 height=50>
```

6.10. PASSING PARAMETERS TO APPLETS

The APPLET tag in HTML allows us to pass parameters to the applet. To retrieve a parameter, we use the **getParameter( )** method. It returns the value of the specified parameter in the form of a **String** object. Thus, for numeric and **boolean** values, it is required to convert the string representations into their internal formats. The example below explains passing parameters:

```
import java.awt.*;
import java.applet.*;
/*
<applet code="ParamApplet" width=300 height=80>
<param name=fontName value=Courier>
<param name=fontSize value=14>
<param name=leading value=2>
<param name=accountEnabled value=true>
</applet>
*/
public class ParamApplet extends Applet
{
String name;
int size;
float lead;
boolean active;

public void start()
{
String pa;
name = getParameter("Name");
if(name == null)
name = "Not Found";
pa = getParameter("Size");
try
 {
   if(pa != null)
   size = Integer.parseInt(pa);
   else
   size = 0;
 }
```

```
 catch(NumberFormatException e)
 {
    size = -1;
 }

 pa = getParameter("lead");

 try
  {
    if(pa != null)
    lead = Float.valueOf(pa).floatValue();
    else
    lead = 0;
  }

 catch(NumberFormatException e)
 {
    lead = -1;
  }

 pa = getParameter("AccountEnabled");
 if(pa != null)
 active = Boolean.valueOf(pa).booleanValue();
 }

 public void paint(Graphics g)
 {
    g.drawString("Name: " + name, 0, 10);
    g.drawString("Size: " + size, 0, 26);
    g.drawString("Leading: " + lead, 0, 42);
    g.drawString("Account Active: " + active, 0, 58);
  }
 }
```

Sample output from this program is shown here:

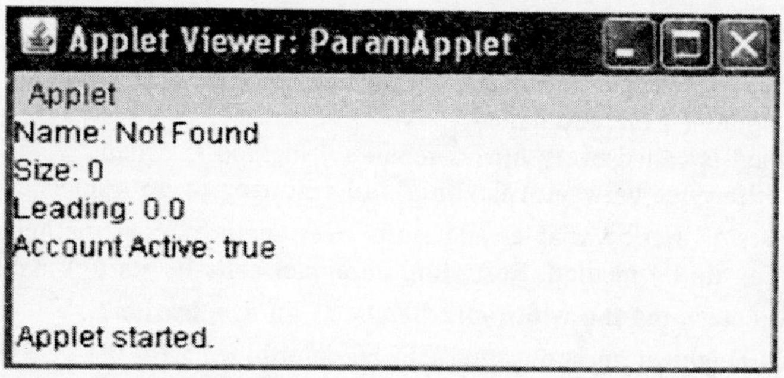

REVIEW QUESTIONS

Q. 1. What are applets? Explain their usage. How are they different from Java programs?
[MDU, May 2012; May 2007]

Ans. An *applet* is a class file that is specially written to display graphics in a web browser. Applets are embedded in web pages using the HTML tag <Applet>. Java applets are downloaded automatically and run by the web browser. *Applets* differ from application programs in the sense that they do not use *main()* program for initiating the execution of code. They do not run independently but from inside a web page that makes an efficient use of HTML tags.

Q. 2. Can Applets communicate with each other?

Ans. At this point in time applets may communicate with other applets running in the same virtual machine. If the applets are of the same class, they can communicate via shared static variables. If the applets are of different classes, then each will need a reference to the same class with static variables. In any case the basic idea is to pass the information back and forth through a static variable. An applet can also get references to all other applets on the same page using the getApplets() method of java.applet.AppletContext. Once you get the reference to an applet, you can communicate with it by using its public members. It is conceivable to have applets in different virtual machines that talk to a server somewhere on the Internet and store any data that needs to be serialized there. Then, when another applet needs this data, it could connect to this same server. Implementing this is non-trivial.

Q. 3. What is the difference between Applet and Application?

Ans. Applets differ from Java applications in the following aspects:

- An applet is a program that has to be run from a web browser while a Java application is a stand-alone program that can run by itself.
- We can use all the features of a web browser that is being used for creating an applet.
- When a user loads a page containing a Java applet in its browser, the browser downloads the applet on the local machine and executes the bytecode on the local system.
- Applets cannot communicate with other servers on the network.
- Applets are restricted from using the libraries from other languages such as C++.
- Applets execute under strict security limitations that disallow certain operations whereas Applications have no inherent security restrictions.
- Applets don't have the main() method as in applications. Instead they operate on an entirely different mechanism where they are initialized by init(),started by start(),stopped by stop() or destroyed by destroy().

Q. 4. How an Applet is initialized?

Ans. An applet is initialized using init( ) method.

Q. 5. What are the steps in Applet lifecycle?

Ans. Various steps are:

- init( )
- start( )
- paint( )
- stop( )
- destroy( )

Q. 6. When is the update( ) method called?

Ans. update( ) method is called every time a repaint( ) method is called.

Q. 7. What is the difference between reloading and restating of an applet?

Ans. Reloading clears all static variables and starts over again by constructing a new instance of the applet by calling init( ) method. Restarting an applet calls its start( ) method.

Q. 8. How can you determine the width and height of an application?

Ans. The width and height of an application can be determined with the getSize( ) method.

❑❑❑

7

CHAPTER

Networking

INSIDE THIS CHAPTER

To implement networking in Java, java.net package classes, that give access to IP address, and a range of methods related to TCP, UDP, and URL. IO stream classes are also used to read and write data from and to communication channel. A virtual-stream is provided by TCP, which supports IP connection across the network. The virtual-stream supports IO stream to transport data from and to either of the connection. There are two important classes that participate in communication. *InetAddress* class provides IP address of target connection. *Socket* class is used to creates TCP connection. Once a connection has been created, we bind both the end socket with their streams, and communication will start by writing and reading data on these streams to and from remote application.

7.1. BASICS OF NETWORKING

Networking means transmitting data from one device to another device on the network, but both devices must use the same protocol for transmission. There is an array of stack, called *protocol stack* that

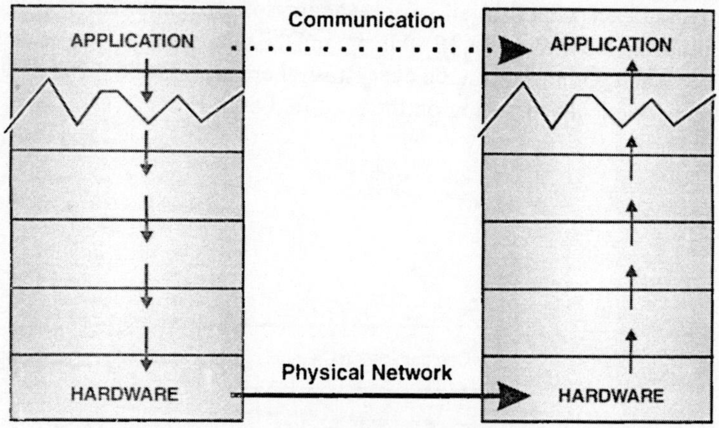

Fig. 7.1. Stack of Layer

involves everything from giving the data to physical device by converting that in voltage pulse to delivering data streams to applications for which it was destined to.

Each layer provides an abstraction and dedicated to a specific fix set of jobs. The intermediate layers also work as an interface between its upward and downward layer, first and the last. The two most famous and important network stacks are given below:

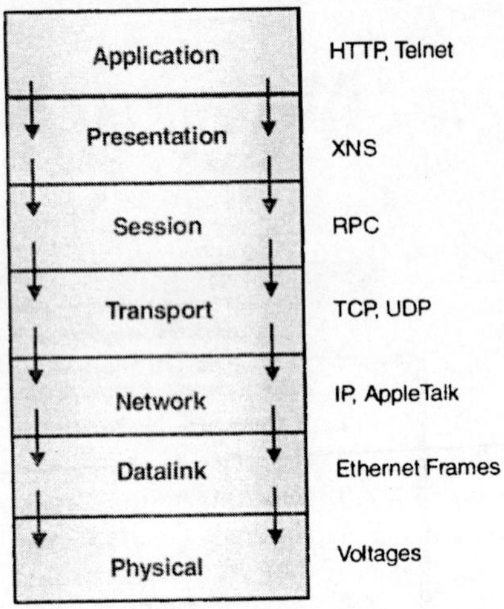

Fig. 7.2. OSI Layer networking stack

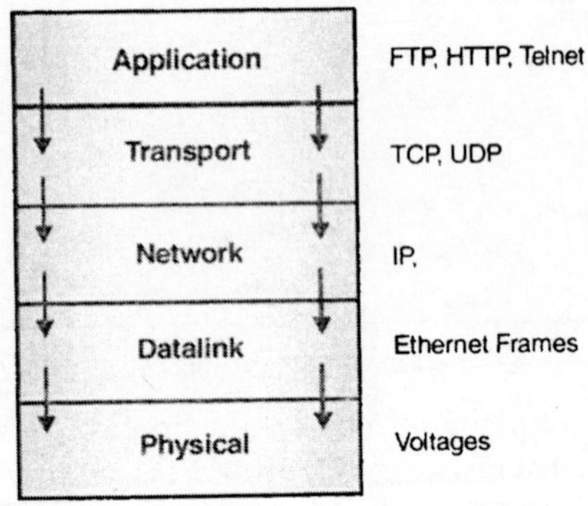

Fig. 7.3. TCP/IP Layer networking stack

Each layer communicates only with its immediate above and below layers. Encapsulation and decapsulation are two mechanism, that embed each layer's packets into the immediately below layer packet, and reverse the process as the data moves up at the receiving machine, respectively. An internet is a bunch of networks of different types of environments. The internet is exclusively the worldwide set of connections of TCP/IP networks which use IP address and protocols under the ultimate command.

7.2. SOCKETS

A *Socket* represents a TCP connection with a machine on network. This class facilitate in establishing the stream-based channel with a remote machine. For TCP/IP communication, there is a need to create a *Socket* with the remote machine. The class Socket creates a socket and automatically established a TCP connection, and throws an exception if it fails. *Socket* address is the combination of two components: one is host id, which can refer to the host name, IP address, and the port number. The port number is an integer value that lies between 1 and 65536. In fact, on every host, there are 655636 different addresses. But there must be a server, and dynamically listening on the specified port.

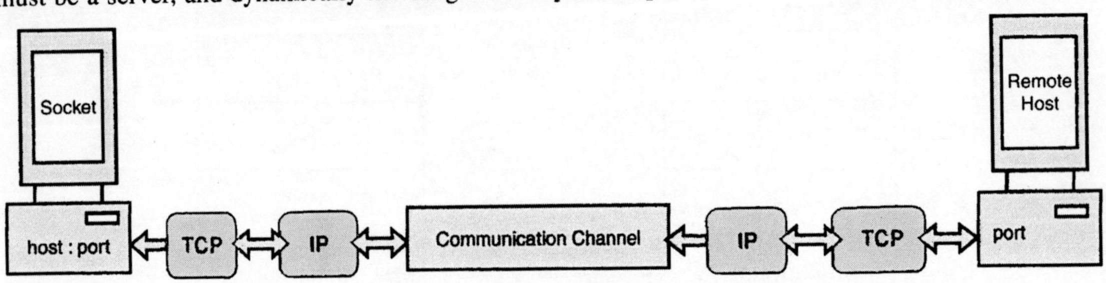

Fig. 7.4. Socket

There are two constructor of the *Socket* class. Creating a *Socket* object automatically connects to the given host and port. As earlier given, that a server actively listen the request, but as connection is refused an *IOException* will be thrown, unknown hostname can be other possibility of connection failure.

Socket(String hostName, int portNumber) throws IOException creates a *Socket* object and connects to the given host on the given port number.

```
Socket sckt = new Socket("mymac", 1889);
```

Socket(InetAddress hostAddress, int portNumber) throws IOException creates a *Socket* object and connect to the given host address on the given port number.

```
InetAddress mac = InetAddress.getByName("127.0.0.1");
Socket sckt = new Socket(mac.getAddress, 1889);
```

OR

```
Socket sckt = new Socket("127.0.0.1", 1889);
```

Socket class methods provide the identification of remote host, port number of remote and local host; also give the stream for communication. First create the Socket object to perform communications across a TCP connection. After creating the socket, bind it with the required or corresponding stream with the help of *getInputStream*() and *getOutputStream*() methods to communicate with the remote server. The client and server, both will have both an *InputStream* and an *OutputStream* for the purposes of communication.

TABLE 7.1. Methods of java.net.Socket class

InputStream getInputStream() throws IOException	Method returns an InputStream that allow stream-based communications for TCP connections. The data written by a host at one end; can be read by remote host from this *InputStream* at other end of the connection. The method actually returns a *SocketInputStream*. The *SocketInputStream* is a nonpublic class.
OutputStream getOutputStream() throws IOException	Method returns an *OutputStream* that allow stream-based communication from TCP connection. The data written to this stream by any machine at one end; will be read by remote host at the other end of connection. This *OutputStream* should be buffered.
void close() throws IOException	The method is used to close the Socket, releasing all network and system resources being used. The data, which have been sent before this the call of this method, will be successfully delivered to the remote end, if there is no network fault.
InetAddress getInetAddress()	Method returns the IP address of the remote host.
int getPort()	Method returns the port number of the remote host to which the Socket is connected.
int getLocalPort()	Method returns the local port number, *i.e.*, the port number of the Socket itself. Every TCP connection in fact consists of IP address and port number of local machine; and IP address and port number of remote machine, which will generally be different. A random and unused port number is assigned to Socket at its creation time.

7.3. INTERNET ADDRESSING

TCP/IP is the architecture, implemented on the internet. The foundation of the TCP/IP is the IP (internetworking protocol). IP works on 3rd layer of the architecture, i.e. network layer. The application of IP is addressing hosts and routing data packets (also called datagram) between them. There are two versions of IP, IPv4 and IPv6. IPv4 contains 32-bit long address, while IPv6 contains 128-bit long address.

In the chapter we will use IPv4 addressing. IPv4 format is xx.xx.xx.xx, where each two consecutive 'x' (xx) separated by dot (.) are one byte, such as 168.196.12.1. Every IP address belongs to any network. Network falls into five different *classes*. The class can be used to determine the number of host in the network of which it is a member.

- **Class A :** 1 – 126 (16M hosts)
- **Class B :** 128 – 191 (65536 hosts each)
- **Class C :** 192 - 223 (256 hosts each)
- **Class D :** 224 – 239 (addresses for multicasting)
- **Class E :** 240 – 255 (addresses reserved for future use)

The class defines the network size of the organization.

An IP datagram contains a header and a protocol data unit (PDU). The header collections of information about fragmentation, length, time to live (TTL), and similar parameters. The PDU contains the encapsulated data, which generally means a TCP segment or UDP packet.

Transmission Control Protocol (TCP) is a connection-oriented protocol that is on top of IP. TCP packets, called *segments*, are encapsulated with IP datagram. Encoding of TCP occurs at the end point of communication channel, i.e. at hosts. Routers inspect the IP packets to route TCP encoding and decoding occur at the hosts which form routing decisions.

TCP provides guaranteed data transmission at the destination host's application because of its connection-oriented features. TCP is not responsible for exact arrival time, or transfer rate at which data will be sent to remote end, only gives surety about delivery of uncorrupted and in order data.

User Datagram Protocol (UDP) is a connectionless transport layer protocol that is on top of IP. It doesn't provide state management and data integrity functions. It also not provides any congestion avoidance, and packet delivery guarantee facility. If any received packed is corrupted it simply throws it away.

7.4. NETWORK CLASSES AND INTERFACE

For networking, java uses java.net package. The classes contained in the **java.net** package are listed below:

Authenticator	InetSocketAddress	SocketImpl
ContentHandler	JarURLConnection	SocketPermission
DatagramPacket	MulticastSocket	URL
DatagramSocket	NetPermission	DatagramSocketImpl
NetworkInterface	URLClassLoader	HttpURLConnection
PasswordAuthentication	URLConnection	InetAddress
ServerSocket URLDecoder	Inet4Address	Socket
URLEncoder	Inet6Address	SocketAddress
URLStreamHandler		

The **java.net** package's interfaces are listed here:

ContentHandlerFactory	SocketImplFactory	URLStreamHandlerFactory
FileNameMap	SocketOptions	DatagramSocketImplFactory

7.5. INET ADDRESS

Computer on internet is identified by IP address, that is currently in two version with different size, in IPv4, it is 32-bit long (four byte) and in IPv6, it is 128-bit long (eight byte). *InetAddress* provides an abstraction to access IP address that is shown in the *figure* given below.

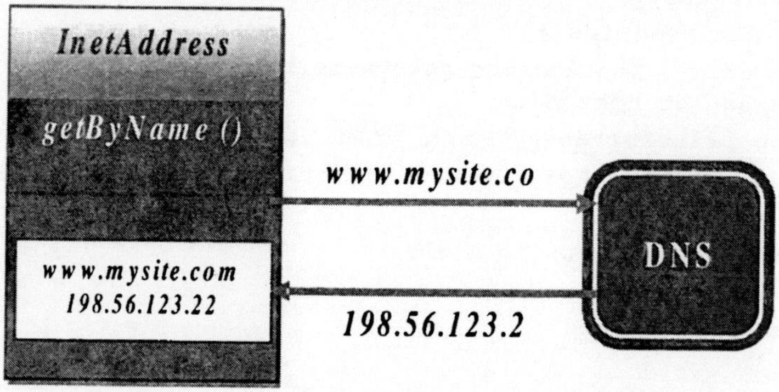

Fig. 7.5. Abstraction provided by InetAddress

There is no constructor of this class. The class provides some static methods to create the instance of the class, i.e. an object.

TABLE 7.2. Methods of java.net.InetAddress class

static InetAddress getLocalHost() throws UnknownHostException	Method returns an object corresponding to the local machine.
static InetAddress getByName(String *host*) throws UnkownHostException	Method returns an object of the specified *host*. The *host* can be a name (mysite.com) or by IP address (198.56.123.22)
static InetAddress[] getAllByName(String *host*) throws UnknownHostException	Methods return an array of objects corresponding to every known IP address for the specified host.

The methods given below provide the information from an InetAddress object :

TABLE 7.3. Methods of java.net.InetAddress class

byte[] getAddress()	Method return an array of bytes corresponding to the IP address, under IPv4 this array will only be of four bytes and for IPv6, it will be eight bytes.
String getHostName()	Method returns the name of the host.

UnknownHostException is subclass of *IOException* and indicates that the named host could not be successfully identified.

Example: Save the file with name "**InetAddressDemo.java**"

```
import java.net.*;
import java.io.*;
public class InetAddressDemo
{
public static void main(String ar[])
{
DataInputStream din = new DataInputStream(System.in);
String ip_name;
try
{
System.out.println("Enter a hostname or IP address :: ");
```

```
System.out.flush();
ip_name = din.readLine();
InetAddress mac = InetAddress.getByName(ip_name); //ip_name can be either
    IP address or host name
System.out.println("Host Name : "+ mac.getHostName());
System.out.println("Host IP : "+ formatIP(mac.getAddress()));
}
catch(UnknownHostException uhe)
{
System.out.println(uhe);
}
catch(IOException ioe)
{
System.out.println(ioe);
}
}
static String formatIP(byte ipadd[])
{
StringBuffer ip = new StringBuffer();
for(int i=0; i< ipadd.length; i++)
{
    if(i > 0)
    ip.append(".");
    ip.append(0xff & ipadd[i]);    //0xff to turn into an unsigned integer
}
return ip.toString();
}
}
```

7.6. FACTORY METHODS

The **InetAddress** class has no visible constructors. To create an **InetAddress** object, we have to use one of the available factory methods. *Factory methods* are simply a convention while the static methods in a class return an instance of that class. Three commonly used **InetAddress** factory methods are shown here.

Method: static InetAddress getLocalHost( ) throws UnknownHostException

Description: It returns the InetAddress object that represents the local host.

Method: static InetAddress getByName(String *hostName*) throws UnknownHostException

Description: It returns an InetAddress for a host name passed to it.

Method: static InetAddress[] getAllByName(String *hostName*) throws UnknownHostException

Description: It returns an array of InetAddresses that represent all of the addresses that a particular name resolves to

7.7. INSTANCE METHODS

The **InetAddress** class also has several other methods, which can be used on the objects returned by the methods just discussed. Some of the most commonly used methods are:

Method: boolean equals(Object *other*)

Description: It returns true if this object has the same Internet address as *other*.

Method: byte[ ] getAddress( )

Description: It returns a byte array that represents the object's Internet address in network byte order.

Method: String getHostAddress( )

Description: It returns a string that represents the host address associated with the InetAddress object.

Method: String getHostName( )

Description: It returns a string that represents the host name associated with the InetAddress object.

Method: boolean isMulticastAddress( )

Description: It returns true if this Internet address is a multicast address. Otherwise, it returns false.

Method: String toString( )

Description: It returns a string that lists the host name and the IP address for convenience.

7.8. TCP/IP CLIENT SOCKETS

TCP/IP sockets are used to implement reliable, bidirectional, persistent, point-to- point, stream-based connections between hosts on the Internet. A socket can be used to connect Java's I/O system to other programs that may reside either on the local machine or on any other machine on the Internet. *Applets can also establish socket connections back to the host from which the applet was downloaded.* There are two kinds of TCP sockets in Java. One is for the servers, and the other is for the clients. The **ServerSocket** class is designed to be a "listener," which waits for clients to connect before doing anything. The **Socket** class is designed to connect to server sockets and initiate protocol exchanges. The creation of a **Socket** object establishes a connection between the client and the server. The constructors that are used to create client sockets are:

Method: Socket(String *hostName*, int *port*)

Description: It creates a socket connecting the local host to the named host and port and can throw an UnknownHostException or an IOException.

Method: Socket(InetAddress *ipAddress*, int *port*)

Description: It creates a socket using a preexisting InetAddress object and a port and can throw an IOException.

A socket can be examined at any time for the address and port information associated with it with the help of following methods:

Method: InetAddress getInetAddress( )

Description: It returns the **InetAddress** associated with the **Socket** object.

Method: int getPort( )

Description: It returns the remote port to which this Socket object is connected.

Method: int getLocalPort( )

Description: It returns the local port to which this **Socket** object is connected.

Once the **Socket** object has been created, it can also be examined to gain access to the input and output streams associated with it. Each of these methods can throw an **IOException** if the sockets have been invalidated by a loss of connection on the Net. The streams are:

Method: InputStream getInputStream( )

Description: It returns the **InputStream** associated with the invoking socket.

Method: OutputStream getOutputStream( )

Description: It returns the **OutputStream** associated with the invoking socket.

7.9. TCP/IP SERVER SOCKETS

The **ServerSocket** class is used to create servers that listen for either local or remote client programs to connect to them on various published ports. **ServerSocket**s are quite different from normal **Socket**s. When we create a **ServerSocket**, it will register itself with the system as having an interest in client connections. The constructors for **ServerSocket** reflect the port number that we wish to accept connections on and how long we want the queue for said port to be. The queue length tells the system how many client connections it can leave pending before it should simply refuse connections. The default is 50. The constructors throw an **IOException** under adverse conditions. The constructors are:

Method: ServerSocket(int *port*)

Description: It creates server socket on the specified port with a queue length of 50.

Method: ServerSocket(int *port*, int *maxQueue*)

Description: It creates a server socket on the specified port with a maximum queue length of *maxQueue*.

Method: ServerSocket(int *port*, int *maxQueue*, InetAddress *localAddress*)

Description: It creates a server socket on the specified port with a maximum queue length of *maxQueue*.

On a multihomed host, *localAddress* specifies the IP address to which this socket binds. **ServerSocket** has a method called **accept( )**, which acts as a blocking call that will wait for a client to initiate communications, and then return with a normal **Socket** which is then used for communication with the client.

7.10. DATAGRAM

Datagrams are the bundles of information that is passed between machines. Once the datagram has been released to its intended target, there is no assurance that it will arrive or even that someone will be there to catch it. When the datagram is received, there is no assurance that it has not been damaged in transit or that whoever sent it is still there to receive a response. Java implements datagrams on top of the UDP protocol by using two classes:

The **DatagramPacket** object is the data container.

The **DatagramSocket** is the mechanism used to send or receive the **DatagramPacket**s.

7.10.1. Datagram Packet

The constructor specifies a buffer that will receive data, and the size of a packet. It is used for receiving data over a **DatagramSocket**. Another form allows us to specify an offset into the buffer at which the data will be stored. A different form specifies a target address and port, which are used by a **DatagramSocket** to determine where the data in the packet will be sent. A still another form transmits packets beginning at the specified offset into the data. The different constructors are:

```
DatagramPacket(byte data[ ], int size)
DatagramPacket(byte data[ ], int offset, int size)
DatagramPacket(byte data[ ], int size, InetAddress ipAddress, int port)
DatagramPacket(byte data[ ], int offset, int size, InetAddress
ipAddress, int port)
```

There are several methods for accessing the internal state of a **DatagramPacket**. They give complete access to the destination address and port number of a packet, as well as the raw data and its length. Some of the most commonly used methods are:

Method: InetAddress getAddress( )

Description: It returns the destination **InetAddress**, typically used for sending.

Method: int getPort( )

Description: It returns the port number.

Method: byte[ ] getData( )

Description: It returns the byte array of data contained in the datagram. It is mainly used to retrieve data from the datagram after it has been received.

Method: int getLength( )

Description: It returns the length of the valid data contained in the byte array that would be returned from the **getData( )** method. This typically does not equal the length of the whole byte array.

7.11. IMPLEMENTING SERVERS

To implement networking in Java, java.net package classes, give access to IP address, and a range of methods related to TCP, UDP, and URL. IO stream classes are also used to read and write data from and to communication channel.

A virtual-stream is provided by TCP, which supports IP connection across the network. The virtual-stream supports IO stream to transport data from and to either of the connection.

There are two important classes that participate in communication. *InetAddress* class is used to provide IP address of target connection. *Socket* class is used to create TCP connection. Once a connection has been created, we bind both ends of a socket with their streams, and communication will start by writing and reading data on these streams to and from remote application.

> **DO YOU KNOW**
>
> A virtual-stream is provided by TCP, which supports IP connection across the network.

Before looking at the implementation of the server, let's have a look on the classes of *java.net* that are used to implement a server:

(I) Class InetAddress

Computer on internet is identified by IP address, that is currently in two version with different size, in IPv4, it is 32-bit long (four byte) and in IPv6, it is 128-bit long (eight byte). *InetAddress* provides an abstraction to access IP address that is shown in the *figure* below.

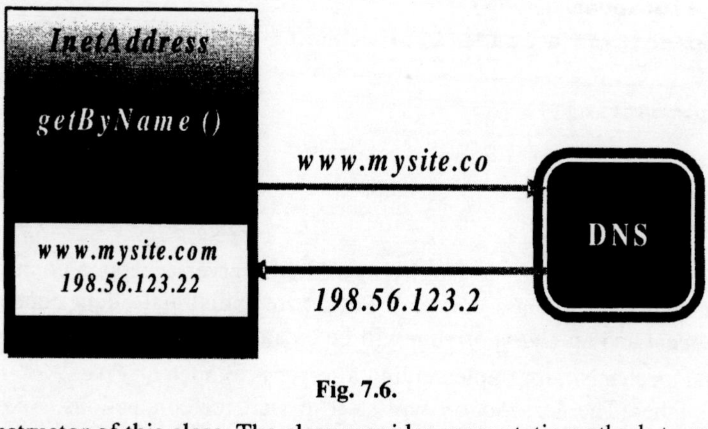

Fig. 7.6.

There is no constructor of this class. The class provides some static methods to create the instance of the class, i.e. an object.

Example: *Save the file with name "InetAddressDemo.java"*

```
import java.net.*;
import java.io.*;
public class InetAddressDemo
{
public static void main(String ar[])
{
DataInputStream din = new DataInputStream(System.in);
String ip_name;
try
{
System.out.println("Enter a hostname or IP address :: ");
System.out.flush();
ip_name = din.readLine();
InetAddress mac = InetAddress.getByName(ip_name); //ip_name can be either
   IP address or host name
System.out.println("Host Name : "+ mac.getHostName());
System.out.println("Host IP : "+ formatIP(mac.getAddress()));
}catch(UnknownHostException uhe){
System.out.println(uhe);
}catch(IOException ioe){
System.out.println(ioe);
}
}
static String formatIP(byte ipadd[])
{
StringBuffer ip = new StringBuffer();
   for(int i=0; i< ipadd.length; i++)
   {
      if(i > 0)
         ip.append(".");
   ip.append(0xff & ipadd[i]);  //0xff to turn into an unsigned integer
   }
return ip.toString();
}
}
```

(ii) Class ServerSocket

The *ServerSocket* class provides the facility by which a server accepts a connections request from clients across the network. The class creates a Socket for individual client connection. After socket creation, an *InputStream* and an *OutputStream* will be created for communicating with client.

The fundamental mechanism for implementing a server is to open a *ServerSocket* with a specific and certain local port number. The *ServeSocket* will start to wait for connections. As client generates the request to connect to this port, the connection will be established.

ServerSocket is constructed by choosing a port on the local machine. This port number should be in the range 1 – 65536; however, ports 1 – 1023 are reserved for system services and can be used only by the machine administrator.

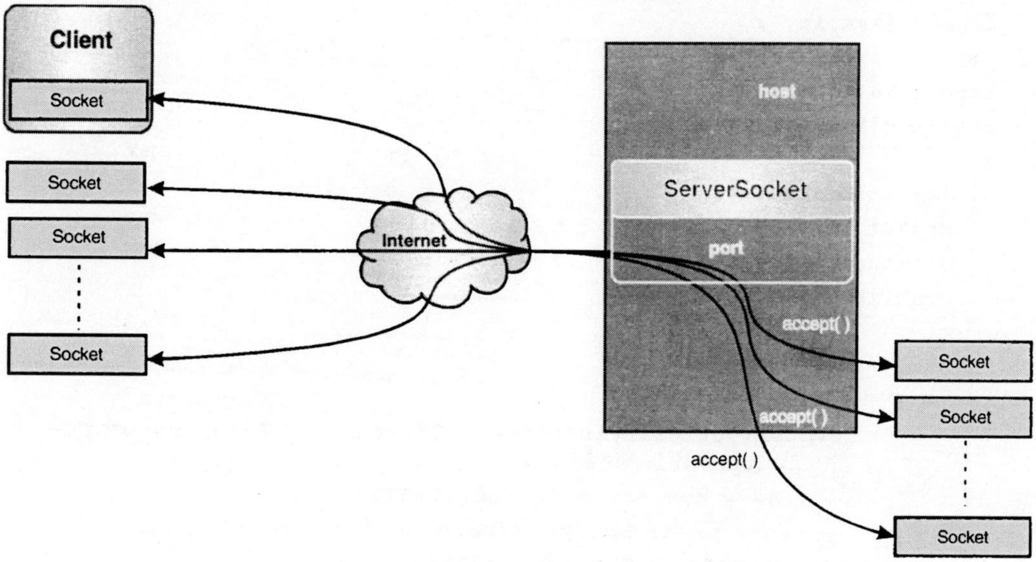

Fig. 7.7. ServerSocket

A server accepts the explicit connection request from a *ServerSocket* to obtain a client's Socket connection. As the *ServerSocket* is created, operating system start to accept the connections from clients, and insert in a queue, and delete one-by-one as the server call the accept() method.

ServerSocket(int portNumber) throws IOException creates a *ServerSocket* object that listens on the specified port *portNumber*.

```
ServerSocket ssckt = new ServerSocket(1889);
```

This server can listen on *port* 1889.

ServerSocket(int portNumber, int countValue) throws IOException creates a *ServerSocket* that listen on the specified port *portNumber* and *countValue* specified the number of outstanding connection requests.

> **DO YOU KNOW**
>
> The *ServerSocket* constructor receive the value of number of connection that will be open and received by the operating system.

```
ServerSocket ssckt = new ServerSocket(1889, 40);
```

All these connection will be stored in a queue. Any connection requests more than given value will be discarded. The default limit of queue is 50 to accept connection request. This server can listen on port 1889, and have 40 outstanding connection requests.

The methods given below allow connection to be accepted and information about the *ServerSocket* to be queried.

TABLE 7.4. Methods of java.net.ServerSocket class

Socket accept() throws IOException	Method waits until a client makes a connection request to the port on which this *ServerSocket* is registered or listening.
void close() throws IOException	Method closes the *ServerSocket* but will not close any connection which have been accepted by *accept*() method and not yet closed, and give instruction to operating system not to accept new client connections. It will maintain existing connected client connections.
int getLocalPort()	Method returns the port on which this *ServerSocket* is listening.

Example: *Save the file with name "MyServer.java"*

```
import java.io.*;
import java.net.*;
import java.util.*;
public class MyServer
{
   ServerSocket ss;
   Socket s;
   DataInputStream dis;
    public MyServer()
   {
     try
      {
          System.out.println("Server Started.  Waiting for CLIENT
               Connection Request ");
               ss = new ServerSocket(1889);
               s = ss.accept(); //Waiting for client connection
 System.out.println("CLIENT Connected");
 dis = new DataInputStream(s.getInputStream());
 serverChat();
 }
 catch(IOException ioe){
       System.out.println(ioe);
       }catch(Exception e){
       System.out.println(e);
       }
    }
    public void serverChat() throws IOException
    {
       String str;
       do{
       str = dis.readUTF();
       System.out.println("CLIENT Message :: " + str);
       }while(!str.equals("stop"));
    }
    public static void main(String ar[])
    {
    new MyServer();
    }
 }
```

(iii) Class Socket

A *Socket* represents a TCP connection with a machine on network. This class facilitate in establishing the stream-based channel with a remote machine.

For TCP/IP communication, there is a need to create a *Socket* with the remote machine. The class Socket creates a socket and automatically established a TCP connection, and throws an exception if it fails. *Socket* address is the combination of two components: one is host id, which refers with the host name, IP address, and the port number. The port number is an integer value that lies between 1 and 65536. In fact, on every host, there are 655636 different addresses. But there must be a server that dynamically listens on the specified port.

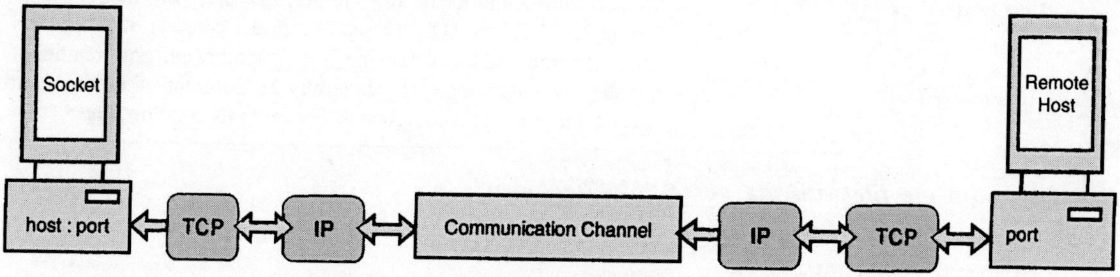

Fig. 7.8. Socket

There are two constructor of the *Socket* class. Creating a *Socket* object automatically connects to the given host and port. As earlier stated, a server actively listens the request, but as connection is refused an *IOException* will be thrown, unknown hostname can be other possibility of connection failure.

Socket(String hostName, int portNumber) throws IOException creates a *Socket* object and connects to the given host on the given port number.

```
Socket sckt = new Socket("mymac", 1889);
```

Socket(InetAddress hostAddress, int portNumber) throws IOException creates a *Socket* object and connect to the given host address on the given port number.

```
InetAddress mac = InetAddress.getByName("127.0.0.1");
        Socket sckt = new Socket(mac.getAddress, 1889);
OR      Socket sckt = new Socket("127.0.0.1", 1889);
```

Socket class methods provide the identification of remote host, port number of remote and local host, and also give the stream for communication.

First create the Socket object to perform communications across a TCP connection. After creating the socket, bind it with the required or corresponding stream with the help of *getInputStream*() and *getOutputStream*() methods to communicate with the remote server. The client and server, both will have both an *InputStream* and an *OutputStream* for the purposes of communication.

TABLE 7.5. Methods of java.net.Socket class

InputStream getInputStream() throws IOException	Method returns an InputStream that allow stream-based communications for TCP connections. The data written by a host at one end; can be read by remote host from this *InputStream* at other end of the connection. The method actually returns a *SocketInputStream*. The *SocketInputStream* is a nonpublic class.
OutputStream getOutputStream() throws IOException	Method returns an *OutputStream* that allow stream-based communication from TCP connection. The data written to this stream by any machine at one end; will be read by remote host at the other end of connection. This *OutputStream* should be buffered.
void close() throws IOException	The method is used to close the Socket, releasing all network and system resources being used. The data, which have been sent before this the call of this method, will be successfully delivered to the remote end, if there is no network fault.

InetAddress getInetAddress()	Method returns the IP address of the remote host.
int getPort()	Method returns the port number of the remote host to which the Socket is connected.
int getLocalPort()	Method returns the local port number, *i.e.*, the port number of the Socket itself. Every TCP connection in fact consists of IP address and port number of local machine; and IP address and port number of remote machine, which will generally be different. A random and unused port number is assigned to Socket at its creation time.

Example: Save the file with the name "MyClient.java"

```
import java.io.*;
import java.net.*;
import java.util.*;
public class MyClient
{
Socket s;
DataOutputStream dout;
public MyClient()
{
try{
    s = new Socket("127.0.0.1", 1889);
    dout = new DataOutputStream(s.getOutputStream());
    System.out.println("Enter the data to send to SERVER");
      clientChat();
    }catch(IOException ioe){
       System.out.println(ioe);
       }catch(Exception e){
       System.out.println(e);
       }
    }
    public void clientChat() throws IOException
    {
    BufferedReader br = new BufferedReader(new
InputStreamReader(System.in));
    String st;
       do{
           st = br.readLine();
           dout.writeUTF(st);
           dout.flush();
       }while(!st.equals("stop"));
    }
    public static void main(String ar[])
    {
       new MyClient();
    } }
```

Compile both file "MyServer.java" and "MyClient.java".

Run *MyServer* program in one command window.

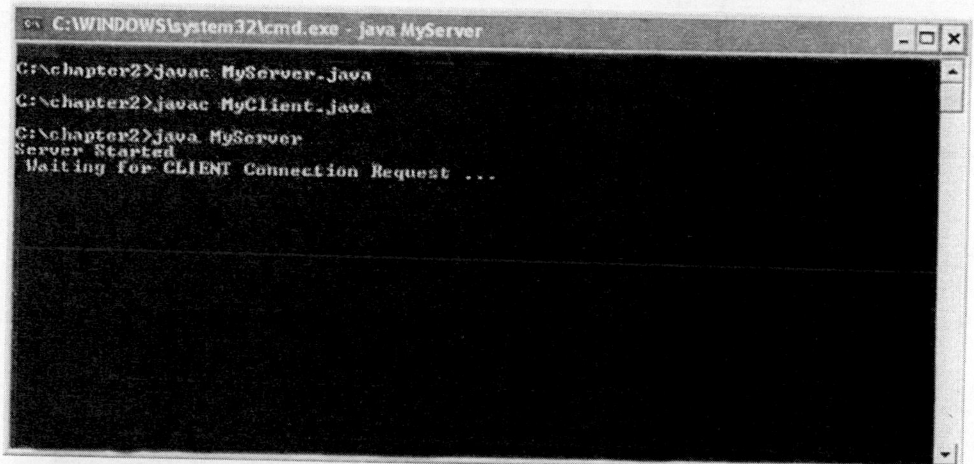

Now server is waiting for the connection request from the client. Here *accept()* method will hold it until any client generate the connection request. As any client generates a request for connection with server, the servers accepts its request and store the client socket reference.

Run the MyClient program in another command window

As request will be generated by client, the client socket reference will be stored in Socket object at server.

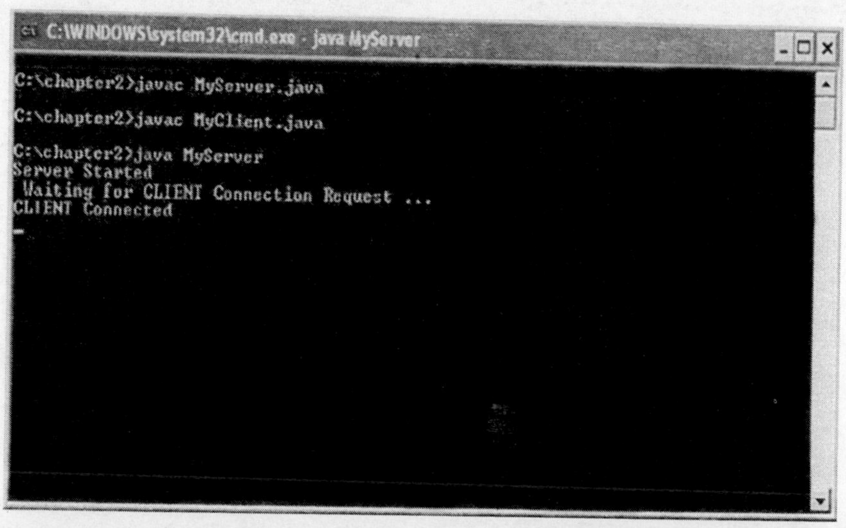

Client sends the data "Hello!" to server. Here client write the data through output stream object, that is bind with socket at server port number.

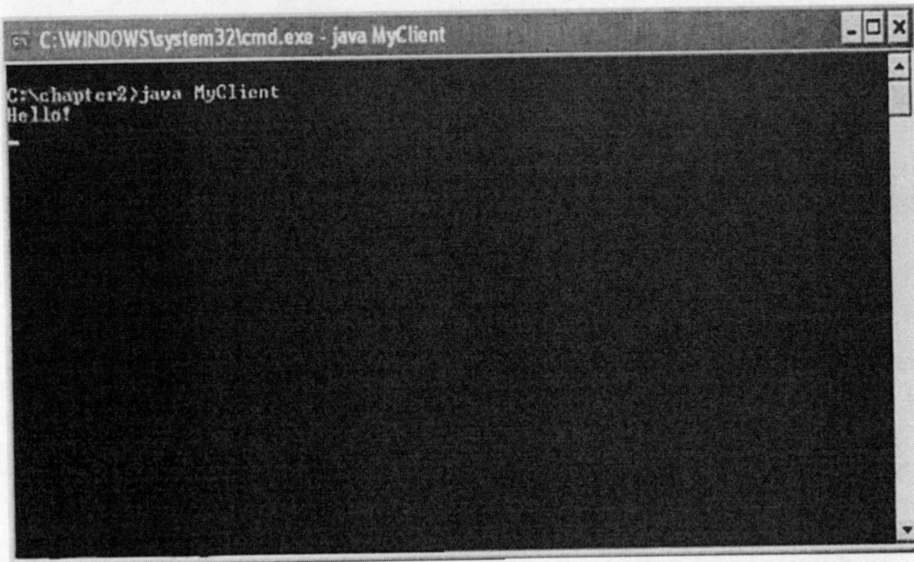

Server receives send data and show with a prompt. In this example, server only read the data from the input stream of client socket.

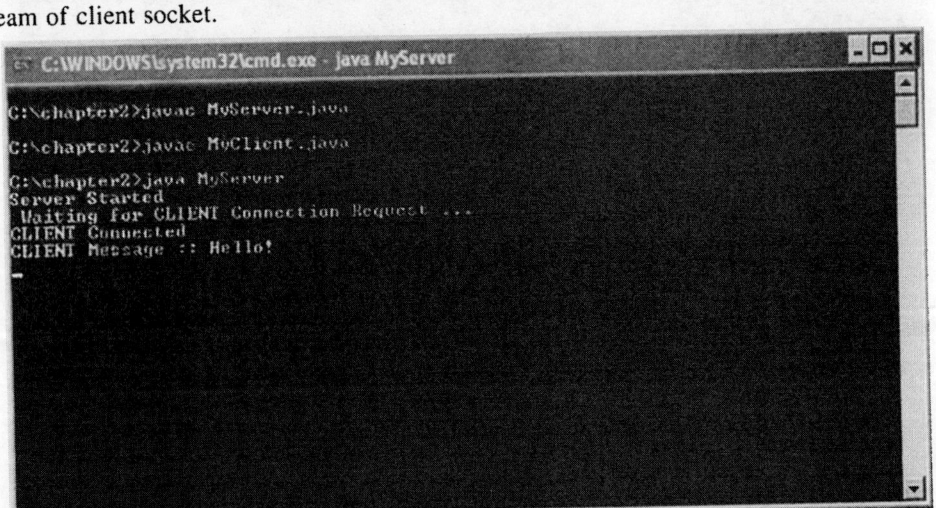

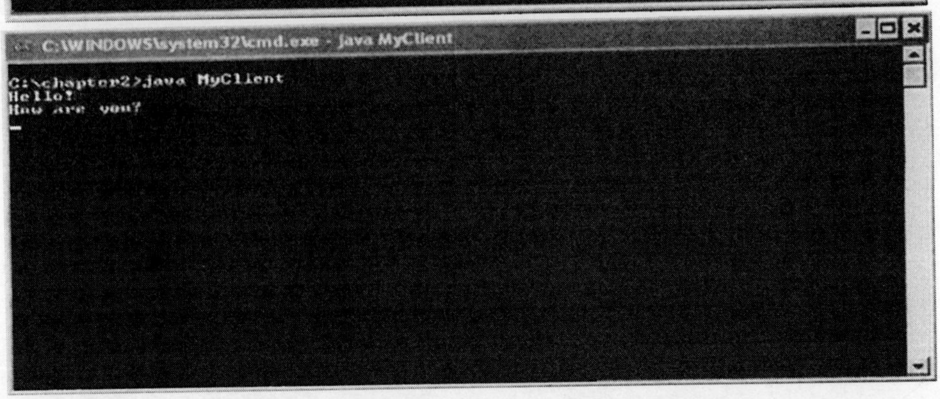

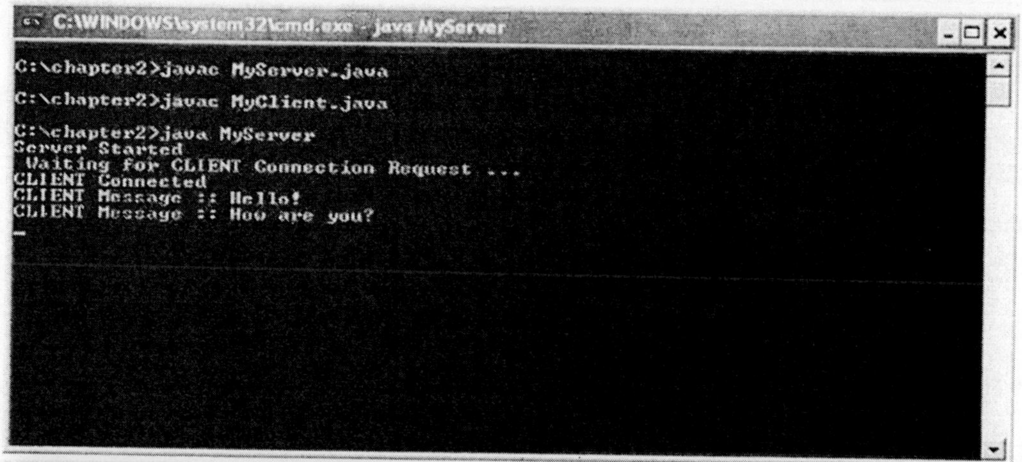

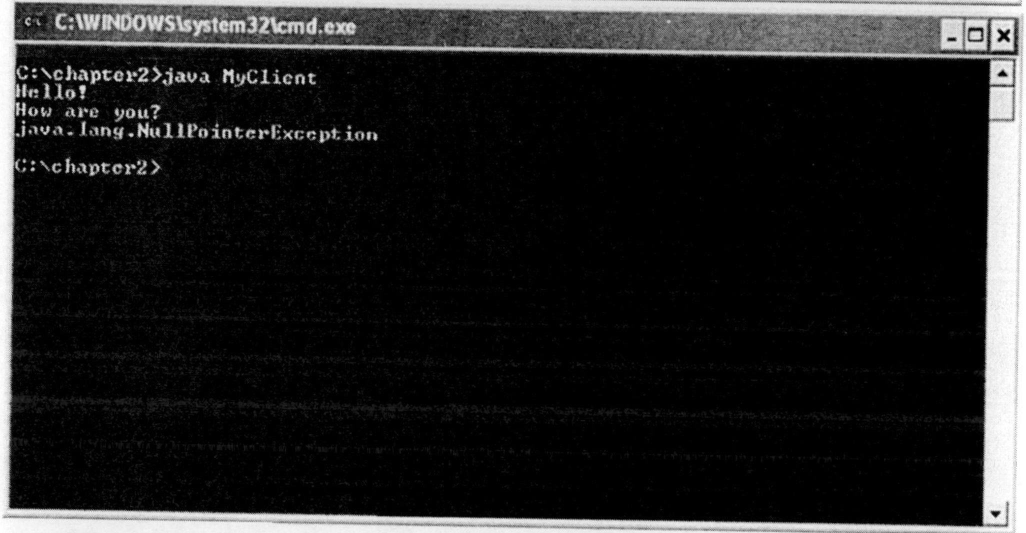

The communication will continue until client gives "stop" or forcefully terminated. As connection terminated *java.net.SocketException* will fire on server machine.

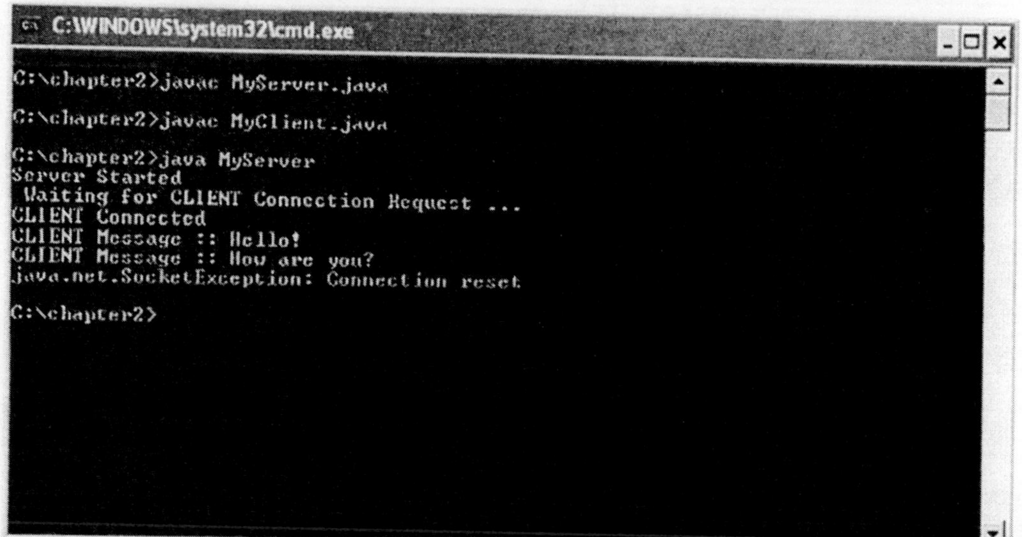

7.12. SENDING E-MAIL

A socket will be created, connected to port 25, which is SMTP port number. Simple Mail Transfer Protocol describes the format in which e-mail messages will be sent. In UNIX OS, *sendmail* daemon was already implemented to support this features.

Example: Sending Mail Using Sockets

```
import java.util.*;
import javax.mail.Message;
import javax.mail.Session;
import javax.mail.Transport;
import javax.mail.internet.InternetAddress;
import javax.mail.internet.MimeMessage;
public class abc {
    public static void main(String[] args) {
        String host = "smtp.gmail.com";
    String user = "anonymous";
    String pass = "not2bgiven";
    String SSL_FACTORY = "javax.net.ssl.SSLSocketFactory";
    String from1 = "anonymous@ymail.com";
    String to1 = "user@ymail.com";
    String subject1 = "java";
    String messageText = "Hello hw r u?";
    Properties props = System.getProperties();
    props.put("mail.host", host);
    props.put("mail.transport.protocol.", "smtp");
    props.put("mail.smtp.port", "465");
    props.put("mail.smtp.socketFactory.class", SSL_FACTORY);
    Session mailSession = Session.getDefaultInstance(props, null);
    Message msg = new MimeMessage(mailSession);
        try{
            InternetAddress  fr = new InternetAddress(from1);
            msg.setFrom(fr);
            InternetAddress[] address = { new InternetAddress(to1) };
            msg.setRecipients(Message.RecipientType.TO, address);
            msg.setSubject(subject1);
            msg.setContent(messageText, "text/html");
            Transport transport = mailSession.getTransport("smtp");
            transport.connect(host, user, pass);
            transport.sendMessage(msg, msg.getAllRecipients());
            transport.close();
            System.out.println("Done Mail");
        }
        catch (Exception err){
            System.out.println("not done mail");
            System.out.println(err);
        }
    }
}
```

7.13. MAKING URL CONNECTIONS

A framework is provided by Java, using java.net package through several interfaces, classes, and abstract classes for handling URL (Uniform Resource Locator). Using this framework, we can easily create sum function of Web browser and can be embedded in any application.

Before discussing the concept in-depth let's have a description about two important classes given below:

(i) Class URI

The term URI stands for Uniform Resource Identifier. It creates an instance of URI, and are used to parse the string, and also providing various methods. Every URL is a URI, but not every URI is a URL. On the basis of conceptual difference between URIs and URLs, two different classes are given with their characteristics and functionalities.

(ii) Class URL

Without URL, browser cannot make out the information about web. URL identifies a resource anywhere on the Internet. The introduction of URLs came up with WWW.

Web is a collection of protocols and different file formats. The URL is combination of four components – the protocol, host name or host IP Address, port number and the file path. Some protocols in use are http, ftp, gopher, and file. But HTTP is most in use and almost everything is done by it.

For example http://java.sun.com/j2se/1.4.2/docs/api/java/net/URL.html indicate that the protocol to use is http (HyperText Transfer Protocol) and that the information resides on a host machine named java.sun.com. The information on that host machine is named /j2se/1.4.2/docs/api/java/net/URI.html and its exact meaning is both protocol dependent and host dependent.

A URL can optionally specify a "port", which is the port number to which the TCP connection is made on the remote host machine. If the port is not specified, the default port for the protocol is used instead. For example, the default port for http is 80.

With the help of URL class's instance some information about the resource can be accessed. The URL always specifies a scheme, called absolute URL. A URL scheme is parsed according to its scheme. A URL is a structure string.

There are following constructor of the URL class;

URL(String protocolName, String hostName, int portNumber, String fileName) throws MalformedURLException creates an absolute URL from combination of all given argument.

URL(String protocolName, String hostName, String fileName) throws MalformedURLException creates an URL but assume the default port number for the URL; the port number depends on the protocol *protocolName*, handled by protocol handler. Internally the port is set to -1.

URL(String str) throws Malformed URLException creates an instance of URL form a complete textual specification, e.g. http://www.nsa.gov:80/index.html

URL(URL urlContext, String str) creates a new URL from an existing URL and a relative textual URLs. E.g. *urlContext* is http://myserver/ and *str* is index.html

Example:

```
URL http_url = new URL("http://www.indianrailway.gov.in");
```

TABLE 7.6. Methods of java.net.URL Class

Method	Description
String getProtocol()	Method returns the protocol name part of the URL as a String, *e.g.*, http, ftp, mailto, etc.
String getHost()	Method returns the hostname as String, which is part of the URL.
int getPort()	Method returns the port number as integer, representing the port number part of the URL.
String getFile()	Method returns the file part of URL as String.
URLConnection openConnection() throws IOException	Method returns an *URLConnection* object, which corresponds to a protocol connection to the URL object. Later, using this class we can access the contents of the URL.
InputStream openStream() throws IOException	Method opens a connection to the URL and returns an object of input stream for reading its contents.
Object getContent() throws IOException	Method brings together the contents of the URL and returns it as an Object type. The actual type is depends on the contents of the URL. If the target URL is an image then for that an appropriate image handled is installed, then object is of type Image, similarly, if the target URL is a text file then appropriate text handler must be there, then object should be returned as String.

Example: Save the file with the name "URLDemo.java"

```
import java.net.*;
public class URLDemo
{
    public static void main(String ar[])
    {
    URL http_url = new URL("http://www.indianrailway.gov.in");
    System.out.println("Protocol Name :: " + http_url.getProtocol());

    System.out.println("Port Number :: " + http_url.getPort());
    System.out.println("Host Name :: " + http_url.getHost());
    System.out.println("File Name :: " + http_url.getFile());
    }
}
```

(III) Class URLConnection

The class *URLConnection* gives some additional information about a web resource. The class provides some methods by which you can send the request to server. The class is actually abstract. There is only constructor to create instance of *URLConnection.*

URLConnection (URL urlString) creates a new *URLConnection* object for the URL *urlString*. This constructor can be called only by a subclass of the *URLConnection* class; and subclass will give the implementations of the various *URLConnection* methods for a specific protocol.

Example:

```
URL http_url = new URL("http://www.indianrailway.gov.in");
URLConnection urlc = new URLConnection(http_url);
```

There is another way to create an instance of *URLConnection*, by invoking the *openConnection*() method of URL class, that return an object of *URLConnection*.

Example:

```
URL http_url = new URL("http://www.indianrailway.gov.in");
URLConnection urlc = http_url.openConnection();
```

When working with a *URLConnection* object, following steps have to be taken:

1. First to obtain an object or *URLConnection*, call the *openConnection*() method of the *URL* class.
2. You can set any request properties, using the given methods, if required.
3. Now connect to the remote resource by calling the *connect*() method.
4. After establishing the connection to the server, you can query for the header information.
5. As a final point, now you can access the all data of resource.

7.14. ADVANCED SOCKET PROGRAMMING

In the following sections, we cover some advanced topic of network programming that happen in real-world programs.

Socket Timeouts

Whenever you try to read data from a socket, the read ( ) call blocks as long as necessary to get enough bytes. By setting SO_TIMEOUT, you can be ensure that the call will never be block for more than a fixed number of milliseconds. When the timeout expires, an *InterruptedException* is thrown, and you should be prepared to catch it, but still the socket is connected. Although this read ( ) call can failed, you can try again to read from the socket again. The next call may succeed.

There are two methods give below to set and get the timeout value.

```
public synchronized void setSoTimeout(int milliseconds) throws
    SocketException
public synchronized int getSoTimeout( ) throws SocketException
```

Example:

```
Socket s = new Socket ();
s.setSoTimeout(10000) ; // time out after 10 seconds
s.connect(...);
```

Timeouts are given in milliseconds. Zero is interpreted as an infinite timeout, and is the default value.

There is one another timeout issue, that is, the constructor

```
Socket (String host, int port);
```

can block for uncertain period of time, until an initial connection is established.

The solution for this problem to make the Socket object in the following way:

```
Socket s = new Socket();
s.connect(new InetSocketAddress(host,port), time_out);
```

Similar option supported for *ServerSocket* is SO_TIMEOUT. SO_TIMEOUT is the amount of time, in milliseconds, that *accept()* waits for an incoming connection before throwing a *java.io.InterruptedIOException* . If SO_TIMEOUT is 0, then *accept()* will never time out. The default is to never time out.

```
public void setSoTimeout(int timeout) throws SocketException
```

The *setSoTimeout()* method sets the SO_TIMEOUT field for the *ServerSocket* object. The countdown starts as *accept()* is invoked. When the timeout expires, *accept()* throws an *InterruptedIOException*. So that, set this option before calling *accept()*; and the timeout value cannot be change while *accept()* is waiting for a connection. The timeout argument must be greater than or equal to zero; if it isn't, the method throws an *IllegalArgumentException*.

Example:

```
try {
ServerSocket serversocket = new ServerSocket(1148);
// block for no more than 20 seconds
server.setSoTimeout(20000);
try {
Socket s = serversocket.accept();
// handle the connection
// ...
} catch (InterruptedIOException e) {
System.err.println("No connection within 20 seconds");
} finally {
serversocket.close( );
}catch (IOException e) {
System.err.println("Unexpected IOException :: " + e);
}
```

Example:

```
// Create a socket without a timeout
try {
    InetAddress addr = InetAddress.getByName("java.sun.com");
    int port = 80;

    // This constructor will block until the connection succeeds
    Socket socket = new Socket(addr, port);
} catch (UnknownHostException e) {
} catch (IOException e) {
}
```

// Create a socket with a timeout

```
try {
    InetAddress addr = InetAddress.getByName("java.sun.com");
    int port = 80;
```

```
        SocketAddress sockaddr = new InetSocketAddress(addr, port);
      // Create an unbound socket
        Socket sock = new Socket();

        // This method will block no more than timeoutMs.
        // If the timeout occurs, SocketTimeoutException is thrown.
        int timeoutMs = 2000;    // 2 seconds
        sock.connect(sockaddr, timeoutMs);
    } catch (UnknownHostException e) {
    } catch (SocketTimeoutException e) {
    } catch (IOException e) {
    }
```

Half-close

Whenever server receives a request from the client application, then it always determines the end of the request. Let's take an example, suppose last request was for writing the data to a file, and now you just want to close the file at the end. But as you close the socket, you'll be disconnected from the server. The half-close is the solution of this problem. Using this, you can close the output-stream of the socket, so that, no more request will goes to server, but input-stream will still open to receive and read the response from the server. There are two methods given below:

public void shutdownInput( ) throws IOException method is used to close or shutdown input stream, cannot read data from the response.

public void shutdownOutput( ) throws IOException method is used to close or shutdown output stream, cannot generate the request to the server

Example:

```
Socket conn = null;
try {
     conn = new Socket("www.indianrailway.gov.in", 80);
     BufferedReader br = new BufferedReader( new
                          InputStreamReader(conn.getInputStream()));
     Writer out = new OutputStreamWriter(conn.getOutputStream( ),
                                              "UTF-8");
     out.write("GET / HTTP 1.0");
     out.flush( );
     connection.shutdownOutput( ); // now socket is half closed and
                  can only read response data
       String str;
       while ((str = br.readLine() ) != null)
        . . .
       } catch (IOException e) {}
       finally {
       try {
             if (conn != null) conn.close( );
```

```
            } catch (IOException e) {
    System.out.println(e); }
    }
```

7.15. CONFIGURING TOMCAT SERVER FOR HANDLING URL CONNECTIONS

A *web application* runs over the Internet. Examples of web applications are google, amazon, ebay, etc. A web application is a *3-tier* (or *multi-tier*) *client-server application*, usually involving a *database*. Most of the web applications run on the HTTP application protocol, with browser as the client to access an HTTP server.

A web database application requires five components, as illustrated below:

- **HTTP Server:** E.g., Apache HTTP Server, Microsoft IIS, Apache Tomcat, and etc.
- **HTTP Client** (Browser): E.g., MSIE, FireFox, Chrome, and etc.
- **Database**: E.g., MySQL, Oracle, DB2, Infomix, MS SQL Server, MS Access, and etc.
- **Client-Side Programs**: could be written in HTML Form, JavaScript, VBScript, Flash, and etc.
- **Server-Side Programs**: could be written in Java Servlet/JSP, ASP, PHP, and etc.

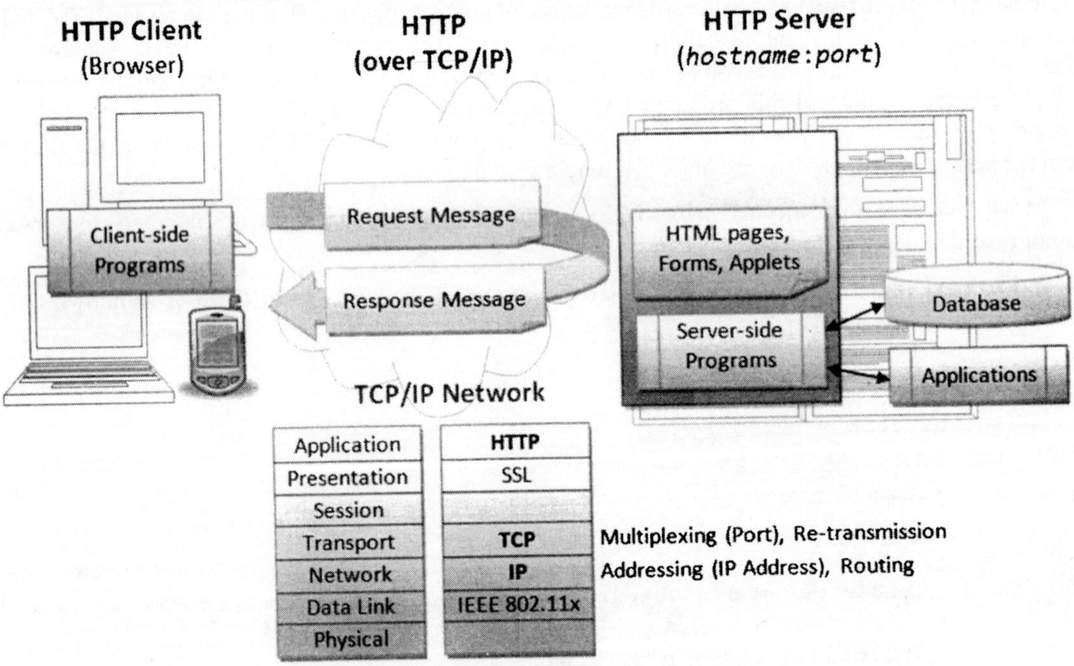

Fig. 7.9. Web Database Application

A user issues a URL to an HTTP server via a web browser to start a web application. The web application first downloads a client-side program (such as an HTML form) into the browser. The user fills up the query criteria in the form. The client-side program sends the query parameters to a server-side program, which queries the database and returns the query result to the client. The client-side program displays the result on the browser.

HTTP is an *asynchronous request-response application-layer protocol*. A client sends a request message to the server. The server returns a response message to the client.

Guidelines to download and install Apache Tomcat Server

➠ **Install JDK and set the PATH and CLASS PATH as given below...**

```
My Computer->properties->advance->Environment Variables-> NEW ->
```

➠ **PATH SETTING**

```
Variable Name : PATH
Variable Value: C:\Program Files\Java\jdk1.6.0_06\bin;
```

➠ **CLASS PATH SETTING**

```
Variable Name : CLASSPATH
Variable Value: C:\Program Files\Java\jdk1.6.0_06\lib;
```

Now Install Apache Tomcat Server according to the steps given below and after installation is completed Update the Class Path as:

➠ **NOW UPDATE CLASS PATH SETTING**

Variable Name: CLASSPATH

Variable Value: C:\Program Files\Java\jdk1.6.0_06\lib;C:\Program Files\Apache Software Foundation\Tomcat 6.0\lib\servlet-api.jar;

NOW Open your Browser and Type : http://localhost:8080/ the Click and Your Tomcat server will run now and will open the Apache Tomcat Page.

Downloading and Installing Tomcat

In this section we will show the process to downloading and installing Tomcat 6 on your computer system. The steps of downloading and installing are easy and you can learn the process very fast. First of all learn the steps of 'How to download Apache Tomcat server on your computer'. Click the link: ***http://tomcat.apache.org/download-60.cgi*** and follow the steps according to your requirement to achieve Tomcat server.

Step 1:

Download the jdk from your intranet's Software Folder and then install it...

Installation of JDK

Before beginning the process of installing Tomcat on your system, ensure first the availability of JDK on your system program directory. Install it on your system if not already installed (because any version of tomcat requires the Java 1.6 or higher versions) and then set the class path (environment variable) of JDK. To set the **JAVA_HOME**

Variable : You need to specify the location of the java run time environment to support the Tomcat else Tomcat server can not run.

This variable contains the path of JDK installation directory.

```
set JAVA_HOME=C:\Program Files\Java\jdk1.6
```

Note: It should not contain the path up to bin folder. Here, we have taken the URL path according to our installation convention.

For Windows OS, go through the following steps:

```
Start menu->Control Panel->System->Advanced tab->Environment
Variables->New->set the Variable name = JAVA_HOME and variable value =
C:\Program Files\Java\jdk1.6
```

Now click on all the subsequent ok buttons one by one. It will set the JDK path.

Step 2:

For setting the class path variable for JDK, do like this:

```
Start menu->Control Panel->System->Advanced tab->Environment
Variables->New->
Set PATH="C:\Program Files\Java\jdk1.6\bin"; %PATH%
```

OR

First, right click on the

```
My Computer->properties->advance->Environment Variables->path.
```

Now, set bin directory path of JDK in the path variable

Step 3:

The process of installing Tomcat 6.0 begins here from now. It takes various steps for installing and configuring the Tomcat 6.0. For Windows OS, Tomcat comes in two forms: .zip file and .exe file (the Windows installer file). Here we are exploring the installation process by using the .exe file. First unpack the zipped file and simply execute the '.exe' file.

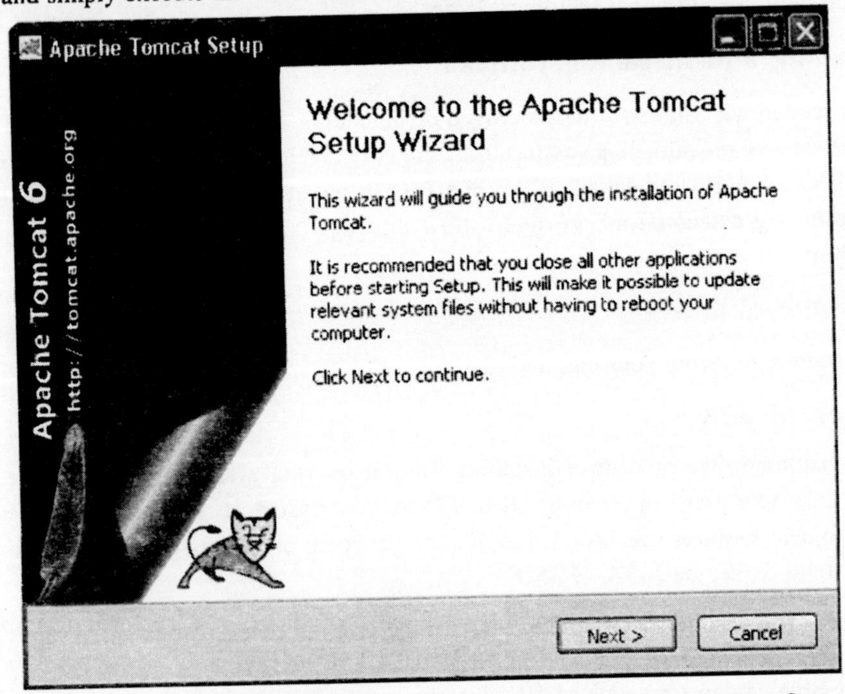

A Welcome screen shot appears that shows the beginning of installation process. Just click on the 'Next' button to proceed the installation process.

Steps 4:

A screen of 'License Agreement' displays.

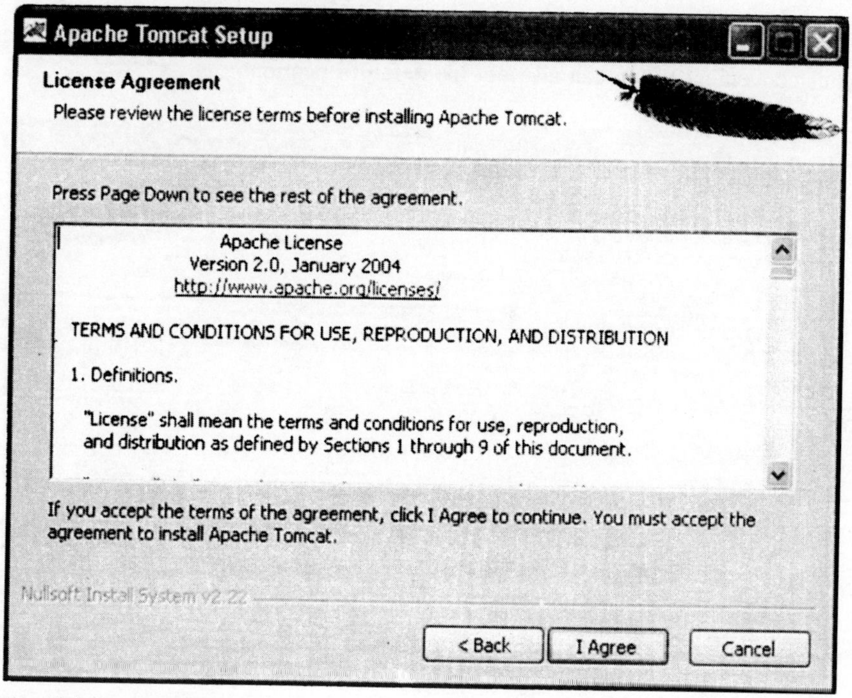

Click on the 'I Agree' button.

Step 5:

A screen shot appears asking for the 'installing location'

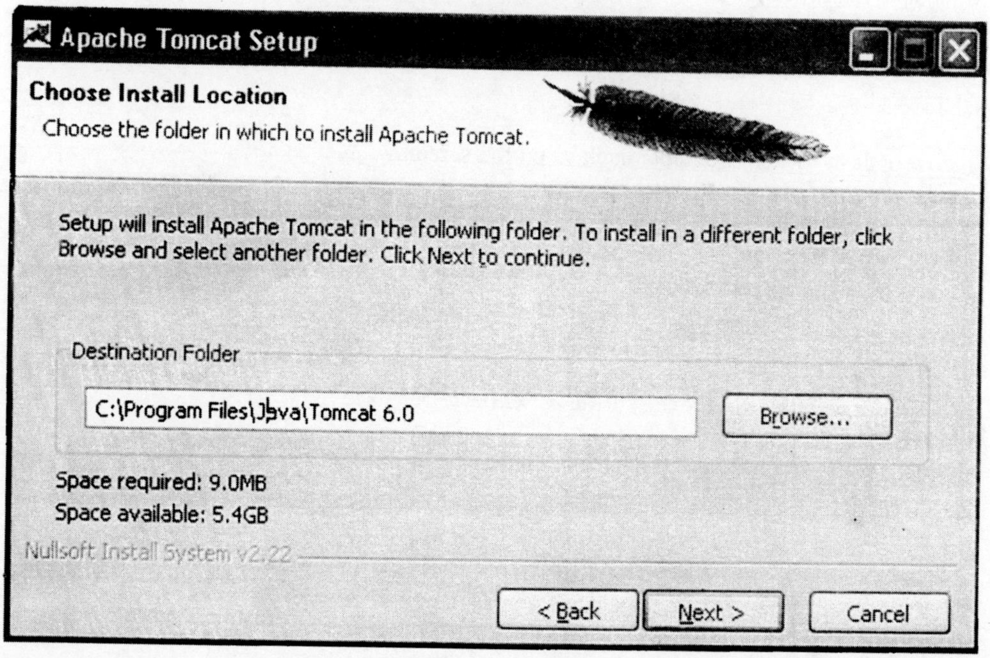

Choose the default components and click on the 'Next' button.

Step 6:

A screen shot of 'Configuration Options' displays on the screen. Choose the location for the Tomcat files as per your convenience. You can also opt the default Location

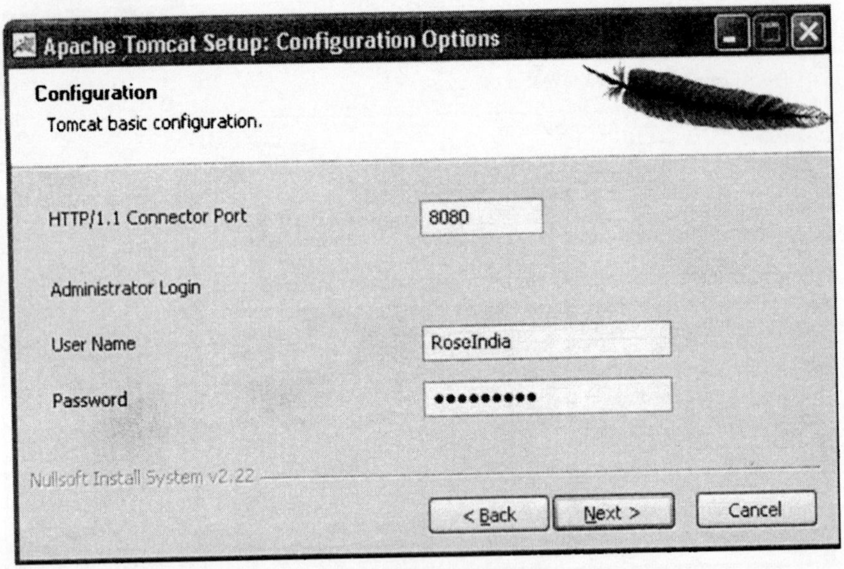

The port number will be your choice on which you want to run the tomcat server. The port number **8080** is the default port value for tomcat server to proceed the HTTP requests. The user can also change the 'port number' after completing the process of installation; for this, users have to follow the following tips.

Go to the specified location as " **Tomcat 6.0 \conf \server.xml** ". Within the server.xml file choose "Connector" tag and change the port number.

Now, click on the 'Next' button to further proceed the installation process.

Step 7:

A Window of Java Virtual Machine displays on the screen

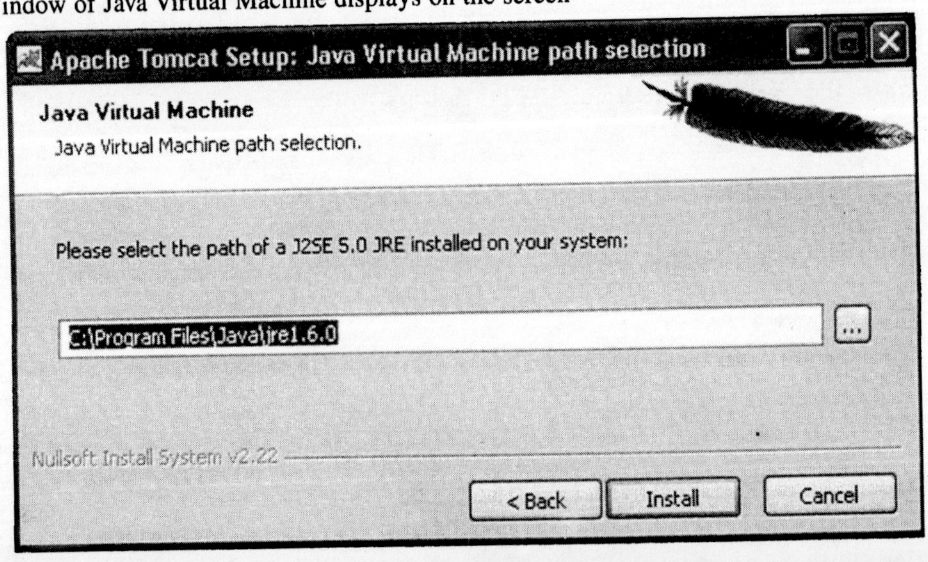

This window asks for the location of the installed Java Virtual Machine. Browse the location of the JRE folder and click on the Install button. This will install the Apache tomcat at the specified location.

Step 8:

A processing window of installing displays on the screen.

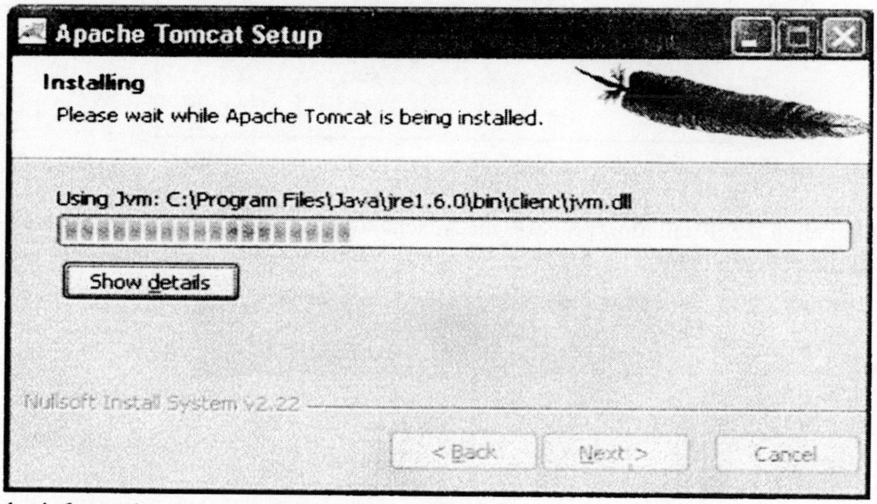

To get the information about installer click on the "Show details" button

Step 9:

A screen shot of 'Tomcat Completion' displays on the screen.

Click on the 'Finish' button.

Step 10:

A window of Apache Service Manager appears with displaying the running process.

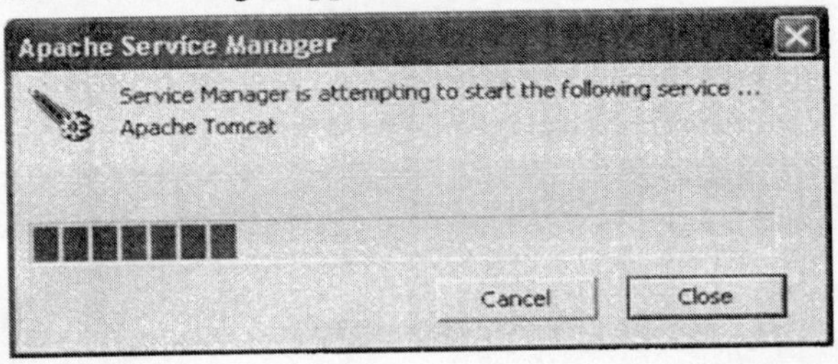

Let the running process goes on.

Step 11:

After completing the installation process, the Apache Tomcat Manager appears on the toolbar panel like shown in the below picture.

Apache Tomcat Manager

Configuring Tomcat Manager

To Configure the Tomcat Manager, there are two ways; either user can configure Tomcat directly from the toolbar panel or can configure it from Control Panel Section.

(i) Configuring from toolbar Panel

To Configure Apache Tomcat web server from the toolbar panel, you have to press 'double click' on the appeared icon.

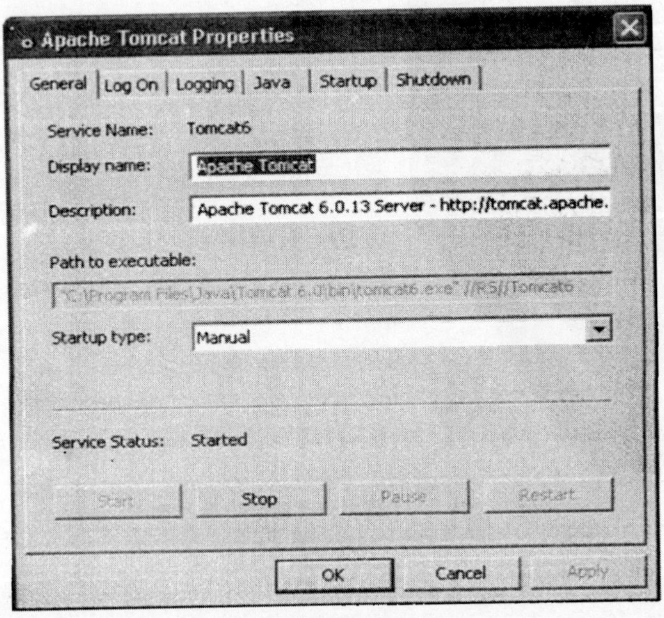

A configured window appears on the desktop. Now, just click on the Startup button. The installation process will be completed.

(ii) Configuration from the Control Panel

To configure the Apache Tomcat Manager, Users will have to follow the follwing steps:

Click on the Start up button — select Control Panel — Administrator Tool — Services — select Apache Tomcat.

The following screen displays on the monitor.

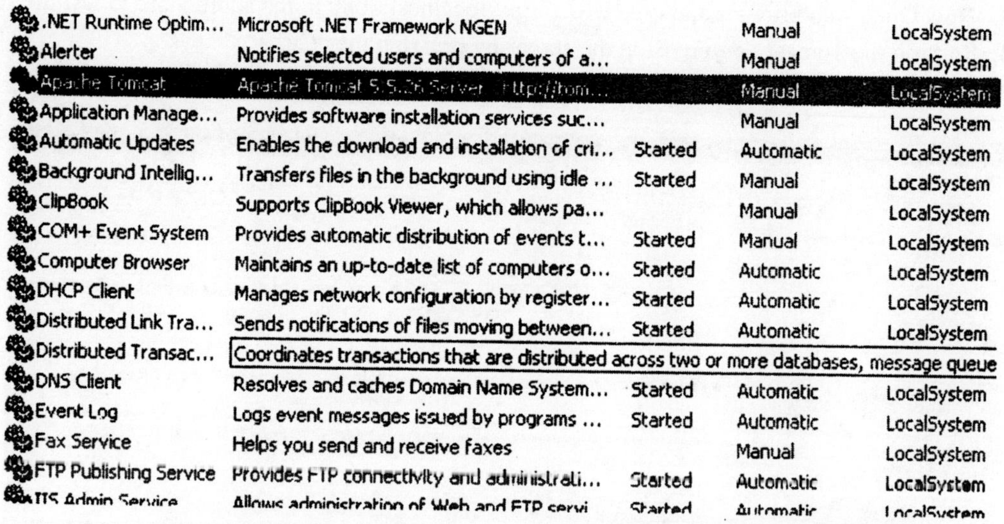

Double click on the Apache Tomcat. The window of Apache Tomcat Properties appears on the screen.

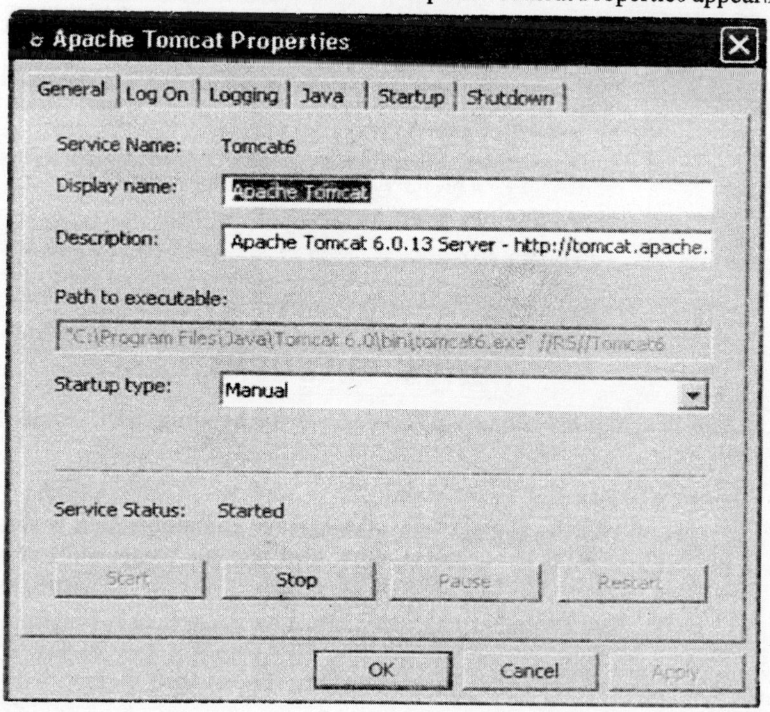

Now, Click on the start up button. The Apache Tomcat is now ready to function.

To operate it, follow the below steps of processing.

Start the Tomcat Server:

1. Start the tomcat server from the bin folder of Tomcat 6.0 directory by double clicking the "tomcat6.exe" file.

   OR

   create a shortcut of this .exe file at your desktop.
2. Now Open web browser and type URL http://localhost:8080 in the address bar to test the server .
3. To Stop the Tomcat Server: Stop the server by pressing the "Ctrl + c" keys.

The screen of Apache Tomcat software looks like this:

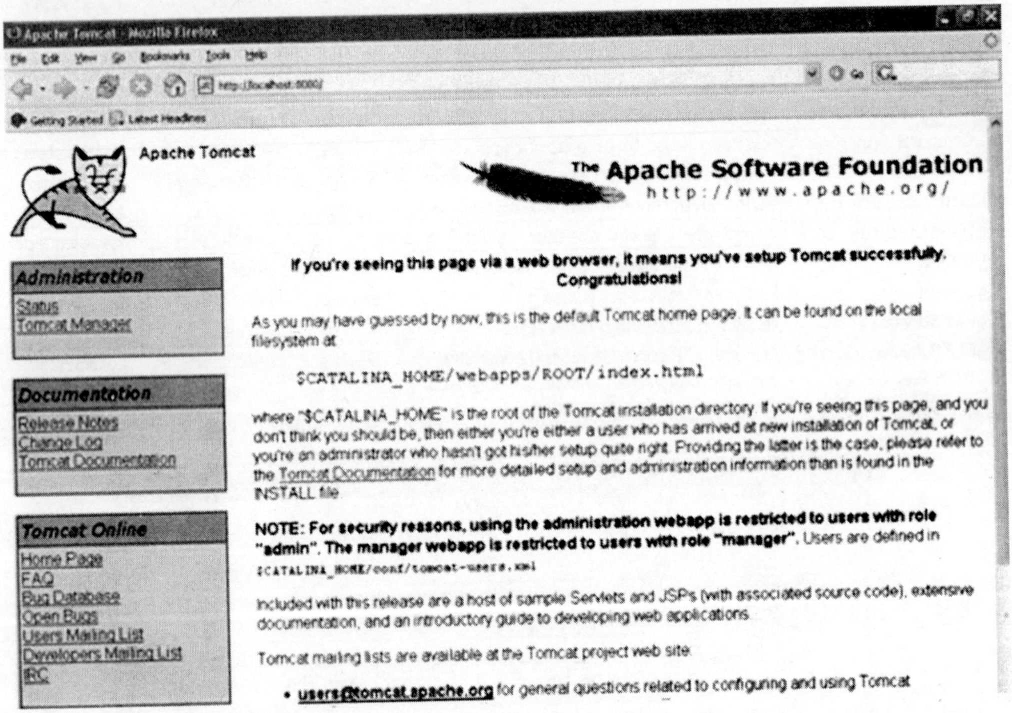

REVIEW QUESTIONS

Q. 1. Explain how to install & configure tomcat server for handling URL connections with example.
[MDU, May 2011]

Ans. *Refer Section 7.15*

Q. 2. With reference to sockets, explain how client/server communication is accomplished in java. Write a program to print the protocol, port, host and file components of a URL.
[MDU, May 2007; Dec. 2007]

Ans. Client and server communication is accomplished by using Socket Progrmming. There are two classes, plays a big role for communication, ServerSocket and Socket. Create an object of ServerSocket class that bind on a fix port number. The method accept() will wait for any client request. As client create an object of Socket type, and bind with a machine on given port number by server, the server accept() method store the client socket reference.

Section 2.1 (ii) Class ServerSocket, and 2.1. (iii) Class Socket covers all the related data for this question.

Program:

```
URL url = new URL("http://www.indianrailway.gov.in:80/index.html");
System.out.println("Protocol ::" + url.getProtocol());
System.out.println("Port Number ::"+url.getPort());
URLConnection conn = url.openConnection();
conn.connect();
Map<String, List<String>> hd = conn.getHeaderFields();
For(Map.Entry<String, List<String>> ent : hd.entrySet())
{
String key = ent.getkey();
for(String value : ent.getValue())
    System.out.println(key+":"+value);
}
```

OR

```
URL url = new URL("http://www.indianrailway.gov.in:80/index.html");
System.out.println("Protocol ::" + url.getProtocol());
System.out.println("Port Number ::"+url.getPort());
URLConnection conn = url.openConnection();
conn.connect();
System.out.println("Content Type::"+conn.getContentType());
System.out.println("Content Length::"+conn.getContentLength());
System.out.println("Content Encoding::"+conn.getContentEncoding());
System.out.println("Creation Date::"+conn.getDate());
System.out.println("Content Expiration::"+conn.getContentExpiration()));
System.out.println("Last Modified::"+conn.getLastModified());
```

Q. 3. Describe implementation of servers. Also explain advanced socket programming with suitable examples. **[MDU, May 2012]**

Ans. Refer 7.11 and 7.14.

Q. 4. Write a program using Socket to receive message from a server computer.

Ans. //Server.java

```
import java.io.*;
import java.net.*;
import java.util.*;
public class Server
{
    ServerSocket ss;
    Socket s;
    DataOutputStream dout;
    public Server()
    {
        try{
```

```
    System.out.println("Server Started \n Waiting for CLIENT Connection
                                                        Request ...");
            ss = new ServerSocket(1889);
            s = ss.accept();        //Waiting for client connection
            System.out.println("CLIENT CONNECTED...");
            System.out.println("Enter the data to send to CLIENT");
            dout = new DataOutputStream(s.getOutputStream());
            serverChat();
        }catch(IOException ioe){
        System.out.println(ioe);
        }catch(Exception e){
        System.out.println(e);
        }
    }
    public void serverChat() throws IOException
    {
    BufferedReader br = new BufferedReader(new
            InputStreamReader(System.in));
    String st;
        do{
            st = br.readLine();
            dout.writeUTF(st);
            dout.flush();
        }while(!st.equals("stop"));
    }
    public static void main(String ar[])
    {
        new Server();
    } }
//Client.java
import java.io.*;
import java.net.*;
import java.util.*;
public class Client
{
    Socket s;
    DataInputStream dis;
    public Client()
    {
        try{
        s = new Socket("127.0.0.1", 1889);
//Bind the DataInputStream object with client socket input-stream
        dis = new DataInputStream(s.getInputStream());
clientChat();
        }catch(IOException ioe){
        System.out.println(ioe);
```

```
        }catch(Exception e){
        System.out.println(e);
        }
    }
    public void clientChat() throws IOException
    {
        String str;
        do{
        str = dis.readUTF();
        System.out.println("SERVER Message :: " + str);
        }while(!str.equals("stop"));
    }
    public static void main(String ar[])
    {
    new Client();
    }}
```

Q. 5. Discuss with example how a client application read a file from server through URL Connection.

[MDU, May 2008; Dec. 2008]

Ans.

```
import java.lang.System,
import java.net.URL;
import java.net.URLConection;
import java.netURLEncoder;
import java.net.MalformedURLException;
import java.io.DataInputStream;
import java.io.IOException;
public class QueryURLApp{
public static void main(String args[]){
try
{
String urlString ="htt://www.jaworski.com/cgi-bin/echo-query";
String extraPathInfo="/this/is/extra/path/information";
String queryString=URLEncoder.encode("Query string with some special
    characters : @#$%?&+");
URL url= new URL(urlString+extraPathInfo+"?"+queryString);
URLConnection connection = url.openConnection();
DataInputStream fromURL = new DataInputStream(url.openStream());
String line;
While((line=fromURL.readLine()!=null)
System.out.println(line);
fromURL.close();
}catch(MalformedURLException ex){
System.out.println(ex);
}catch(UnknownServiceException use){
System.out.println(use);
}catch(IOException ioe){
```

```
System.out.println(ioe);
}
}
}
```

Q. 6. What is Socket programming? Write a program to illustrate client-server Communication.

Ans. Socket programming is a mechanism by which two computers can communicate with each other. Standard protocols are in used for the communication. To communicate with any machine there is requirement of two components, one is address of machine on network, that can be IP or machine name, and other is port number, on which the remote machine listen the request. For socket programming, we create a server, which is bind with a port number. Client machine first connect to the server with its address on network and port number to a socket. The request is generated form the client to server established the connection. The server accepts the request and stores the client socket reference. Then for communication to each other streams are bind with the socket object. The data will write to and read from these stream objects on either end.

In very common words Socket is a one end point of two communication link between two running programs running on network. A socket is bound to a port number, IP address and protocol, so that TCP layer can identify the application that data is destined to be sent.

Steps to reading from and writing to a Socket

(*i*) Open a socket

(*ii*) Open an I/O stream to a socket

(*iii*) Read from and write to the stream according to the service protocol

(*iv*) Close the stream

(*v*) Close the socket

The example of client-server communication is discussed in last answer.

Q. 7. Describe how a server sends data to client? **[MDU, Dec. 2008]**

Ans. Server sends information to clients after starting the server program; it waits for some client to attach to its port. We chose port number for example 8189, which is not used by any of the standard services.

Q. 8. What information is needed to TCP Socket classes?

Ans. The Local System's IP address and Port Number. And the Remote System's IPAddress and Port Number.

Q. 9. What are the two important TCP Socket classes?

Ans. Socket and ServerSocket.

ServerSocket is used for normal two-way socket communication. Socket class allows us to read and write through the sockets.getInputStream() and getOutputStream() are the two methods available in Socket class.

Q. 10. When Malformed URLException and UnknownHostException throws?

Ans. When the specified URL is not connected then URL throw MalformedURLException and If InetAddress method getByname and getLocalHost are unabletoresolve the host name they throw an UnknownHostException.

❑❑❑

8

CHAPTER

Database Networking

Account balance, userid, password, telephone number, etc. are daily use information; each has some implicit meaning, the value called *data*. These are the facts and can be recorded. A collection of related data is called *database*, and the system used to manage it is called *Database Management System* (DBMS). Data is an atomic entity that is stored in DBMS and become the information after pull together.

INSIDE THIS CHAPTER

As the use of computer is growing, there is a major impact of Database system to handle it, called Database Management System, on everyday life in modern society. A DBMS offer strength to industry solutions to manage the data and uses different algorithm to assure efficient interaction in a secure way with stored data in the database.

Many DBMSs are available in the market, most of which are implemented on the basis of widely accepted relational database model. A database model gives an explanation about the organization of data in a database. Relational database model supports normalization, i.e. data is grouped in the form of tables. A relational database is thus a collection of *Tables*.

Once a database is created, at least containing one table, then the following operations can be performed on the tables of the database:

- *Save data into table*
- *Retrieve data from table*
- *Update data of table*
- *Manipulate data of table*
- *Delete data from table*

DO YOU KNOW

A database model gives an explanation about the organization of data in a database.

There are a number of vendors in market that provides DBMS packages such as Oracle, DB2, Sybase, and Microsoft Access.

A component that gives access to the database to Java application is JDBC. JDBC is used to connect enterprise component to any data source. JDBC is the standard API to work on data source, retrieve the data in tabular format from any kind of data source. A data source can be relational database, spreadsheet, or flat file. In this chapter we'll discuss how Java data objects are used to create a communication links with a data source and SQL is used to create a message (Query), later sent to DBMS, the message contains the information about an operation like retrieve data, update data, delete data and other manipulations.

In most of the applications, information repository has to be accessed and manipulated by the application (implemented with J2SE or J2EE). Here, we'll discuss about the architecture of JDBC data sources, SQL queries related to some operations (update, insert, delete) on database and updates, and advanced concepts like prepared statements, batch updates, and metadata. Finally, we'll look at the built-in features of the JDBC API.

JDBC API allows the programmer to write an application with the separation of the underlying architecture and operating environment. JDBC in a nutshell performs all the database operations related methods, connecting methods, process database results and also support heterogeneous environment. It provides an interface that has intelligence how to communicate with a proprietary database. JDBC is same as ODBC (Open Database Connectivity) standard that facilitates the programmer by giving the interoperable function through JDBC-ODBC bridges.

> **DO YOU KNOW**
>
> JDBC is a unified standard API that facilitates you to access any data source in tabular format.

8.1. INTRODUCTION OF JDBC

Using JDBC, it is easy to send SQL statements to any relational database. In other words, with the JDBC API, it isn't necessary to write one program to access a Sybase database, another program to access an Oracle database, another program to access an Informix database, and so on. One can write a single program using the JDBC API, and the program will be able to send SQL statements to the appropriate database. And, with an application written in Java programming language, one doesn't have to worry about writing different applications to run on different platforms. The combination of Java and JDBC lets a programmer write it once and run it anywhere. Java, being robust, secure, easy to use, easy to understand, and automatically downloadable on a network, is an excellent language basis for database applications.

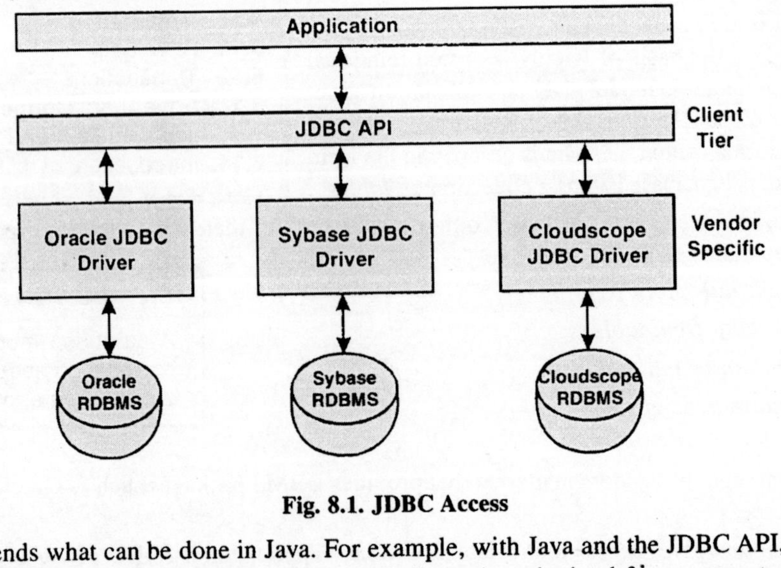

Fig. 8.1. JDBC Access

JDBC extends what can be done in Java. For example, with Java and the JDBC API, it is possible to publish a web page containing an applet that uses information obtained from a remote database. Or an enterprise can use JDBC to connect all its employees to one or more internal databases via an intranet. With more and more programmers using the Java programming language, the need for easy database access from Java is continuing to grow. Once a database is created, at least containing one table, then the following operations can be performed on the tables of the database:

- *Save data into table*
- *Retrieve data from table*
- *Update data of table*
- *Manipulate data of table*
- *Delete data from table*

There are a number of vendors in market who provide DBMS packages such as Oracle, DB2, Sybase, and Microsoft Access.

8.2. DATABASE CONNECTIVITY/ DESIGN OF JDBC

Java provides JDBC product components that are a part of Java Development Toolkit. The first component is **JDBC Driver Manager** that acts as the backbone of JDBC architecture and connects Java applications to the correct JDBC Driver. We will discuss about JDBC architecture and JDBC Drivers in the proceeding sections. Another component is the **JDBC Driver Test Suite** that provides a surety that the JDBC drivers will run our program. The last component is **JDBC-ODBC Bridge** that allows ODBC drivers to be used as JDBC drivers.

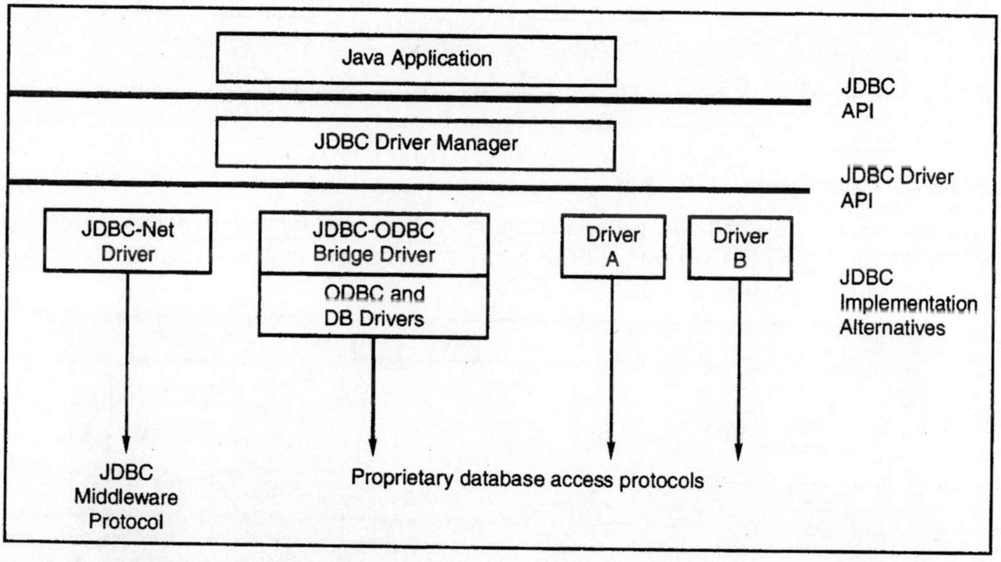

Fig. 8.2. JDBC Driver Implementations

8.3. JDBC DRIVERS

JDBC Driver works as middleware layer between the data source and the application, either written through J2SE OR J2EE.

There are four types of JDBC Drivers that are discussed below:

1. JDBC-ODBC Bridge plus ODBC driver

The JavaSoft bridge product provides JDBC access via ODBC drivers. Note that ODBC binary code, and in many cases database client code, must be loaded on each client machine that uses this driver. As a result, this kind of driver is most appropriate on a corporate network where client installations are not a major problem, or for application server code written in Java in three-tier architecture.

JDBC-ODBC Bridge – Type 1: It uses the Open Database Connectivity (ODBC) as the prime interface to data source (Database). Initially ODBC was developed for Microsoft Windows platform to give API standard to communicate with SQL, but later upgraded to provide SDKs for other platform as well. In JDBC-ODBC Bridge the client makes the JDBC calls. The type 1 driver provides a bridge between the JDBC API and the ODBC API. The bridge is responsible to convert the JDBC calls to the corresponding ODBC calls, and finally request to data source using ODBC libraries, that must be required on client tier.

J2SE contains the classes for the JDBC-ODBC Bridge and so that there is no need to install any other packages. Though, you have to create Data Source Name (DSN) using ODBC Manager.

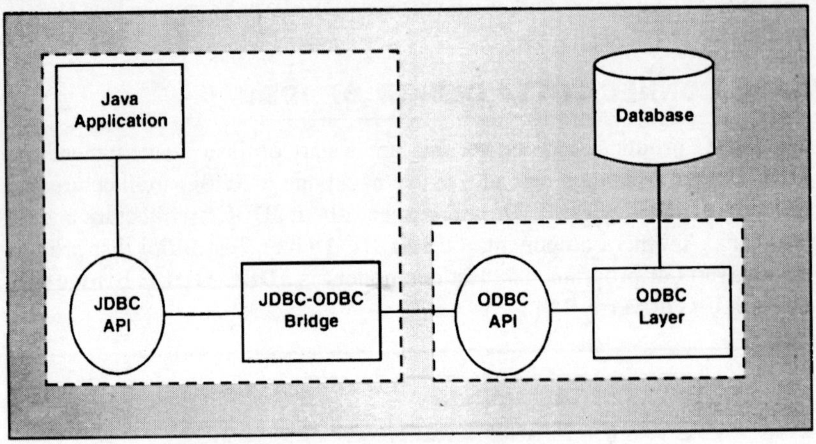

Fig. 8.3. Type 1 Driver

2. Native-API partly-Java driver

This kind of driver converts JDBC calls into calls on the client API for Oracle, Sybase, Informix, DB2, or other DBMS. Note that, like the bridge driver, this style of driver requires that some binary code be loaded on each client machine.

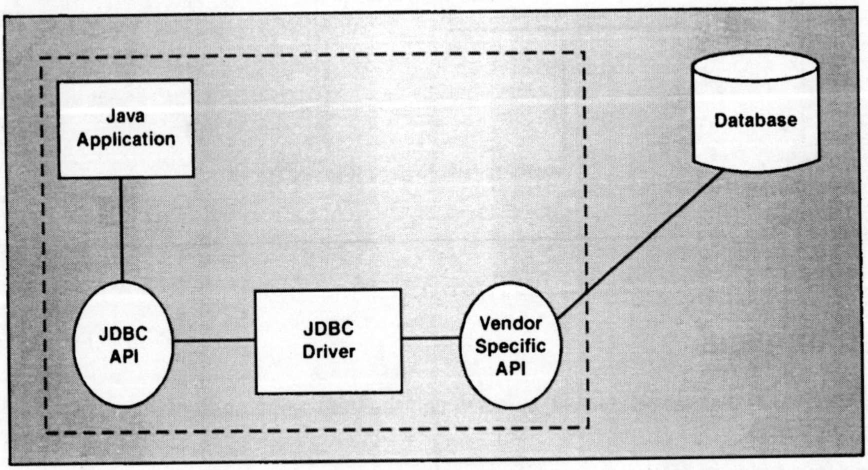

Fig. 8.4. Type 2 Driver

Native-API Partly Java Driver – Type 2: Type 2 driver works on a combination of vendor-specific and Java implemented APIs to give to communicate with data source. Here, the client makes the JDBC calls and these calls are interpreted into vendor-specific APIs, finally from vendor-specific APIs to database to perform the operations on it. Database process the request and send the results back through the vendor-specific API, which will be forwarded back to the JDBC driver. Now, here JDBC driver will convert the response (results) to the JDBC standards and send to Java application. The driver can be used where is the need of high-performance and large transaction volume.

3. JDBC-Net pure Java driver

This driver translates JDBC calls into a DBMS-independent net protocol, which is then translated to a DBMS protocol by a server. This net server middleware is able to connect its pure Java clients to many different databases. The specific protocol used depends on the vendor. In general, this is the most flexible JDBC alternative. It is likely that all vendors of this solution will provide products suitable for intranet use. In order for these products to also support Internet access, they must handle the additional requirements for security, access through firewalls, and so forth, that the Web imposes.

JDBC-Net Pure Java Driver – Type 3: Type 3 driver works as middleware server between a numbers of Java clients to multiple vendor databases. Type 2 is a multitier and vendor-neutral driver. To connect with database server, client (Java client application) uses a server component in-between the client and database that works as a gateway for multiple database servers. The client application sends the request to intermediate data access server, by making the JDBC call through a JDBC driver. The intermediate data access server completes the request to the Database using native driver.

Different protocols are used at different level of communication. Client and intermediate server use different protocol, and based on the vendor of middleware server, similarly middleware sever connect and communicate with database server by using different native protocol. There is no need of any driver at the client machine. The JDBC call will be generated through RMI Call to the JDBC driver that is stored in the middle tier. It is used to provide solution and is suitable for internet use. To support these applications, there is need of internet access, requirements for security, firewalls etc.

DO YOU KNOW

The JDBC call will be generated through RMI Call to the JDBC driver that store in the middle tier.

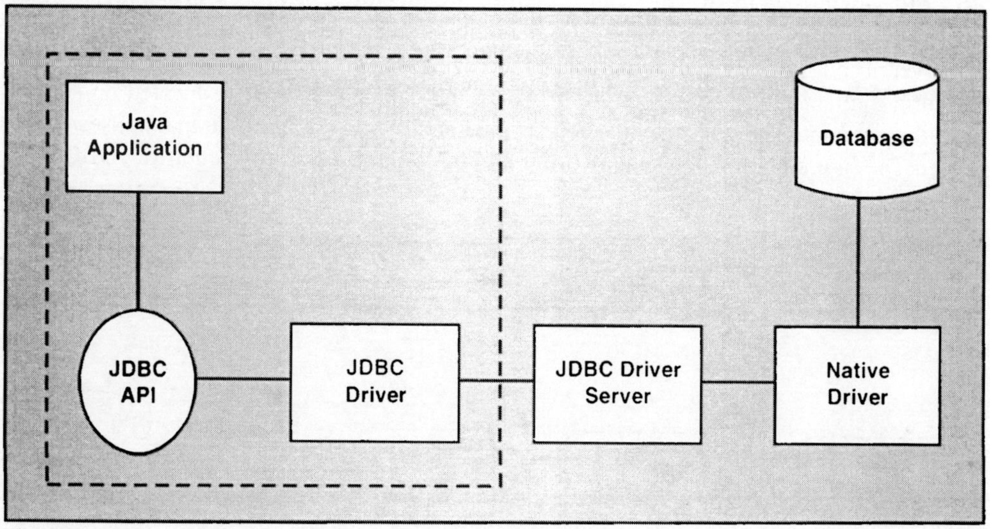

Fig. 8.5. Type 2 Driver

4. Native-protocol pure Java driver

This kind of driver converts JDBC calls directly into the network protocol used by DBMSs. This allows a direct call from the client machine to the DBMS server and is an excellent solution for intranet access. Since many of these protocols are proprietary, the database vendors themselves will be the primary source. Several database vendors have these in progress.

Native-protocol Pure Java Driver – Type 4: Type 4 drivers converts JDBC API request directly to networking protocols of vendor-specific. By making the socket connection with database they communicate with database. There is no need of other libraries or any middleware interface to deploy it.

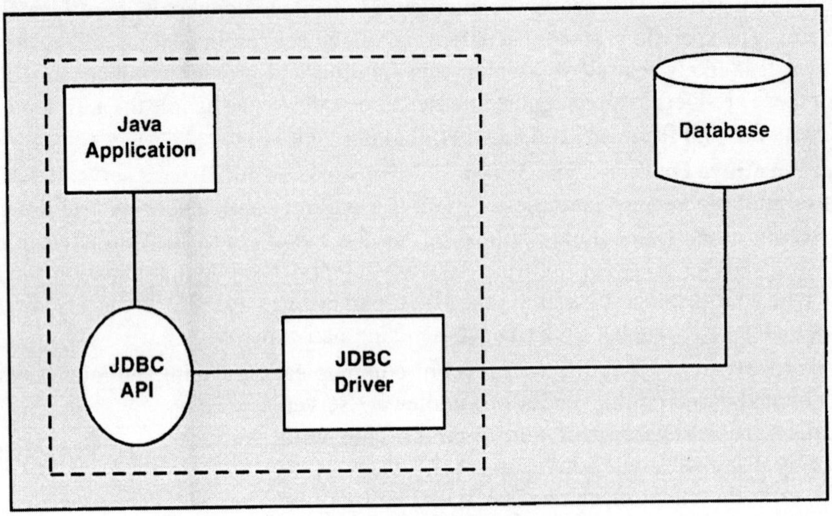

Fig. 8.6. Type 4 Driver

8.4. JDBC APPLICATION ARCHITECTURE

JDBC API supports both two-tier and three-tier architectures or processing models for database access. In the two-tier model, a Java application communicates directly to the data source with the help of JDBC driver. The user's commands are sent to the database and the response is sent back to the user. This may form a client/server configuration in which the data source may be located on some another machine in the network. The user machine acts as the client while the machine that houses the data acts as the server. The figure below shows the two-tier and three-tier processing models of JDBC.

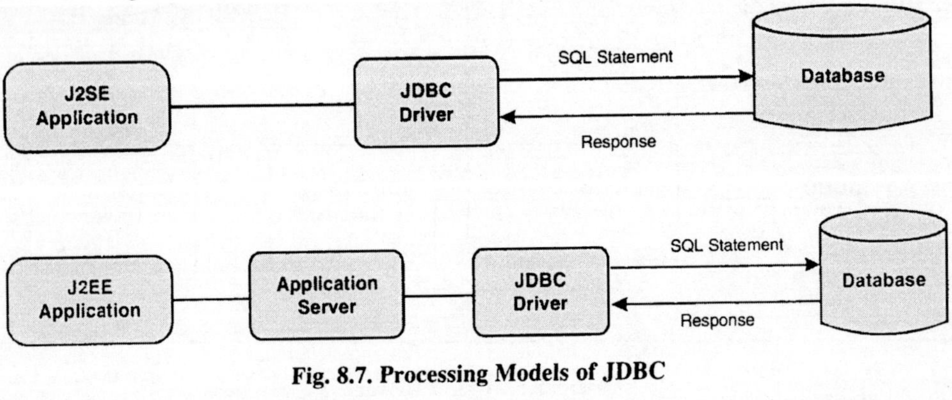

Fig. 8.7. Processing Models of JDBC

In the three-tier architecture, the commands are sent to the middle-tier which sends the command to the data source. The data source processes the commands and sends the result back to the middle-tier which then sends them to the user. Middle-tier makes it possible to maintain control over access and the kinds of updates that can be made to corporate data. It also simplifies the deployment of applications. The three tier architecture provides performance advantages.

8.5. RELATIONAL DATABASES

A database is a means of storing information in such a way that information can be retrieved from it easily. A *relational* database presents information in tables with rows and columns. A table is referred to as a *relation*. A Database Management System (DBMS) handles the way data is stored, maintained, and retrieved. In the case of a *relational database*, a Relational Database Management System (RDBMS) performs these tasks. The following terms are associated with RDBMS that perform discrete tasks for accessing tables in the simplest and efficient manner. RDBMS is the simplest method for accessing data.

Integrity Rules

Integrity means accuracy. Relational tables follow certain integrity rules to ensure that the data they contain remains accurate and accessible. The rules are as follows:

1. First, the rows in a relational table should all be distinct. If there are duplicate rows, there can be problems resolving which of two possible selections is correct.
2. A second integrity rule is that column values cannot be repeating groups or arrays.
3. A third aspect of data integrity involves the concept of a *null* value. A database has to take care of situations where data may not be available. A null value indicates that a value is missing. It does not equate to a blank or zero. A blank is considered equal to another blank, a zero is equal to another zero, but two null values are not considered equal.
4. When each row in a table is different, it is possible to use one or more columns to identify a particular row. This unique column or group of columns is called a *primary key*. Any column that is part of a primary key cannot be null. This rule is referred to as *entity integrity*. The Table below illustrates some of these relational database concepts:

Student Table

Student-ID	First_Name	Last_Name	Date_of_Birth	Marks
534	Ram	Verma	28-AUG-87	87
589	Manan	Sood	30-MAY-85	78
596	Munish	Sharma	06-APR-85	89
675	Shivendra	Tripathi	28-AUG-86	75
743	Swati	Verma	26-AUG-84	92
859	Bhanu	Priya	10-NOV-91	86

In this student table, there are five columns: Student-ID, First_Name, Last_Name, Date_of_Birth and Marks. There are six rows, each representing a different student. The primary key for this table would be Student-ID because each one is guaranteed to be different. It would also be possible to use First_Name and Last_Name because the combination of the two also identifies just one row. Using the last name alone would not work because there are two students with the last name "Verma." In this particular case the first names are all different, so one could conceivably use that column as a primary key, but it is best to avoid using a column where duplicates could occur.

8.6. STRUCTURED QUERY LANGUAGE

As we know the meaning of the database, i.e. database is a methodology of storing data in it that later it can be fetched from it. A type of database called *relational database, discussed above,* presents data in the form of tables (combination of rows and columns). The database system supports relational database called Relational Database System (RDBMS).

Table Example: Employee

EmpID	FirstName	LastName	DOB	Salary
1001	Mariya	Surkozi	22-AUG-45	339834
1009	Alice	Haden	12-NOV-47	534877
1011	Bob	Paul	10-JAN-69	569849
1088	Rosy	Paul	14-JUL-74	236652
1090	Anna	Kurnikova	11-SEP-81	258498
1099	Rahina	Washington	28-DEC-57	484599

The above table showing some record of table *EMPLOYEE* with attributes EmpID, FirstName, LastName, DOB, CarNumber. All the SQL command example will be based on the table *EMPLOYEE.*

There are varies operations on data:

- Writing to database
- Extract from database
- Modification of data
- Deletion of data

For all significant operations on data there is a language called Structure Query Language (SQL), supported by all RDBMS. SQL Commands are divided into the following categories:

1. Data Manipulation Language (DML)

DML commands work on data, and perform operations like, retrieving it; manipulate it. A brief discussion about the more common DML commands are as follows:

(a) **SELECT:** A flexible and simple statement use to select rows with some attributes (columns) from the given table name.

Syntax:

```
    select [column1,column2,...] from [TABLENAME] where [criteria_1 {AND/
OR} criteria_2 . . . {AND/OR} . . . criteria_2]
```

For example :

select * from Employee

Employee				
empid	**firstname**	**lastname**	**salary**	**DOB**
1001	Mariya	Surkozi	339834	8/22/1945
1009	Alice	Haden	534877	11/12/1947
1011	Bob	Paul	569849	1/10/1969
1088	Rosy	Paul	236652	7/14/1974
1090	Anna	Kurnikova	258498	9/11/1981
1099	Rahina	Washington	484599	12/28/1957

select EmpID, FirstName, DOB from Employee

Employee		
EmpID	**FirstName**	**DOB**
1001	Mariya	8/22/1945
1009	Alice	11/12/1947
1011	Bob	1/10/1969
1088	Rosy	7/14/1974
1090	Anna	9/11/1981
1099	Rahina	12/28/1957

select * from Employee where Salary > 500000

Employee				
empid	**firstname**	**lastname**	**salary**	**DOB**
1009	Alice	Haden	534877	11/12/1947
1011	Bob	Paul	569849	1/10/1969

(*b*) **INSERT :** It use to insert data into the table as a record. Here we've to give the attribute name (if store in selected attribute) and value (data) to store in database.

Syntax:

```
insert into [TABLENAME] [columnsname1, columnname2,....] values
    [value1, value2,...]
```

For example :

insert into Employee values (1110,'Sujain','David','12-JAN-78',578989)

Employee				
empid	**firstname**	**lastname**	**salary**	**DOB**
1001	Mariya	Surkozi	8/22/1945	339834
1009	Alice	Haden	11/12/1947	534877
1011	Bob	Paul	1/10/1969	569849
1088	Rosy	Paul	7/14/1974	236652
1090	Anna	Kurnikova	9/11/1981	258498
1099	Rahina	Washington	12/28/1957	484599
1110	Sujain	David	1/12/1978	578989

insert into Employee (EmpId, FirstName, DOB, Salary) values (1111,'David','21-MAR-78',473554)

Employee				
empid	**firstname**	**lastname**	**salary**	**DOB**
1001	Mariya	Surkozi	8/22/1945	339834
1009	Alice	Haden	11/12/1947	534877
1011	Bob	Paul	1/10/1969	569849
1088	Rosy	Paul	7/14/1974	236652
1090	Anna	Kurnikova	9/11/1981	258498
1099	Rahina	Washington	12/28/1957	484599
1110	Sujain	David	1/12/1978	578989
1111	David		3/21/1978	473554

(*c*) **DELETE :** The command is use to delete record (row) from the table. We can give some criteria using where clause to give selected record.

Syntax :

```
delete from [TABLENAME] where criteria N
```

delete from Employee where Empid=1099

Employee				
empid	**firstname**	**lastname**	**salary**	**DOB**
1001	Mariya	Surkozi	8/22/1945	339834
1009	Alice	Haden	11/12/1947	534877
1011	Bob	Paul	1/10/1969	569849
1088	Rosy	Paul	7/14/1974	236652
1090	Anna	Kurnikova	9/11/1981	258498
1110	Sujain	David	1/12/1978	578989
1111	David		3/21/1978	473554

delete from Employee where FirstName= 'David'

Employee				
empid	**firstname**	**lastname**	**salary**	**DOB**
1001	Mariya	Surkozi	8/22/1945	339834
1009	Alice	Haden	11/12/1947	534877
1011	Bob	Paul	1/10/1969	569849
1088	Rosy	Paul	7/14/1974	236652
1090	Anna	Kurnikova	9/11/1981	258498
1110	Sujain	David	1/12/1978	578989

(*d*) **UPDATE :** It's a simple SQL command used to update or modify the value of any attribute (column) of records on the base of given criteria.

Syntax :

```
update [TABLENAME] set [column1 = expression1], [column2 =
expression2], … [columnN = expression] where criteria1 [AND/OR] … criteriaN
```

update Employee set salary= salary + salary* 0.10 where Empid=1111

Employee				
empid	**firstname**	**lastname**	**salary**	**DOB**
1001	Mariya	Surkozi	8/22/1945	339834
1009	Alice	Haden	11/12/1947	534877
1011	Bob	Paul	1/10/1969	569849
1088	Rosy	Paul	7/14/1974	236652
1090	Anna	Kurnikova	9/11/1981	258498
1110	Sujain	David	1/12/1978	636888

2. Data Definition Language (DDL)

DDL commands use to define the database, like creating the table, modifying the table structure, creating views; and indexing the database. Let's have a look on more common use of DDL commands as follows:

(*a*) CREATE TABLE
(*b*) DROP TABLE
(*c*) ALTER TABLE

8.7. JDBC INSTALLATION

Registering an ODBC data source

To configure the ODBC data source implement the following steps:

1. Open the control panel from Windows settings and start the application "ODBC Data Sources [32bit]"

NOTE

In Windows 2000, the Control Panel does not contain the pointer to the "ODBC Data Sources". Using Help, search on ODBC and then select the topic "Using Data Sources". From there, select "Data Sources" to bring up the ODBC Data Source Administrator window.

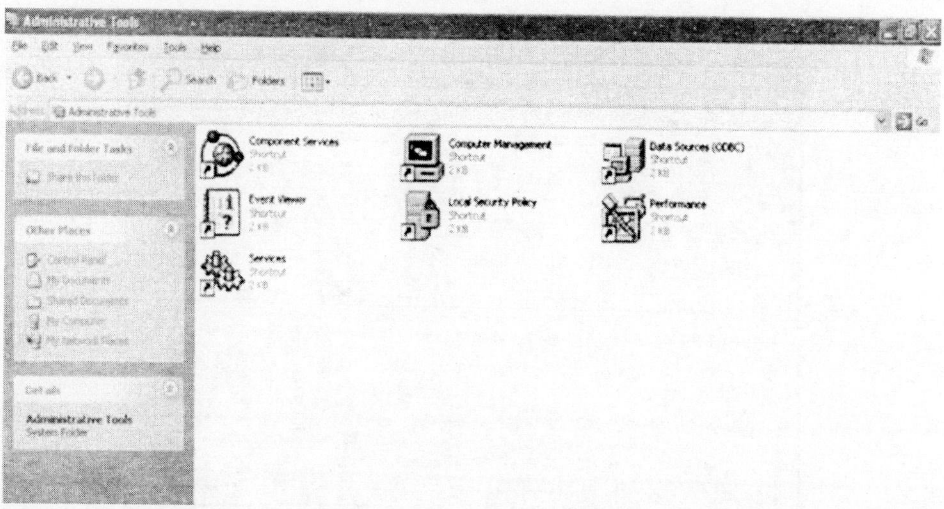

2. Select the Add button to add a new data source.

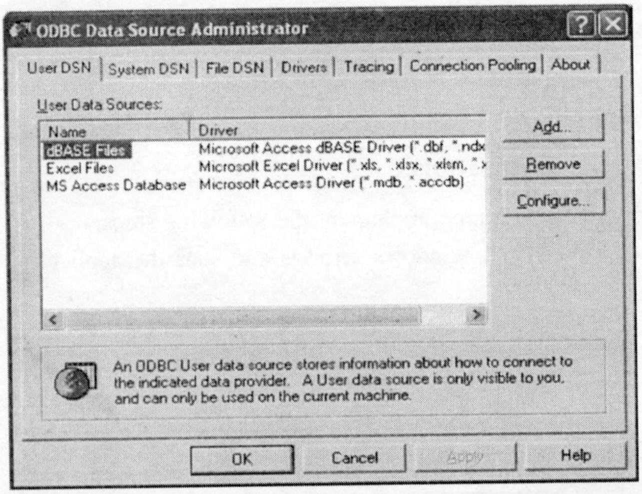

3. Select the appropriate driver for your database and click the Finish button.

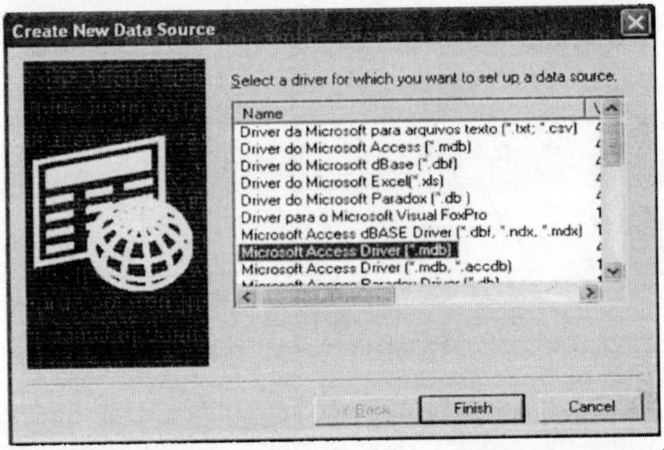

4. Enter a string in the Data Source Name that you want to use to register your database, and click the Select... button.

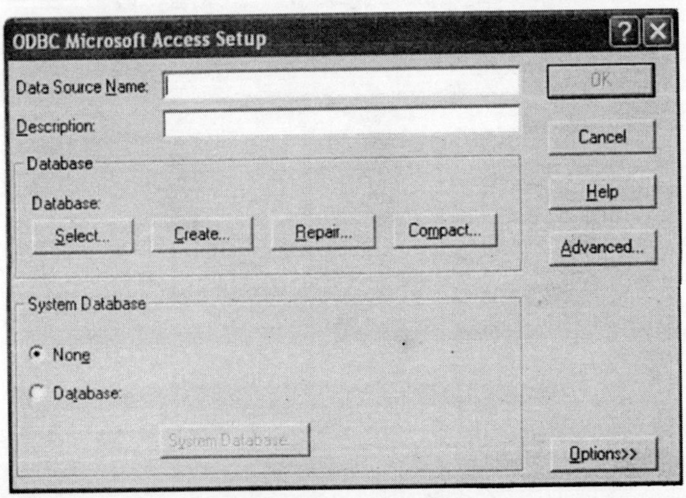

5. Find the actual file of your database and click OK when you're done.

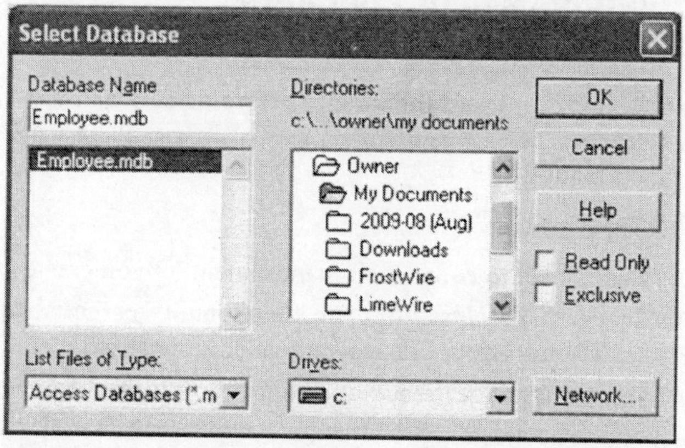

6. Click on OK

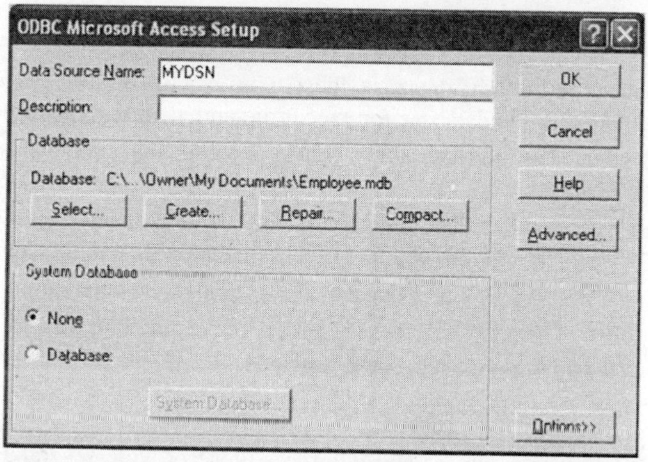

7. Click on OK

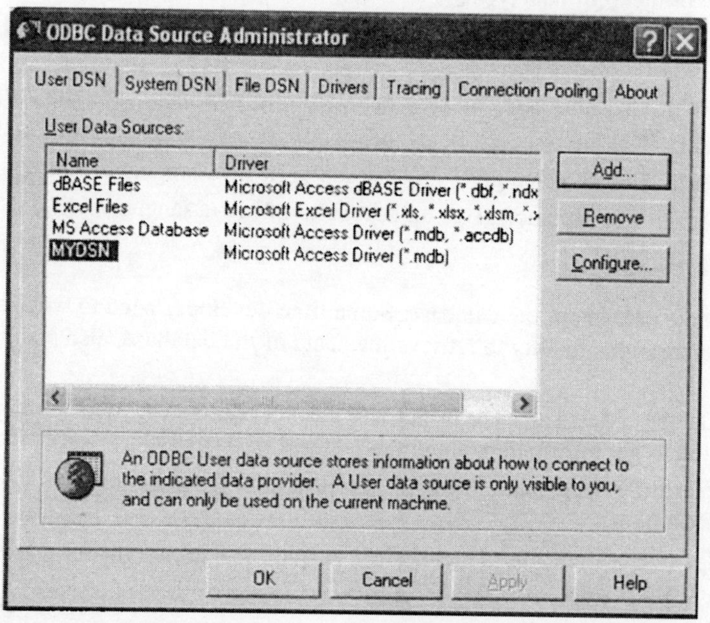

8.8. BASIC JDBC PROGRAMMING CONCEPTS [INTRODUCTION TO JAVA.SQL PACKAGE]

Let's have some concise look on java.sql package, before go in-depth to write the Java application with database connectivity code. The classes in the java.sql package can be separated into the subsequent groups based on their functionality:

1. Management of Connection

The classes and interface related to connection management let you create a connection to the data.

- java.sql.DriverManager: The *class* offer the functionality essential for running one or more database drivers, individual driver connects to specific database.
- java.sql.Driver: The *interface* abstracts the connection protocol of specific-vendor.
- java.sql.Connection: It abstracts the interaction with the database. Sending SQL statement to database, read the results of operations is done by this *interface*.

2. Access Database

After creating connection with the database, these types of classes and interfaces allow you to send SQL commands (statements) to the database for execution, and also read the response (result).

- java.sql.Statement: The interface allow you to execute SQL commands above the connection and access the results.
- java.sql.PreparedStatement: The interface used to give parameterized SQL Statement, include place holders (as "?"), later on which will be replaced with actual values.
- java.sql.CallableStatement: The interface allows you to execute stored procedure.
- java.sql.ResultSet: The interface store the results of execute SQL SELECT commands. To access the results row-by-row, it provides some methods.

3. Type of Data

In java.sql package there are classes and interfaces those provide several Java data types so as to correspond to some of the SQL data type.

- java.sql.Array: The interface gives an abstraction in Java language with ARRAY collection of SQL data types.
- java.sql.Blob: The interface gives an abstraction in Java language with SQL type BLOB
- java.sql.Clob: The interface gives an abstraction in Java language with SQL type CLOB
- java.sql.Data: The interface gives an abstraction in Java language with SQL type DATE
- java.sql.Time: The interface gives an abstraction in Java language with SQL type TIME

4. Metadata of Database

Metadata is some data about the database. Some time developer need to work on metadata. So that JDBC API also supports this facility to retrieve metadata of the database, also parameters to statements; and result too.

- java.sql.DatabaseMetadata: The instance of this interface can be obtained by java.sql.Connection. This is used to get information about the features of database.
- java.sql.ResultSetMetadata: Methods are provided by this interface to access metadata of java.sql.ResultSet.
- java.sql.ParameterMetadata: The interface permits you access the database types of parameters in prepared statements.

5. Exceptions and Warnings

The collection of classes support the runtime exception and warning related to JDBC.

- java.sql.SQLException: Handle all type of exception of JDBC related, database level, driver level exceptions and error levels.
- java.sql.SQLWarnings: It represents database warnings.
- java.sql.BatchUpdateException: Handles the exception for batch updates.

For writing the code to connect with data base and work on database with the SQL commands the developer has to perform the following steps.

6. Loading the Database Driver

The concept of a URL in JDBC is same as used other application like internet. In basic concept of JDBC URLs, in application more then one database can be in use, each will be accessed via a different database driver. In such a situation, it is necessary every driver must be uniquely identified.

To communicate with the data source, you must load a driver. Here we're taking the example of *JdbcOdbcDriver*. The syntax to load the driver is given below:

Syntax:

```
(i) Class.forName( "DriverClassName");
      OR
       Class.forName("DriverClassName").newInstance();
```

Example:

```
try{
Class.forName ("sun.jdbc.odbc.JdbcOdbcDriver");
}catch(ClassNotFoundException e)
{
System.out.println("Unable to load the Driver"); }
```

```
(ii) java -D jdbc.drivers=DriverClassName JavaApplication
```

The above command (*ii*) specified the driver from the command line, using a system property *jdbc.drivers*, but it's necessary the driver must be in *CLASSPATH* at the time of compilation.

The DriverManager class has ***three variations*** of a static method *getConnection*() to establish the connections as follows:

public static Connection getConnection(String urlString) throws SQLException return a *Connection* object to a driver from the set of registered JDBC driver. The *urlString* must be in form of *jdbc:subprotocol:subname.*

Example:

```
Connection conn = DriverManager.getConnection("jdbc:odbc:MYDSN");
```

public static Connection getConnection(String urlString, java.util.Properties prop) throws IOException, two arguments are passed to it, *utlString* as same as above, and *prop* is an object of *Properties* type, contains the additional information required for the specified database.

Example:

```
String URL = "jdbc:cloudscape:Movies";
Properties p = new Properties();
```

```
p.put("autocommit","true");
p.put("create","true");
Connection conn = DriverManager.getConnection(URL,p);
```

public static Connection getConnection(String urlString, String userName, String password), this variant receives the *userName* and *password* as the argument to connect with the database.

Example:

```
String urlString = "jdbc:odbc:MYDSN";
String userName = "admin";
String password = "adminadmin"
Connection conn = DriverManager.getConnection(urlString,
                        userName, password);
```

7. Opening Connections or Establishing a Connection

DriverManager class defines object to connect application to a JDBC Driver. It has a method getConnection() used to find a connection to a database. It receives a URL, the URL contains protocol, sub-protocol, sub-name, and other attribute as well, and return the connection object.

A Connection object represents a context. SQL statements are fired on it and return the results. You can create more than one connection in an application with one database, can have many connection with many databases.

URL Syntax: jdbc: <subprotocol>:<subname>

Each driver supports it's own sub-protocol, and each sub-protocol has its own syntax, varies on the base of number of attributes.

Example:

```
try{
String url =  "jdbc:odbc:MYDB";
Connection conn = DriverManager.getConnection(url, "USER1",  "MYNAME")
}catch( SQLException e ){
    System.out.println( "Error in Connection " + e ); }
```

In the given example, *url* contains the protocols name : jdbc, the sub-protocol : odbc, and data-source name : MYDB is a DSN. USER1 is the username to access the database and MYNAME is the password.

The ***java.sql.Connection*** has many public methods, we'll see the use of other later in this chapter but all three of them are used to create different type statements as follows:

8. Creating Statement

Once the connection has been established with data-source, we can interact with database. *Connection* object are used to execute SQL statements by creating *Statement, PreparedStatement,* or *CallableStatement. These* objects are abstract SQL commands.

Once we have created the object on one of these statements, we can execute the statement and read the results, though the result can be stored in *java.sql.ResultSet* object if the SQL command is SELECT. The methods are given below to create either type of these statements.

Statement createStatement() throws SQLException

The method creates a *Statement* object, which use to send the SQL command to the database.

```
Statement createStatement(int result_Set_Type, int result_Set_Concurrency)
                                        throws SQLException
```

In it *resultSetType* and *resultSetConcurrency* arguments apply to *ResultSet* object, which is created by executing queries. For the first argument there are three different constant *ResultSet.TYPE_FORWARD_ONLY, ResultSet.TYPE_SCROLL_INSENSITIVE* and *ResultSet.TYPE_SCROLL_SENSITIVE* for different purposes discuss later in the same chapter. For the second argument there are two constants ResultSet.CONCUR_READ_ONLY, ResultSet.CONCUR_UPDATABLE

The no-argument *createStatement()* returns a forward-only scrollable and read-only ResultSet.

Example:

```
Statement stmt = conn.createStatement();
OR
Statement stmt = conn.createStatement(
        ResultSet.TYPE_SCROLL_INSENSITIVE,
        ResultSet.CONCUR_READ_ONLY);
```

public PreparedStatement prepareStatement(String sqlString) throws SQLException

It's a pre-compiled SQL statement that avoids the repeated compiling of the same SQL command. It is used to create parameterized statements in a way that the parameters will be placed at run time. It is useful to create statements where data values are not always as character string. It's superclass is *Statement*.

Example:

```
PreparedStatement prstmt = conn. prepareStatement("INSERT INTO" + "EMPLOYEE
(empid, firstname, lastname, salary)"+              VALUES  ?, ?, ?, ?, ?)");
```

To set the value at place holder i.e. at '?' (Question Mark) there are number of setXXX(int placeHolderIndex, dataTypeXXX data) method of each data type value.The *placeHolderIndex* starts from 1. For the *prstmt,* the value will be set to the placehoder in the following manner:

```
prstmt.setInt(1, 1078);
prstmt.setString(2, "Holzner");
prstmt.setString(3. "Steve");
prstmt.setInt(4, 567439);
pstmt.executeUpdate();
```

public CallableStatement prepareCall(String sqlString) throws SQLException

```
    public CallableStatement prepareCall(String sqlString, int
resultSet_Type,  int resultSet_Concurrency)  throws SQLException
```

The purpose of this interface is to create statements for executing stored procedures.

Example:

```
String queryString = " { CALL LastAttendanceRecord (?) }";
CallableStatement cstmt = conn.prepareCall(queryString);
```

9. Executing the Querying

After creating the statement of either type, execute the query on the database. To execute the query on database we use executeQuery(), executeUpdate(), and execute() method of Statement interface according to the need. Before exploring the logic in database programming, a discussion on execute(), executeQuery() and executeUpdate() method of Statement interfaces is required:

Here we'll take the table **EMPLOYEE** *for demonstartion* **executeUpdate()** method is used to execute SQL INSERT, UPDATE, and DELETE statements, and returns number of rows updated by the either of these SQL statement.

The method of *Statement* interface will receive a sqlString as argument that is a SQL statement has to execute. In *PreparedStatement* interface, the SQL statement is given at its creation time, so that the method passes no argument. The prototype is given in the corresponding method description table.

Example: Using the Statement object

```
Statement stmt;
Connection conn;
String url = "jdbc:odbc:MYDSN";
try
{
   Class.forName("sun.jdbc.odbc.JdbcOdbcDriver");
   conn = DriverManager.getConnection(url);
    String qryStr = "UPDATE Employee set Salary = 708767 where
                      EmpId = 1011";
   stmt = conn.createStatement();
   int rows = stmt.executeUpdate(qryStr);
  System.out.println("Update Rows : " + rows);
}catch(ClassNotFoundException cnfe){
System.out.println(cnfe);
}catch(SQLException se){
System.out.println(se);
}
```

In the above code *stmt.executeUpdate(qryStr)* will execute the query string given in *qryStr* and returns the number of updated records count to *rows*. Here the value of *rows* will be one only, because only one record of employee of whom *EmpId* is '1011' will be updated in *Employee* table.

Example : Using PreparedStatement Object

```
PreparedStatement prstmt;
Connection conn;
String url = "jdbc:odbc:MYDSN";
try
{
   Class.forName("sun.jdbc.odbc.JdbcOdbcDriver");
   conn = DriverManager.getConnection(url);
    String qryStr = "UPDATE Employee set Salary = Salary + Salary * 0.10
where                        EmpId = ?";
   prstmt = conn.prepareStatement(qryStr);
  prstmt.setInt(1, 1090);
  prstmt.executeUpdate();
 }catch(ClassNotFoundException cnfe){
System.out.println(cnfe);
}catch(SQLException se){
System.out.println(se);
}
```

In the above code *prstmt.executeUpdate()* will execute the query string given in *qryStr* that is already passed at the creation time of *PreparedStatement* object. The statement *prstmt.setInt(1,1090)* will set the value of first placeholder as *'1090'*, as the statement *prstmt.executeUpdate()* executes, only one record of employee of whom *EmpId* is '1090' will be updated in *Employee* table. The method *executeUpdate*() of PreparedStatement will never return any value.

Example:

```
stmt = conn.createStatement();
stmt.executeUpdate("INSERT INTO EMPLOYEE (EmpId, FirstName, LastName,
        Salary) VALUES (1212, 'Aman','Rai',676545)");
stmt.executeUpdate("INSERT INTO EMPLOYEE (EmpId, FirstName, LastName,
        Salary) VALUES (1046, 'Neomi','Lio',548975)");
stmt.executeUpdate("INSERT INTO EMPLOYEE (EmpId, FirstName, LastName,
        Salary) VALUES (1045, 'David','Fernandis',549598)");
executeQuery() method execute SQL SELECT statements, and store the
        results of exeution in ResultSet object.
```

ResultSet contains the data returned by executing Java method with SQL statement. It also provides methods for conversion of data between SQL and Java. ResultSet also acts as an iterator to move tuple by tuple.

To retrieve the value from any column, we have to use getXXX(int columnIndex) or getXXX(String columnName), where XXX is the data type in Java that corresponds to the datatype in SQL. The mapping of Java datatype with SQL datatype, is given at the last of this chapter.

Let the columnName is FIRSTNAME of Employee table and of type VARCHAR, then its value can be retrieve by the following statement, where rst is an object of ResultSet:

```
System.out.println(rst.getString("FIRSTNAME");
OR
String fname = rst.getString("FIRSTNAME");
```

Example: Using the Statement object

```
public void showAllData() throws SQLException{
Connection conn=null;
try{
Class.forName("sun.jdbc.odbc.JdbcOdbcDriver");
conn= DriverManager.getConnection("jdbc:odbc:MYDSN");
}catch(ClassNotFoundException cnfe){
System.out.println(cnfe);
}
Statement stmt =  conn.createStatement();
ResultSet rst = stmt.executeQuery("select * from Employee");
while(rst.next())
{
System.out.println(rst.getString("EMPID")+ ", "
    +rst.getString("FIRSTNAME")+ ", " +rst.getString("LASTNAME")+ ", "
    +rst.getInt("SALARY"));
    }
 }
```

In the above code the statement *stmt.executeQuery("select * from Employee")* returns the result of the SQL statement and store it in ResultSet object *rst*. The while loop will continue till *rs.next()* return *true*, and *rs.next()* return true until last record traversed. The *getString("EMPID")* will return the value of EMPID of current row , similarly with other getXXX() methods.

Example : Using PreparedStatement Object

```
import java.sql.*;
import java.io.*;
public class JDBCDemo1
{
public static void main(String ar[]){
Connection conn=null;
PreparedStatement pstmt=null;
try{
Class.forName("sun.jdbc.odbc.JdbcOdbcDriver");
conn= DriverManager.getConnection("jdbc:odbc:MYDSN");
pstmt=conn.prepareStatement("SELECT EmpID, FirstName, LastName, Salary
    from EMPLOYEE where EmpID =?");
pstmt.setString(1, ar[0]);
ResultSet rst = pstmt.executeQuery();
if(rst.next())
{
System.out.println(rst.getString("EMPID")+ ", "
    +rst.getString("FIRSTNAME")+ ", " +rst.getString("LASTNAME")+ ", "
    +rst.getInt("SALARY"));
}
else
System.out.println("No record found");
}catch(ClassNotFoundException ce){
System.out.println("Error "+ce);
}catch(SQLException se){
System.out.println("Error"+se);
} } }
```

execute() method execute the SQL statement given to it by String. It returns true if the statement returns a result set, otherwise false. Method *getUpdateCount()* returns number of updation in table, and *getResultSet()* will return result set that come out after executing the SQL statement.

The method of *Statement* interface will receive a sqlString as argument, that is a SQL statement has to execute, in *PreparedStatement* interface, the SQL statement is given at its creation time, so that the method pass no argument, but both returns *ResultSet* object The prototype is given in the corresponding method description table.

Example:

```
public class JdbcExecuteStatement {
        public static void main(String args[]) {
        ~~ Connection con = null;
                Statement st = null;
                String url = "jdbc:mysql://localhost:3306/";
```

```
                String db = "komal";
                String driver = "com.mysql.jdbc.Driver";
                String user = "root";
                String pass = "root";
                try {
                    Class.forName(driver);
                   con = DriverManager.getConnection(url + db, user, pass);
                    st = con.createStatement();
                String sql = "create table Abc(no integer,name varchar(10))";
                    st.execute(sql);
                } catch (Exception e) {
                    System.out.println(e);
                }
            }
        }
```

8.9. SCROLLABLE AND UPDATABLE RESULT SETS

Sometimes there is a need to move in forward direction as well as in backward direction. The next() method of *ResultSet* allows to iterate in forward direction only. i.e. by default the *ResultSet* object is forward-only scrollable. As we traversed the ResultSet, the next() will return false when it has reached the last record, now no further details can be retrieved.

New JDBC 2.1 API enhanced the old version and functionalities are categories in following section:

(*a*) **Scrollable :** The scrollable result set allows to move the cursor in either direction, and also able to support absolute positioning of the cursor.

(*b*) **Scroll Sensitive:** There are two types of sensitivity of result set: sensitive and in-sensitive. In-sensitive result set give a static snapshot, while sensitive result set give dynamic view that supports the changes made to the data after the query has been executed.

(*c*) **Updatable:** By default the resultsets are read-only in nature, i.e. the contents of the result set can not be changed. Now it supports updateable resultsets.

Before we create an object of Statement or PreparedStatement let's have a look on the three types of resultset of java.sql.ResultSet as follows:

TYPE_FORWARD_ONLY	Set the result set is not scrollable.
TYPE_SCROLL-INSENSITIVE	Set the result set is scrollable but not sensitive to database changes.
TYPE-SCROLL_SENSITIVE	Set the result set is scrollable and sensitive to database changes.

Scrollable results means to move cursor in forward and backwards direction, and also allow moving to absolute and relative position.

Scroll sensitive resultset means the change will be reflected in the resultset and database after the execution of query while insensitive resultset give a static snapshot of query results.

Updatable results allow the updation in resultset as well as in underlying database. By default resultset is read-only.

To obtain the result set of scrollable type, the Statement object has to be creating using the following syntax;

Syntax:

```
Statement stmt = conn.createStatement(scroll_type, concurrency_type);
PreparedStatement pstmt = conn.prepareStatement(command, scroll_type,
                                                    concurrency_type);
```

Example:

```
PreparedStatement pstmt = conn.prepareStatement("select * from
                                   Employee where salary > ?",
                              ResultSet.TYPE_SCROLL_INSENSITIVE,
                                   ResultSet.CONCUR_UPDATABLE);
```

By default, resultsets are created in read-only mode. That is, the contents of the reasultsets can not be changed. But enhanced version of JDBC (JDBC 2.1 API) provides updateable resultsets. As we perform update operations on resultset, it updates the corresponding original data.

There are two constants of java.sql.ResultSet to show whether the resultset will be read-only or updateable are given below:

CONCUR_READ_ONLY	Set the result set cannot be used to update the database.
CONCUR_UPDATABLE	Set the result set can be used to update the database.

Example:

```
Statement stmt = conn.createStatement (
                  ResultSet.TYPE_SCROLL_INSENSITIVE,
                  ResultSet.CONCUR_UPDATABLE);
```

Example: Code is to update the set of attribute of Employee Table and print the records in reverse order

```
Statement stmt = conn.createStatement(
                  ResultSet.TYPE_SCROLL_INSENSITIVE,
                  ResultSet.CONCUR_UPDATABLE);
ResultSet rst = stmt.executeQuery("select * from employee where empid=
'E1001'");
if(rst.next())
{
rst.updateString("address", "new defence colony, south ex-1");
rst.updateString("phonon","9898980001");
rst.updateRow();
}
rst = stmt.executeQuery("seletct * from employee");
rst.last();
while(rst.previous())
{
System.out.println("Emp ID :: " + rst.getString("empid"));
System.out.println("Emp First name :: " + rst.getString("firstname"));
System.out.println("Emp Last name :: " + rst.getString("lastname"))
}
```

Example: Code is to insert data into Employee Table

```
Statement stmt = conn.createStatement(
                        ResultSet.TYPE_SCROLL_INSENSITIVE,
                        ResultSet.CONCUR_UPDATABLE);
ResultSet rst = stmt.executeQuery("select * from employee");
rst.moveToInsertRow();
```

```
rs.updateString("empid", "E1009");
rs.updateString("firstname", pubid);
rs.updateString("lastname", lastname);
rs.updateDouble("P", price);
rs.insertRow();
rs.moveToCurrentRow();
```

8.10. METADATA

Meta means "Data About Data". As any database is created, or any table is created in a database, then database system store some information about all these for the internal use. If there is any need then we can retrieve from system file or the tools given by database system. For example, metadata describes that table structure, like column name, size, type etc.

JDBC also provides two interfaces by which we can get the information about the structures of a database and its relations. Supporting methods are useful to write the tool to interact with database. To explain the need of these interfaces let's take one example: only the database name and its URL is given, now the developer has to write the code to extract the stored data in the tables. So before writing the code the developer must know all tables name, their structures, etc. In that these interfaces will provide help to the developer to collect the information about the database and their components.

Let's have a look on the JDBC interfaces that facilitate the developer to work on metadata of the database and resultset. There are two interfaces as following:

- **DatabaseMetaData:** The interface java.sql.DatabaseMetaData is used to provide information about the database itself. Vendor to vendor the differences are found. For example, it gives data or information about the version, table name list, and other function too.

The object of *DatabaseMetaData* can be get from *getMetaData*() from Connection object.

Example:

```
DatabaseMetaData mtd = conn.getMetaData();
Sysetm.out.println("Major Version :: "+mtd.getJDBCMajorVersion());
Sysetm.out.println("Minor Version :: "+mtd.getJDBCMinorVersion());
Sysetm.out.println("Maximum Connection :: "
                        +mtd.getMaxConnections());
/*Print name of all tables of the database*/
ResultSet mrst = mtd.getTables(null, null, null, new String[] { "TABLE" });
while(mrst.next())
System.out.println(mrst.getString(3));
mrst.close();
```

- ResulstSetMetaData: The interface java.sql.ResultSetMetaData is used to provide the information about the characteristics ResultSet object. For example, using this developer can collection the name of each column of the corresponding table of which ResultSet object has the reference.

  The object of *ResultSetMetaData* can be get from *getMetaData*() from ResultSet object.

Example:

```
public void get_Meta_Data()
{
ResultSet rst = stmt.executeQuery("select * from Employee");
ResultSetMetaData metad = rst.getMetaData();
System.out.println("COLUMN NAME\t COLUMN LABLE \t COLUMN  TYPE");
```

```
for(int i=1; i<= metad.getColumnCount(); i++){
System.out.print(metad.getColumnName(i)+ "\t" +
                                metad.getColumnType(i)+"\t"+
                                metad.getColumnLabel(i));
}
}
```

In the above example, *metad* is an object of ResultSetMetaData, created by ResultSet object *rst.getColumnCount()* will return the number of columns as a result of corresponding ResultSet. The *getColumnName()* returns the name of the column on ith index, and *getColumnType*() return the data type of the column as integer value. These types are constants and define in the *java.sql.Types* interface.

8.11. ROW SETS

There is a drawback of scrollable result set, because database connections must be open throughout the interaction. Row set is a connectionless state with database, and extends ResultSet interface. Connection less means that after fetching the data from database, it disconnects the connection.

Row sets support the features to move the query result to another tier of an application or to any other supporting device.

Properties of RowSet

The javax.sql.RowSet interface provides a set of JavaBeans properties, those can be set at design time. The following are some of the commonly required parameter:

- URL
- Data source name
- User name
- User password
- Transaction isolation
- Command

Events of RowSet

The javax.sql.RowSet object generates three types of events discuss following:

There are three types of implementation of RowSet:

- **Cached rowsets** (sun.jdbc.rowset.CachedRowSet) : A CachedRowSet support disconnected operation. It contains all data from a resultset because it extends ResultSet. The developer can use it similarly as result set.

Whenever any command is going to execute, then command opens the connection, issues a query, store the result in rowset, and then disconnects the connections. But there is one drawback of rowset that it doesn't support large query results.

If the database is updated after populating the rowset, it leads to data inconsistency. At the run time, the references check the database values with rowset value. If both are same then it is replaced with new value otherwise SyncProviderException is thrown.

The CachedRowSet object is suitable in at least following two situations:

- The client application holds the results for long time interval.
- The client has neither capacity nor the resources for connecting to a database.
- **JDBC rowsets** (sun.jdbc.rowset.JdbcRowSet): It is a connected rowset. It provides JavaBeans type layer. The sun.jdbc.rowset.JdbcRowSet class is connected, not serializable and nor is it cloneable.

- **Web rowsets** (sun.jdbc.rowset.WebRowSet): For web applications, the webrows is used. The Cached RowSet use XML to communicate with other components over HTTP protocol.

8.12. TRANSACTIONS

Transaction is a set of statements. As the transaction is completed or done successful, then it can be committed. If any error occurs during transaction execution with any one of them, then it can be rollback to the last committed state.

Data integrity is an important issue of any database, so as to maintain it we group the related commands into one transactions.

For Example: Transfer money from one account to another account.

By default, a database connection is in auto-commit mode. As any SQL command is executed, it commit the database as soon as possible, and cannot be rollback.

Concepts of Transaction Processing

Before starting explanation of related APIs, let's have look at concepts associated with distributed transactions:

1. Transaction demarcation
2. Transaction context and propagation
3. Resource enlistment
4. Two-phase commit

Methods related to Transaction's operations

setAutoCommit() –

conn.setAutoCommit(false);

The above command turn off autocommit mode.

commit() –

After executing executeUpdate() method one or more then one times commit() method is called.

Example:

```
Statement stmt = conn.createStatement();
        stmt.executeUpdate("insert into Employee values
               (1110,'Sujain','David', '12-JAN-78',578989)");
             stmt.executeUpdate("insert into Employee values
                  (1121,'Edward','Luis', '21-DEC-78',545676)");
    conn.commit();
```

rollback()

If any error occurred then

```
conn.rollaback();
```

can be called, and every updates will be roll back to last commit point.

Save Points

You can gain finer-grained control over the rollback process by using save points. Creating a save point marks a point to which you can later return without having to return to the start of the transaction.

Example:

```
        Statement stmt = conn.createStatement();
        stmt.execute("insert into Employee values (1221,'Tom','Glib',
                            '09-FEB-76',496876)");
        Savepoint sp = conn.setSavepoint();
    stmt.execute("insert into Employee values (1231,'Jerry','Messa',
                        '19-MAY-74',597869)");
    if(error)
    conn.rollback(sp);
    conn.commit();
```

You can assign or give the a name to save point.

Syntax:

```
        Savepoint st = conn.setSavepoint("firstSavePoint");
```

After done with the save point , it must be release using the following syntax:

```
        stmt.releaseSavepoint(st);
```

Batch Updates

Apart from scrollable resultsets, JDBC support for batch updates. The features facilitate, to use update statements (INSERT, UPDATE, OR DELETE) in a single request to the datasource. In batch update mechanism more than one commands is collected and submitted as a batch. All the commands have to be added to batch exception SELECT because it returns ResultSet object. To add command to batch and to execute the batch there is following two commands:

- *addBatch(command)* : It used to add command in batch. The commands will execute then ther order of execution will be same as, the command was added in batch.
- *executeBatch()* : To execute all the added command, this method is used. It submit entire batch to datasource, but one statement will execute at a time. It return the updated row count the table,

Example:

```
    Statement stmt = conn.createStatement();
    boolean flag = conn.getAutoCommit();
    if(flag)
    conn.setAutoCommit(false);
    stmt.addBatch("INSERT INTO EMPLOYEE (EmpId, FirstName, LastName, Salary)
                    VALUES (1212, 'Aman','Rai',676545)");
    stmt.addBatch("INSERT INTO EMPLOYEE (EmpId, FirstName, LastName, Salary)
                    VALUES (1046, 'Neomi','Lio',548975)");
    stmt.addBatch("INSERT INTO EMPLOYEE (EmpId, FirstName, LastName, Salary)
                    VALUES (1045, 'David','Fernandis',549598)");
    int a[] = stmt.executeBatch();
    conn.commit();
    conn.setAutoCommit(flag);
```

8.13. ADVANCED CONNECTION MANAGEMENT (JNDI AND DATA SOURCE)

The JDBC can be used at small scale application that is for small organization. In that database connection can be setup with the help of properties files. These database properties file will contains some information about the database that is require to makes the connection. But for large scale application it will not be so efficient.

In an enterprise environment, JDBC is deployed in a different manner. The database management will work with the mechanism of Java Naming Directory Interface (JNDI). All the properties are not stored in a properties file, but all the details are stored in a directory. Using the directory, centralized management can be possible of user name, passwords, database names, and URLs of JDBC.

The following code is used to establish the database connection:

```
Context jndi_Context = new InitialContext();
DataSource data_Source = (DataSource) jndi_Context.lookup("java:comp/
env/jdbc/ad_java");
Connection con = data_Source.getConnection();
```

Here *DataSource* is used instead of *DriverManager*. *DataSource* interface is defined in *javax.sql* standard extension package. It provides some advance services with simple JDBC connectivity. It can execute distributed transaction on multiple databases simultaneously. So data source is a better option than *DriverManager*.

A data source is considered as a shareable network resource, and can be retrieved from a JNDI service. Any object can be *bind* with a name to JNDI service from any application. The application that binds objects with names in the JNDI service, and the application that lookup object by the name, are on remote side. JDBC APIs provides methods to build the database resources based on this approach.

Let's have a look on how to create a data source and later how to retrieve data source object.

Create a Data Source

```
XDataSource xyz = new XDataSource(…);
//Set the properties for the datasource
try {
Context ctx = new InitialContext();
ctx.bind("jdbc/mydb",xyz);
}catch(NamingException ne){}
```

XDataSource is a class that implements javax.sql.DataSource. In the above code the data source is bind with a name 'jdbc/mydb". Any application can access the data source by retrieving it from JNDI services.

Retrieving a DataSource object

```
try{
Context ctx = new InitialContext();
DataSource dataSource = (DataSource) ctx.lookup("jdbc/mydb");
}catch(NamingException ne){}
```

There are following advantages for data source approach:

- There is no need for each client application to initialize JDBC drivers.
- The client application need not be know the details about the driver.
- It provides location independency.

An enterprise application server will deal with more clients. As the number of clients increase there is need of a technique that can give efficient services to all clients. In case of distributed system, each client will create its own object, initialize it, and the JVM has to keep track of each and every object, so that no longer used object can be garbage-collected. In the same way server receive the request from more then one client for same type of object. Every time server creates object for each client, and initializes it. It increases the overhead on the server.

The solution for minimizing the overhead of server for creating and managing the object is *Object Pooling*. It is the technique for organization and reusability of the objects. Connection objects are most frequent access objects on the server. So we create a pool of these objects. As any request come from client to create a connection object, server allots one connection object from the pool to the client thread, after performing all the jobs, the connection object is returned to pool. Before allocating the pool object, the server checks the authentication and other security credentials.

By distribution a set of connections along with number of clients, instead of creating them when required, we can distribute and improve the load on resources and therefore the responsiveness of the application will be better.

8.14. INTRODUCTION OF LDAP

The Lightweight Directory Access Protocol (LDAP) was developed in the early 1990s as a standard directory protocol. LDAP is now probably the most popular directory protocol and, because JNDI can access LDAP.

"LDAP defines the mechanism by which clients should be able access data on the serve. It does not however; give details how the data should be stored on the server."

LDAP uses the hierarchical tree structure to organize the data, called a Directory Information Tree (DIT). In DIT each *leaf* is an entry, and the first entry is called *root entry.*

An entry contains a Distinguished (DN) and attributes-value pairs related to it. The DN is a unique name by which entry will be identified, and it works as unique key of relational database table. It comprises the information how it is related to rest of the DIT, same as the pathname shows how the file is related to rest of file system.

The path to a file on a system start root to file. A DN reads right to left for example:

```
uid=sujjain, oi=people, o=theweb.com
```

The leftmost part of a DN is called a Relative Distinguished Name (RDN his example would be uid=scarter.

LDAP attributes often use mnemonics as their names. Some of the more common LDAP attributes are following.

LDAP Attribute	Definition
Cn	Common name
Sn	Surname
Givename	First Name
Uid	User ID
Dn	Distinguished Name
Mail	E-mail address

An attributes can have one or more values. For example, a user can have more than one e-mail address. Attribute values can be text or binary data and are referred as name-value pairs.

Attributes also have matching rules. These rules tell the server how it should decide whether a particular entry is a match or not for a given query. The possible matching rules are:

Matching Rules	Meaning
DN	Attribute is in the form of a Distinguished Name.
Case-Insensitive String (CIS)	Attribute can match if both the query value and the attribute's value are equals, regardless of case.
Case-Sensitive String (CSS)	Attribute can match if both the query value and the attribute's value are equals, including the case.
Integer	Attribute match is determined by numbers.
Binary	Attribute matches if both the query value and the attribute's value are the same binary value.

LDAP servers have other following benefits:

- They support reference pointers to other LDAP directories where data is stored. So that just by one client request a single LDAP server could search millions of entries.
- Reliability and speed is improved by replication of LDAP data.
- LDAP also has a exceptionally strong security model that contains
  - ACLs to protect data inside the server.
  - Secure Socket Layers (SSL)
  - Transport Layer Security (TLS)
  - Simple Authentication and Security Layer (SASL) protocols.

Application Areas of LDAP

There are basic three basic applications of LDAP in Java today;

- Access Control
- White Pages Services
- Distributed Computing Directory

Access Control

All application defines when user has permission to access them. As any user selects the application to start it, the required credentials are scan to match. Most applications like in between these two extremes. Access control can be broken into following two part:

- *Authentication* is determining who can work on specific part of the complete application software. There are various authentication mechanism.
- *Authorization* is determining what certain operations are allowed to do with the application module for that user is authenticated.

White Page Services

White pages services are services that provides a facility to look up any user based on attributes contained in their entries. For example, you can look up Mariya's e-mail address, and obtain the telephone number of the Accounts office, the building number of Sales Department, etc. they are called *white pages* because this type of information you search in the white pages of US telephone books.

Distributed Computing Directory

According to the current environment and current organization need the distributed networking programming become the most common solution, when different modules resides separately on different machine (JVM) of an application. To implement this approach Java provides following three distributed architecture:

- Remote Method Invocation (RMI)
- Common Object Request Broker Architecture (CORBA)
- Enterprise Java Beans (EJB)

LDAP Operations

We must first obtain a references and a network connection to the LDAP server, before we can perform any type of operation on an LDAP server. Then we must also give the authentication credentials. Some LDAP servers allow some type of anonymous access, with the support of ACL.

Standard LDAP Operations

- Connect to the LDAP server
- Bind to the LDAP server
- Perform a sequence of LDAP operations
  - Search the server
  - Add a new entry
  - Modify an entry
  - Delete an entry
- Disconnect from the LDAP server

REVIEW QUESTIONS

Q. 1. Write a function to create a database connection and retrieve all the records from a table that has name="RAHUL". **[MDU, May 2007; Dec. 2007]**

Ans.

```
public class DemoPrg
{
Connection conn;
Statement stmt;
ResultSet rst;

    public DemoPrg() throws IOException
   {
      try{
        Class.forName("sun.jdbc.odbc.JdbcOdbcDriver");
        conn = DriverManager.getConnection("jdbc:odbc:MYDSN");
        stmt = conn.createStatement();
        rst = stmt.executeQuery("select
      empid,firstname,lastname,salary from Employee
      where firstname = 'RAHUL'");
        System.out.println("EMPID\tFIRST_NAME\tLAST_NAME\tSALARY\n");
        while(rst.next())
         {
        System.out.print(rst.getInt("empid")+'\t');
        System.out.print('\t'+rst.getString("firstname")+'\t');
        System.out.print('\t'+rst.getString("lastname")+'\t');
        System.out.print("\t"+rst.getInt("salary")+"\n");
         }
```

```
        }catch(ClassNotFoundException cnfe){
    System.out.println(cnfe);
        }catch(SQLException sqle){
    System.out.println(sqle);
        }
    }
     public static void main(String ar[]) throws IOException
     {
        new DemoPrg();
      }  }
```

Q. 2. What is the difference between static and dynamic query? Explain in reference to prepared statements. **[MDU, May 2007]**

Ans. A static query is the one that is defined statically with the help of annotation before the entity class. A dynamic query is the one whose strings are provided at run-time.

Q. 3. Explain various JDBC programming concepts. **[MDU, May 2008]**

Ans. The classes in the java.sql package can be separated into the subsequent groups based on their functionality:

- Management of Connection
- Access Database
- Type of Data
- Metadata of Database
- Exceptions and Warnings

Q. 4. Describe LDAP and Metadata with example. **[MDU, May 2012; May 2009; Dec. 2009]**

Ans. LDAP defines the mechanism by which clients should be able access data on the serve. It does not however; give details how the data should be stored on the server. Meta means "Data About Data". As any database is created, or any table is created in a database, then database system store some information about all these for the internal use. If there is any need then we can retrieve from system file or the tools given by database system. For example, metadata describes that table structure, like column name, size, type etc.

Q. 5. What is JDBC? Explain the different JDBC drivers. Explain the steps of java to database communication in details. **[MDU, Dec. 2007]**

Ans. The Design of JDBC given all explanation about the JDBC with their Drivers Type. For writing the code to connect with data base and work on database with the SQL commands the developer has to perform the following steps.

- Loading the Database Driver
- Opening Connections or Establishing a Connection
- Creating Statement
- Executing the Querying

All these steps are given in details under *Basic JDBC Programming Concepts.*

Q. 6. Write a program to connect to a database query and display the results. **[MDU, May 2008; Dec. 2008]**

Ans. Refer programs in the chapter.

Q. 7. Explain the following:

(*a*) **Metadata**

(*b*) **Design of JDBC** **[MDU, May 2012]**

Ans. Refer 8.1 and 8.10.

Q. 8. How can you load the drivers?

Ans. Loading the driver or drivers you want to use is very simple and involves just one line of code. If, for example, you want to use the JDBC-ODBC Bridge driver, the following code will load it:

For example,

```
class.forName("sun.jdbc.odbc.jdbcodbcdriver");
```

Your driver documentation will give you the class name to use. For instance, if the class name is jbdc.DriverXYZ, you would load the driver with the following line of code:

For example,

```
class.forName("jdbc.DriverXYZ");
```

Q. 9. How can you create JDBC statement?

Ans. A statement object is what sends your SQL statement to the DBMS. You simply create a Statement object and then execute it, supplying the appropriate execute method with the Sql statement you want to sent. For a SELECT statement, the method to use is executeQuery. For statements that create or modify table, the method to use is executeUpdate.

For example, I takes an instance of an active connection to create a Statement object. In the following example, we use our Connection object con to create the Statement object stmt :

```
Statement stmt = con.createStatement();
```

Q. 10. What are the different types of statement?

Ans. There are three types of statement :

1. Statement(Use createStatement Method)
2. Prepared Statement(Use prepareStatement method)
3. Callable Statement(Use prepareCall)

Q. 11. What is the usage of PreparedStatement? Discuss with an example. **[MDU, Dec. 2008]**

Ans. It Execute precompiled sql queries with or without parameters as:

```
PreparedStatement prepareStatement(String sql)
returns a new PreparedStatement object. PreparedStatement objects are
precompiled SQL statements.
```

❑❑❑

9

CHAPTER

Distributed Objects

In the era of Internet, there is a need of distributed applications.

For example, in client-server architecture, client and server are located at different location, now client generate a request through an object on server that contains information about request. The server object assembles the information about the request, perform the operations and get back to the client with the response.

INSIDE THIS CHAPTER

POINTS TO REMEMBER

Distributed application means objects can be located anywhere, and can communicate with each other through standard protocols across the network.

For the fundamental communication technique, the Java supports sockets programming, for communication between two hosts. But limitation of socket programming, both hosts have to engage in application level protocols for encoding and decoding of message at time of communication *i.e.*, exchanging the information. The implementation of such protocols is a tedious job and there is a possibility of errors.

The goals for supporting distributed objects in the Java programming language are:

- Support seamless remote invocation on objects in different virtual machines
- Support callbacks from servers to applets
- Integrate the distributed object model into the Java programming language in a natural way while retaining most of the Java programming language's object semantics
- Make differences between the distributed object model and local Java platform's object model apparent
- Make writing reliable distributed applications as simple as possible
- Preserve the type-safety provided by the Java platform's run-time environment
- Support various reference semantics for remote objects; for example, live (non-persistent) references, persistent references, and lazy activation

- Maintain the safe environment of the Java platform provided by security managers and class loaders.

Underlying all these goals is a general requirement that the RMI model is both simple (easy to use) and natural (fits well in the language). RMI allows invoking a running object in a JVM from another running object in another JVM.

9.1. CLIENT SERVER ARCHITECTURE

In a network, client computers generate a request, and translate it to an intermediate transmission channel format, and send it to server across a network.

POINTS TO REMEMBER

The server converts channel format to its own application format for which the request was destined, executes the request and get back to client with a response.

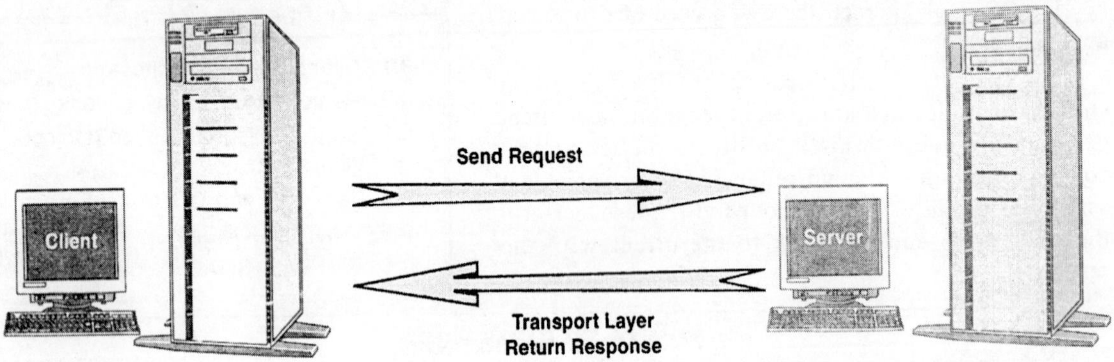

Fig. 9.1. Client Server Architecture

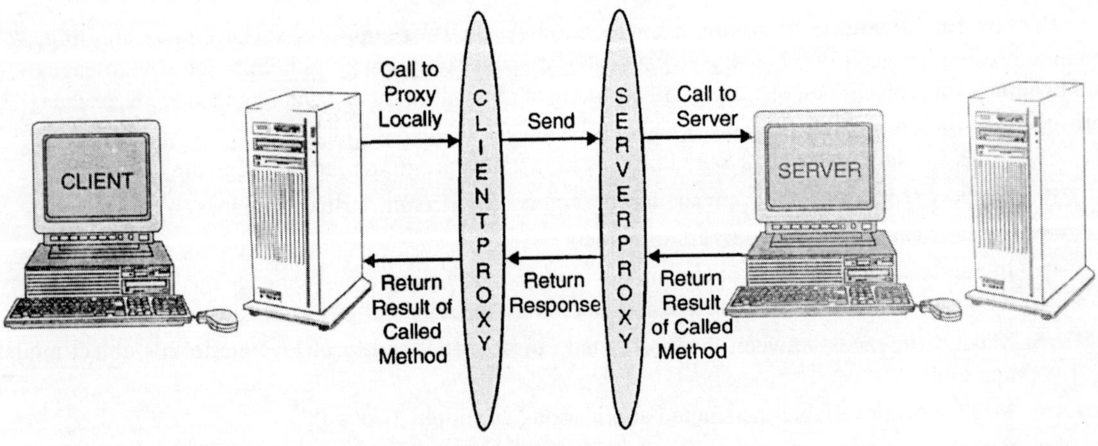

Fig. 9.2. Client-Server with Proxy Application

Implementation of such type of approach needs a huge coding and not sure that will be error free. You have to set a design of definite standard transmission format between client and server. Even you have to write the code for exchanging and translation for request and response data into transmission format and to transmission format respectively.

There is feeling of need a mechanism that facilitates the above task with less effort. Programmer must not worry about format conversion. Proxy application is the solution of the problem. One proxy has to install on client machine and other one has to install on server machine. The client proxy will communicate with server, and vice versa.

How do the proxies communicate with each other? That depends on the implementation technology there are three common choices:

- Remote Method Invocation (RMI) allows the remote method call between distributed object located on different JVM.
- Common Object Request Broker Architecture (CORBA) allows the remote method call between distributed object of any programming language. It uses the Inter-ORB and IIOP protocol communication between objects.
- Simple Object Access Protocol (SOAP) also allows the same mechanism as above, but it uses XML for transmission of request and response.

9.2. REMOTE METHOD INVOCATIONS

POINTS TO REMEMBER

Remote Method Invocations (RMI) frameworks provides a mechanism by which a Java object running in one JVM calls method of another Java object running in another JVM.

It provides object-to-object communication regardless of their location. This allows an application to access and make the call to object method located remotely.

In RMI jargon, the client object makes a remote call to the remote object located at server machine, called server object. The computer running the application or a process that calls the remote method is called client, and the computer running the object that processes the call request from client, is called server. If the server generates a remote call to an object residing on another machine (JVM) then the server becomes the client and the remove computer will become server.

If we want two Java applications executing within different virtual machines to communicate with each other there are a couple of other Java-based approaches that can be taken besides RMI:

- Sockets
- Java Message Service (JMS)

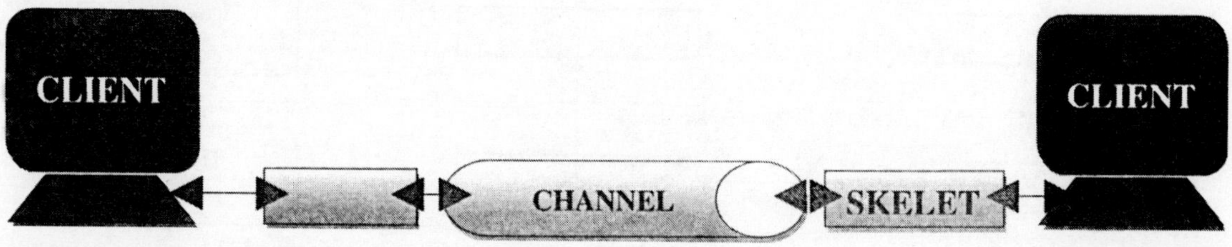

Fig. 9.3. Remote Method Invocation

Sockets have already been discussed in Chapter 7. RMI is different with JMS in such a way that, the object stays local at one JVM, but in JMS the object (message) pass through across the network from one JVM to another JVM.

Architecture of RMI

Now, it is clear, what the use of RMI is. Transparency is the one of the most aspect of RMI design. A method invocation on a remote object has the same syntax as a method invocation on a local object. The RMI architecture supports to create a distributed application, in which one object call the method of remote object.

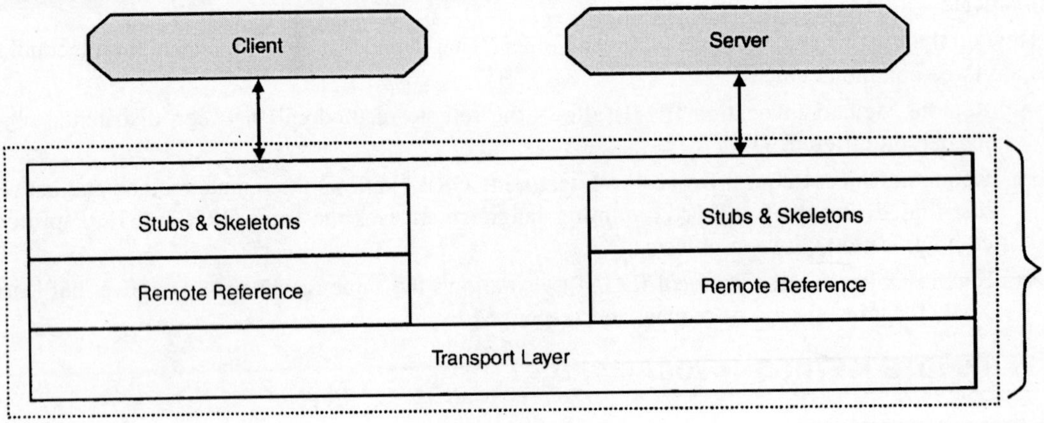

Fig. 9.4. RMI System

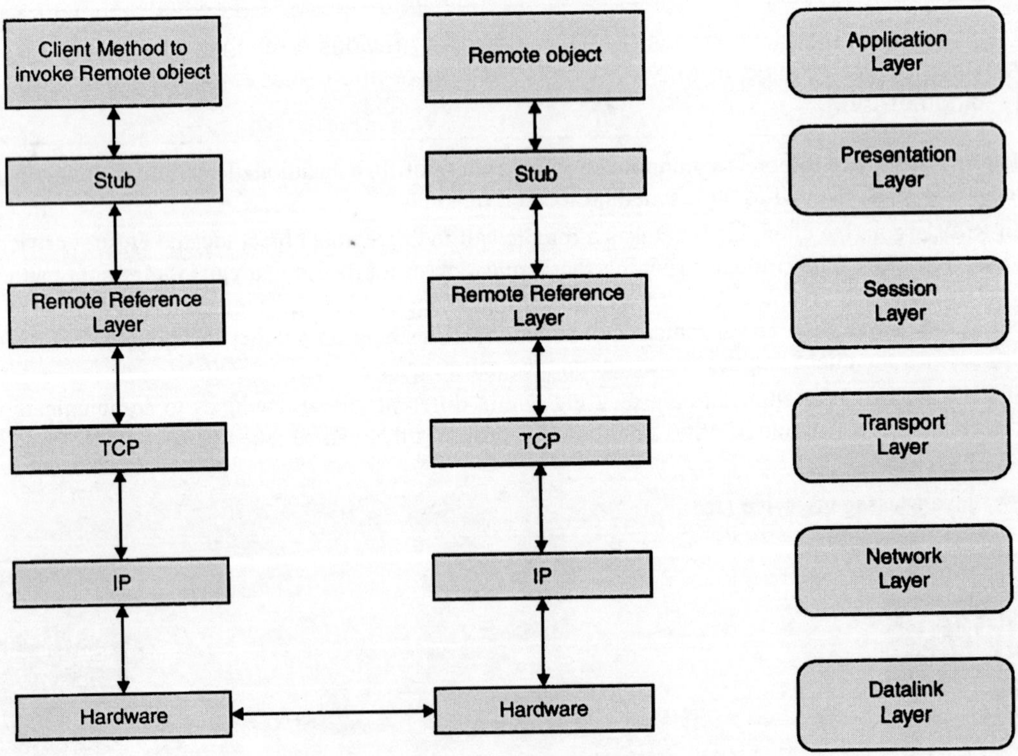

Fig. 9.5. *RMI Layered Architecture*

Specifically, in RMI, the definition of a remote service is coded using a Java interface. The implementation of the remote service is coded in a class. Therefore, the key to understanding RMI is to remember that *interfaces define behavior* and *classes define implementation.*

The RMI architecture is high-level of abstraction. It is constructed using three layer architecture. The layered architecture gives a transparency for application architecture, mean implementation gives flexibility.

The RMI is constructed using the following three abstract layers:

- ***The Stub-Skeleton Layer*:** The responsibility of this layer is to intercept and redirect the client request for method calls to remote object.
- ***The Remote Reference Layer*:** It interprets and manages reference, that are made by client to the remote object.
- ***The Transport Layer*:** It provides the connectivity between two machines.

The Stub-Skeleton Layer

Stub and Skeleton are two proxy applications; those are attached with client and server application respectively. The stub-skeleton layer gives the location transparency. Both of these work as an interface. The application of this layer is to transfer data to the Remote Reference Layer via marshaling and unmarshaling. Marshaling is the process that converts the data or objects being transferred to remote computer into a byte stream format, and unmarshaling is the reverse process, convert the byte stream data to an application format. This conversional done via object serialization.

DO YOU KNOW

The stub-skeleton layer gives the location transparency. Both of these work as an interface.

In the RMI use of the Proxy pattern, the stub class plays the role of the proxy for the remote service implementation.

POINTS TO REMEMBER

The skeleton is a helper class that is generated by RMI to help the object communicate with the stub across the RMI link.

The skeleton carries on a conversation with the stub; it reads the parameters for the method call from the link, makes the call to remote service implementation object, accepts the return value, and then writes the return value back to the stub.

***Role of Stub*:**

POINTS TO REMEMBER

The stub is an object at client-side that interacts with remote object located at server-side. The stub has the same interface and methods as the remote object has. When the client calls a stub method implicitly, the stub forwards the request to the remote object via the skeleton, under RMI infrastructure.

The events are executed by the stubs are in sequence, and are given below:

(1) Initiates a connection with the remote JVM.

(2) Marshals the client method call and its parameter to the remote JVM.

(3) Waits for the result.

(4) Unmarshals the response or any exception returned from remote JVM.

(5) Return the value to the caller.

***Role of Skeleton*:** Server machine have this skeleton object, which contains all the information about the remoteness. The events are executed by the skeleton are following:

(1) Unmarshals the requested parameters
(2) Invokes the method of which client generated the request through remote reference layer
(3) Marshals the response of the called method, and returned back to the caller (client).

The Remote Reference Layer

The Remote Reference Layer provides Java Remote Method Protocol (JRMP), which is responsible to handle remote object using *java.rmi.server.RemoteRef* object. The stub objects use the ***invoke(Remote, Method, Object[], long)*** method in *RemoteRef* to make the method call. The *RemoteRef* object understands how to intercept and interpret the remote services.

The Transport Layer

The RMI transport layer is responsible to make the connection between the client JVM and the server JVM, yet it has to face networking obstacles. The layer is designed on TCP-based. All connections use TCP/IP are called stream-based network connections. Java Remote Method Protocol (JRMP) is a wire level protocol, works on top of TCP/IP.

Distributed Object Applications

RMI applications consist of two separate sub-application programs: a server and a client. A server application jobs includes remote object creation, making reference to accessible remote objects, and wait for client request for remote object. A client application acquire a remote reference of any number of remote object from the server, and then invokes then remotely. The communication will be there in both directions.

Such an application is sometimes referred to as a distributed object application.

Distributed object applications need to:

- Locate remote objects
- Communicate with remote objects
- Load class bytecodes for objects

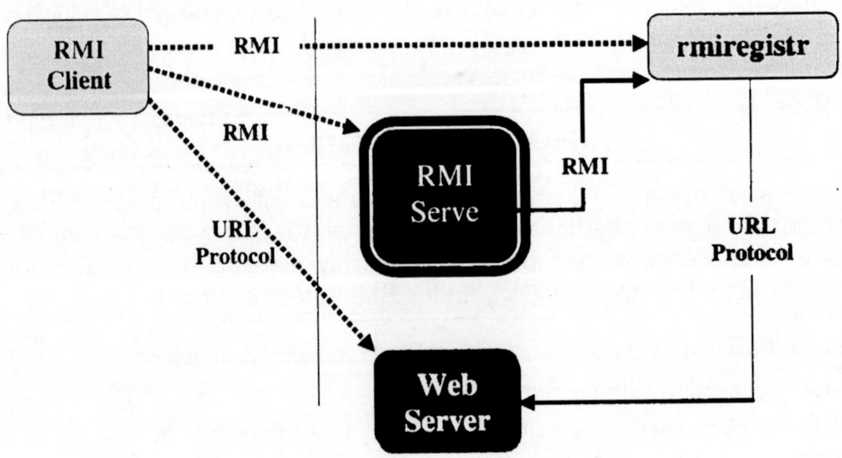

Fig. 9.6. RMI with Distributed Environment.

It is depicted in figure above RMI registry is used to obtain the reference to a remote object. The server calls that registry to register the remote object. Later the client will lookup the remote object reference from the registry, and generates remote method call to the server. RMI loads the bytecode (class) of the any URL protocol.

Dynamic Class Loading

Sending the objects, parameters, return values and exceptions from one JVM to another JVM is done by using serialization mechanism. For that stream call are there, so that the class definition files can be loaded at the receiver.

POINTS TO REMEMBER

When values are return for a remote method call, and are unmarshalled to real corresponding object at the receiving JVM, the class definitions are required to be loaded for all types of objects in the stream.

RMI provides a capability for dynamically loading the class definitions of the type of object is passed as parameters and return values. It also includes the dynamic downloading of remote stub.

To provide this facility the RMI uses special subclasses of *java.io.ObjectOutputStream* and *java.io.ObjectInputStream* for the marshal streams. It is used for marshalled and unmarshalled RMI parameters and return values.

9.2.1. Setup for Remote Method Invocation

RMI supported application requires more setup rather than any standalone program. The application developed with RMI mechanism will have two applications, client application that run on client machine and server application that run on server machine. The necessary details must be located on client and server machine with their related interfaces. There is also a lookup mechanism to locate the server object.

Any RMI application would be a collection of few classes and interfaces, and obviously use some statements to registers and locate the server object. RMI facilitate to create stub and skeleton class, and programmer has to least bother about how to transfer the data to the remote side.

How does RMI Implemented and Works?

First step, defines an interface called *RemoteInterface* which declares public methods. In these method signatures, parameters and return values can be of any type. Second step, define a class called *Implementation Class* or *Remote Object Class* which implements *Remote Interface* and some other class and interfaces for its local uses. Third step, to create *Stub Class* and *Skeleton Class*, which are generated automatically using *rmic* (RMI Compiler), a tool that comes with JDK, stored in *jdk\bin* directory. The *Stub Class* functionality is to translate remote method call into network communication channel format and passing parameter. The *Skeleton Class* is a class at server virtual machine, that accepts data from network connection and translates into actual method call. The data transferred on the channel is handled by *Object Streams* automatically.

The final step, registered the remote object with a naming service that allows client to address the object by name. The remote object, which is running on a remote machine, will be addressed with the combination of two components, one is machine address and another is name of the particular remote object.

When a client request to make call to a remote object, it first searches for the object in the naming service. This returns a remote reference to the object. It prevents the object from garbage collection when there are still remote references to it.

The *Stub Class* will be subsequently downloaded if it is not available locally. *Stub Class* has all the details necessary to invoke the remote object. RMI implicitly take care of low level communication work. The client can invoke the remote object method call directly. These calls will actually being passed to the stub. And the stub has all the detail, how to setup the communication, transmitting the call and parameters passing to the skeleton. The skeleton then makes the actual method call through *Remote Reference Layer*.

The **value** returns as a response from the method will send back to client stub, which then returns the results to the client application. The response can be either correct results or an exception.

Point to be noted here is that all the interfaces must be implemented and specific class files must be **generated.** It also means that only those remote methods can be invoked, those are declared in the *RemoteInterface*. With these limitations, RMI framework is tremendously powerful technology in which a network application can be developed just by using a method call.

Creating Distributed Applications by Using RMI

The following steps have to be taken to implement any distributed application.

First. ***Implementation of RemoteInterface***

```
import java.rmi.*;
public interface Calculator extends java.rmi.Remote {
    public long add(long a, long b) throws RemoteException;
    public long sub(long a, long b) throws RemoteException;
    public long mul(long a, long b) throws RemoteException;
    public long div(long a, long b) throws RemoteException;
}
```

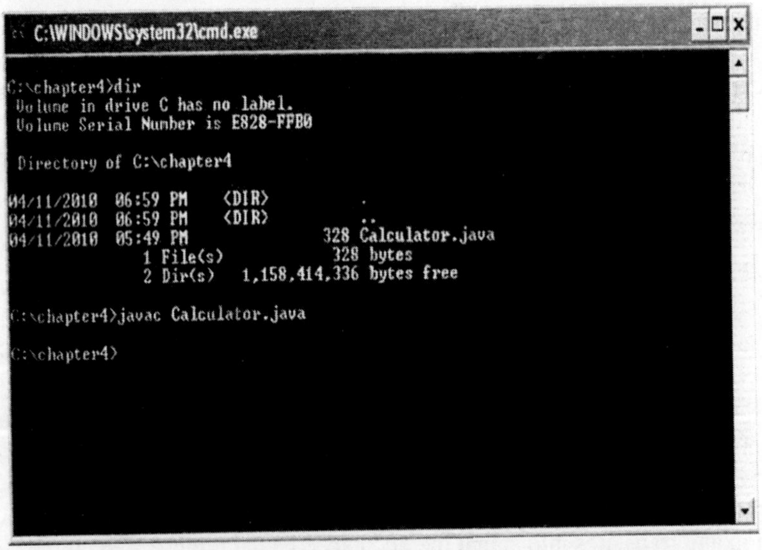

Second. ***Implementation of RemoteClass***

```
import java.rmi.*;
public class CalculatorImpl extends java.rmi.server.UnicastRemoteObject
    implements Calculator {

    public CalculatorImpl() throws java.rmi.RemoteException {
        super();    }

    public long add(long a, long b) throws java.rmi.RemoteException {
        return a + b;   }
```

```
        public long sub(long a, long b) throws java.rmi.RemoteException;{
            return a - b;    }

        public long mul(long a, long b) throws java.rmi.RemoteException;{
            return a * b;    }

        public long div(long a, long b) throws java.rmi.RemoteException;{
            return a / b;    }
    }
```

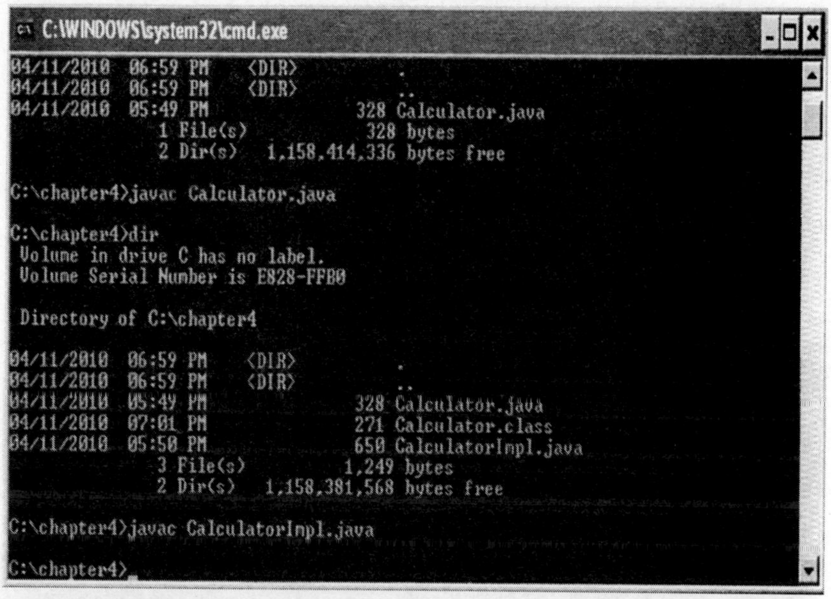

Third. ***Generation of Stub class***

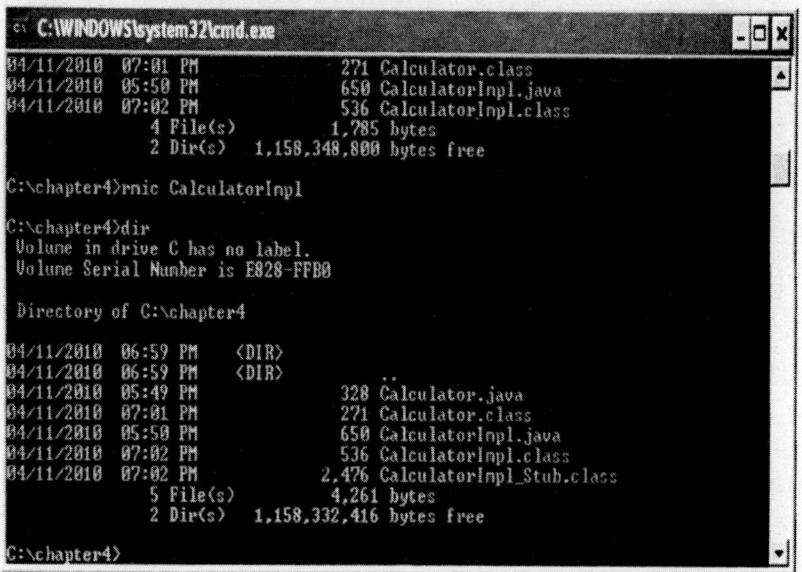

Fourth. *Implementing Server Class and Binding RemoteClass object*

```
import java.rmi.Naming;

public class CalculatorServer {

   public CalculatorServer() {
     try {
       Calculator c = new CalculatorImpl();
       Naming.rebind("rmi://localhost:1099/CalculatorService", c);
     } catch (Exception e) {
       System.out.println("Trouble: " + e);
     }
   }
   public static void main(String args[]) {
     new CalculatorServer();
   }
}
```

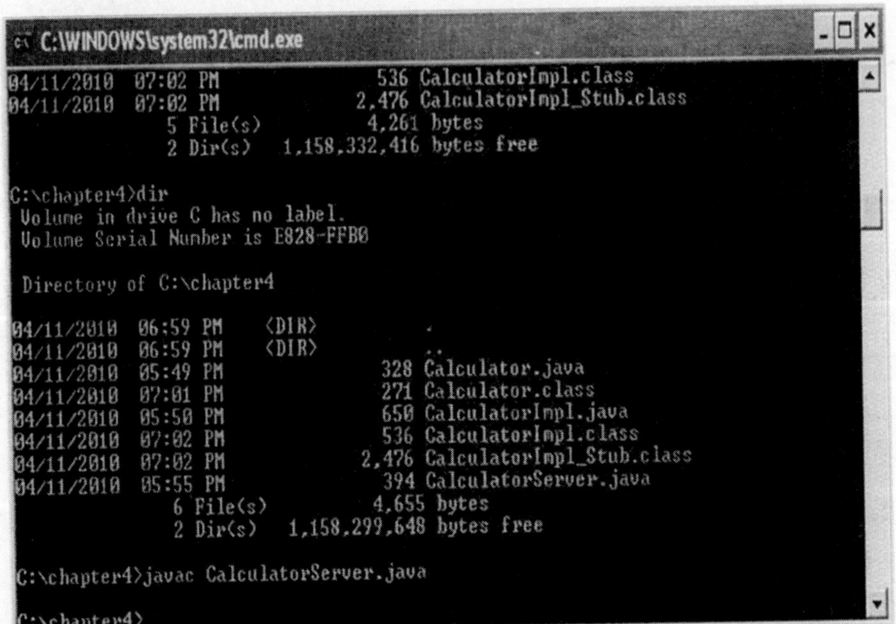

Fifth. *Implementing Client Class and Locating RemoteClass object*

```
import java.rmi.Naming;
import java.rmi.RemoteException;
import java.net.MalformedURLException;
import java.rmi.NotBoundException;
```

```
public class CalculatorClient {

    public static void main(String[] args) {
        try {
            Calculator c = (Calculator) Naming.lookup("rmi://localhost/
                                        CalculatorService");
             System.out.println( c.sub(4, 3) );
             System.out.println( c.add(4, 5) );
             System.out.println( c.mul(3, 6) );
             System.out.println( c.div(9, 3) );
        } catch (MalformedURLException murle) {
             System.out.println();
             System.out.println("MalformedURLException"
                                        + murle);
        } catch (RemoteException re) {
             System.out.println();
             System.out.println("RemoteException" + re);
        }  catch (NotBoundException nbe) {
             System.out.println();
             System.out.println("NotBoundException" + nbe);
        }  catch (ArithmeticException  ae) {
             System.out.println();
             System.out.println("java.lang.ArithmeticException" + ae);
        }
    }
}
```

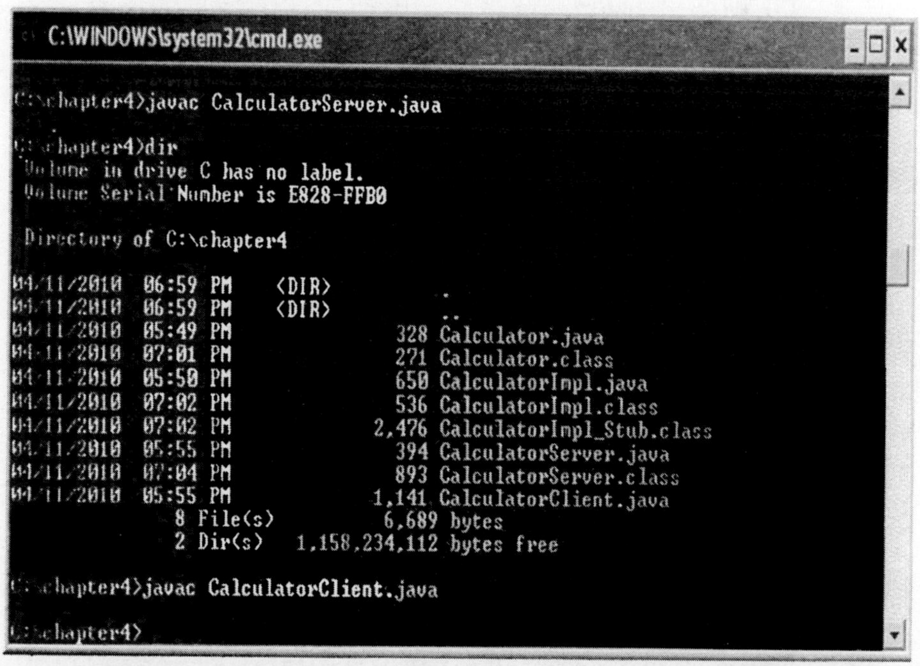

Sixth. ***Deployment***

Deploying an application that uses RMI can be tricky because so many things can go wrong and the error message that you get when something does go wrong are so poor. We have found that it really pays off to stage the deployment locally.

The server machine contains all files that are needed to run the server. You have to move these files to the machine running the server process. In our example, the server contains the following files:

```
CalculatorServer.class
CalculatorImpl.class
Calculator.class
```

The client machine contains the files that are needed to start the client. There are as follows:

```
CalculatorImpl_Stub.class
Calculator.class
CalculatorClient.class
```

Seventh. ***Start RMI Registry***

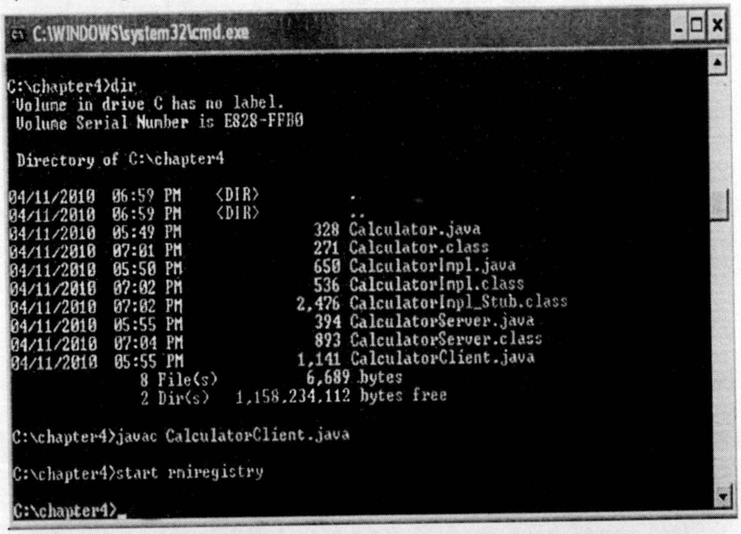

Eighth. *Run Server Program*

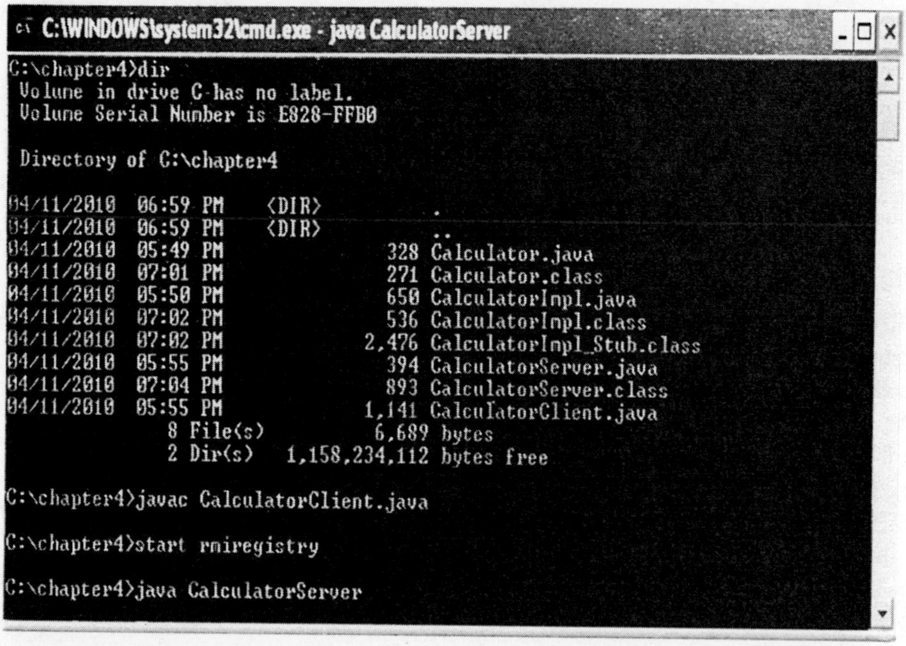

Ninth. *Run Client Program*

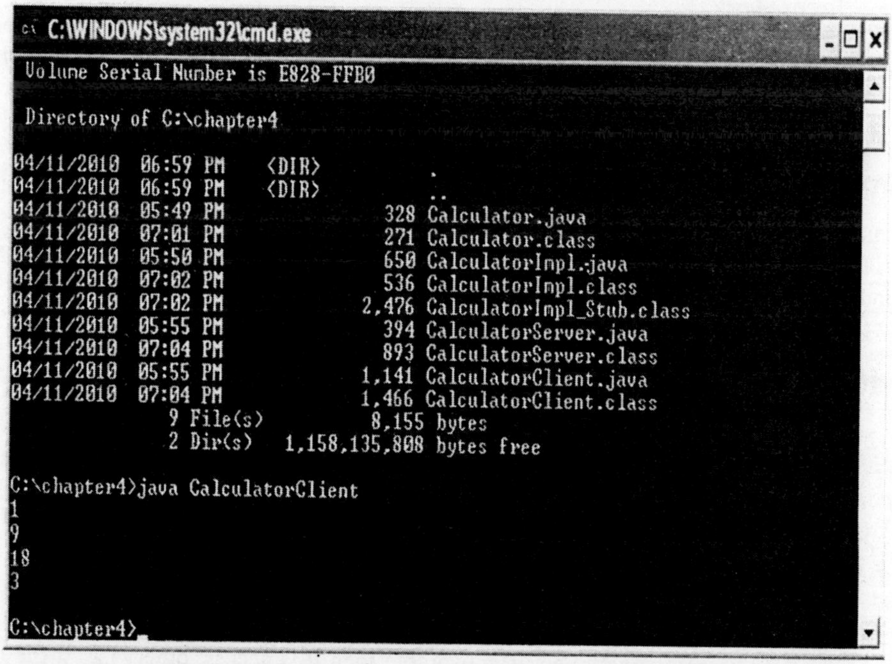

9.2.2. Parameter Passing in Remote Methods

In RMI, whenever there is a remote method call, the any types of parameters can be passed to the remote method. The parameter can be of primitive type, can be of object type or it can be a remote object.

If primitive data type is passed as parameter to a method, then JVM copies the value of the variable and pass copies to the method. But if any object has to pass then a copy of its reference will be made from the heap memory, and placed on the stack, then this copy is passed anywhere.

Method uses the copy of the reference to access the object and its member. If there is any method call on the reference, then the original object move to another state. Any alteration in reference will not affect to the original object, but if there is any modification in original object, must reflect the reference.

So that we have to take care of all these problem, and keep in mind in which format the parameter is passing to the remote method.

Primitive Parameters

Whenever any primitive data type is passed to the remote method as parameter, then it passes by value. A copy of the value of the primitive data type variable is sent to the remote method, in the similar way the value is returned back from the remote method.

Object Parameters

Objects are resides in heap memory. The heap can not be share between two JVM. So that the whole object will be sent to the remote JVMs, not the reference of an object. The object is actually passed by value. In the similar way the remote method returns an object, that is, the copy of the object will returned.

POINTS TO REMEMBER

RMI can send the references of object using the concept of serialization. In serialization the object will be transformed into a linear format that can be transmit over the network.

At the receiving end of the communication channel the serialized object will be de-serialized. The object being passed across the network must implement the java.io.Serializable or the java.io.Externalizable interface.

Remote Parameters

It is something different, as compare to pass a primitive data type parameter or object as parameter. The object has to pass as remote parameter from either end (server or client), it has to register in RMI registry, and any program, either client or server will access through the RMI registry program. It's an essentially feature, because used in remote callbacks.

9.3. SERVER OBJECT ACTIVATION (OBJECT ACTIVATION)

It was observed, that the remote objects are instance of the java.rmi.UnicastRemoteObject. The remote objects are accessible at all the time, but some time there is a situation when no client is executing, and there is no call for remote objects. At this instant of time, the resources used by the remote machine on server are high. It increases overhead on the JVM and runtime environment, and as well as create a bottleneck condition for JVM. For the solution of this problem and Object Activation concept was introduced.

The concepts allocate the remote objects to be executed as the demand is generated. As any request comes for any remote object, if the object is activatable then it is access, otherwise, if the object is not currently in use, the corresponding JVM initiates the object execution. For the first time execution of the remote object is postponed for a time being. But all these activation and initialization are transparent to the client.

There is an activation model to provide this facility, and the model has few components with certain set of actions, are as following:

- **Activator:** It is an important component. Remote object activation and starting an instance of JVM on the server whenever the need is generated is the responsibility of the activator. It keeps track of all the information that is required for activation and starting the instance.
- **Activation Group:** It is a complete and separate instance of the JVM that involves creating instances groups of object belongs to its groups and simultaneously sent the details to monitor about the active and passive states of groups. There can be multiple activation groups.
- **Activation Monitor:** It is created when a groups is active. And the responsibility or job of the activation monitor is to keep track of the state (active or passive) of the group and any object belongs to the group.
- **Activation System:** The system provides the functionality to register as the groups creates, and activatable object to be activated within that group. It works near to activator, as to activate the group or object, both will be registered via the activation system. Activation monitor will get the status about the state of the group and objects form it.

TABLE 9.1. Activation Model Contains Following Classes and Iterfaces

Entity	Implementation	Implemented As
Activatorjava.rmi.activation.	Activator	Interface
Activation Group	java.rmi.activation.ActivationGroup	Abstract Class
Activation Monitor	java.rmi.activation.ActivationMonitor	Interface
Activation System	java.rmi.activation.ActivationSystem	Interface

The activation technique uses identifiers and descriptors. Always keep following point:

- Every activatable object has an ID and a descriptor
- Every activatable object will be a member of an activation group, and group will has an ID.

API's discussion of Activation Object

The Activation Group

An activation groups is created explicitly as follows:

```
public static ActivationGroup createGroup(ActivationGroupID gid,
                    ActivationGroupDesc gdesc, long incarnationNum)
```

where:

(*i*) *gid* is the activation group identifier

(*ii*) *gdesc* is the activation group descriptor

(*iii*) *incarnationNum* is the activation group incarnation number

ActivationID

As the group is registered with the activation system, ActivationID is implicitly assigned. It consist two pieces of information:

(*i*) A remote reference for the object activator

(*ii*) An object unique identifier.

Activation Descriptor

The activation descriptor contains the information related to activation of an object:

(*i*) Activation group identifier

(*ii*) The name of class being activated

(*iii*) The location of the class

(*iv*) The marshaled object that contains data for initialization, but this object is optional

There is four constructor to create an object of Activation Descriptor.

```
ActivationDesc(ActivationGroupID groupId, String classname, String
classLoation, MarshaledObject data, Boolean restart)

ActivationDesc(ActivationGroupID groupId, String classname, String
classLocation, MarshaledObject data)

ActivationDesc(String classname, String classLocation,
MarshaledObject dataObject, boolean restart)

ActivationDesc(String classname, String classLocation,
MarshaledObject dataObject)
```

How Activation Works

- A stub is generated for an activatable object. It contains all details about the object, activation identifier and remote reference type of the object.
- The stub holds the activation identifier, and make the call to the activator to activate the object for the associated identifier.
- As activator gets the request to activate an object, it will locate the activation descriptor and activation group of the corresponding object. If the group does not exist, then activator starts an instance of a JVM to creates an activation group for the object, and forward this activation request with details to the group.
- The activation groups loads the class and instantiates the object

Building Objects Activatable

(*i*) Create the Remote Interface

(*ii*) Create the Object Implementation

(*iii*) Register the Object with the System

(*iv*) Compile the file then generates the stub and skeletons for the remote object.

(*v*) Start the activation system using rmid utility

Inactive the Object

As we know the activation allows an object to be started on demand, but it will not deactivated any object automatically. *Activatable.inactive(ActivationID id)* is the method by invoking it the object will be deactivated. It does not unregistered the object.

In deactivation process, a temporary state of the object will be change, later activation system will restart the object. But *Activatable.unregister(ActivationID id)* is used for un-registering the object, and it will be permanently remove from activation system.

9.4. JAVA IDL AND CORBA

Java IDL (Java Interface Definition Language) is a technology design for distributed objects; those are communicating with each other and running on different platform across the network. It is similar to RMI but with one advantage is that there is interaction among the object written in any programming language like C, C++ etc, including Java.

POINTS TO REMEMBER

Java IDL provide this functionality only because it is based on Common Object Request Brokerage Architecture (CORBA).

CORBA is defined as distributed object model. The language supports the CORBA will have its own IDL mapping.

Object Request Broker (ORB) is a class library that allows low-level communication between the Java IDL application and CORBA supported application.

IDL-Java Bridge Data Type Mapping

When data moves between IDL and a Java object, IDL automatically converts variable data types. The following table maps how Java data types correlate to IDL data types.

TABLE 9.2. Java to IDL Data Type Conversion

Java Type (# bytes)	IDL Type
boolean (1)	Integer
byte (1)	Byte
char (2)	Byte
short (2)	Integer
int (4)	Long
long (8)	Long64
double (8)	Double
java.lang.String	String
Arrays of the above types	IDL array of the same dimensions (from 1 to 8 dimensions) and corresponding type.
java.lang.Object	DL array of primitives or IDL array of IDLjavaObjects
Null object	IDL Null object

TABLE 9.3. IDL to Java Data Type Conversion

IDL Type	Java Type (# bytes)
Byte	byte (1)
Integer	short (2)
Unsigned integer	short (2)
Long	int (4)
Unsigned long	int (4)
Long64	long (8)
Unsigned Long64	long (8)
Float	float (4)
Double	double (8)
String	java.lang.String
Arrays of the above types	Java array of the same dimensions and corresponding type
IDLjavaObject	Object of corresponding Java class
Arrays of objects	Java array of the same dimensions, consisting of corresponding Java proxy objects
Null object	Java null

TABLE 9.4.

IDL Type	Java Type
module	package
boolean	boolean
char, wchar	char
octet	byte
string, wstring	java.lang.String
short, unsigned short	short
fixed	java.math.BigDecimal
enum, struct, union	class
interface (non-abstract)	signature interface and an operations interface, helper class, holder class
interface (abstract)	signature interface, helper class, holder class
constant (not within an interface)	public interface
constant (within an interface)	fields in the Java signature interface for non-abstract, or the sole Java interface for abstract
exception	class
Any	org.omg.CORBA.Any
type declarations nested within interfaces	"scoped" package
typedef	helper classes
pseudo objects	pseudo interface
readonly attribute	accessor method
readwrite attribute	accessor and modifer methods
operation	method

The Java IDL Development Process

(*i*) Define the remote interface
(*ii*) Compile the remote interface
(*iii*) Implement the server
(*iv*) Implement the client
(*v*) Start the applications

9.5. WHAT IS CORBA?

The Common Object Request Broker Architecture (CORBA) is a industry standard and the technique by the multiple application written in different languages, and running of multiple computers, either on same platform or different platform, can work together. For this functionality Object Management Groups (OMG) is given by it. CORBA allows normalizing of the method-call semantic from one platform to another platform, *i.e.*, it's interpret the call from one platform to another platform by giving some standard way.

CORBA uses an Interface Definition Language (IDL), that gives an interface that object will use to interact with the outside world. CORBA provides the mapping from IDL to another certain implementation language. It supports standard mapping C++, Java, Ada, C, Lip, COBOL, Python, Ruby.

The applications interact with other application object using ORB. The ORB will be initialized in application, will access an internal Object Adapter which will maintains reference counting, reference instantiation policies, object lifetime policies, etc. The object adapter is responsible for instance registration of code classes, these classes are generated by compiling the IDL code. That code translates the high-level interface definition into the language-specific class that is used by user application and into OS used by user application.

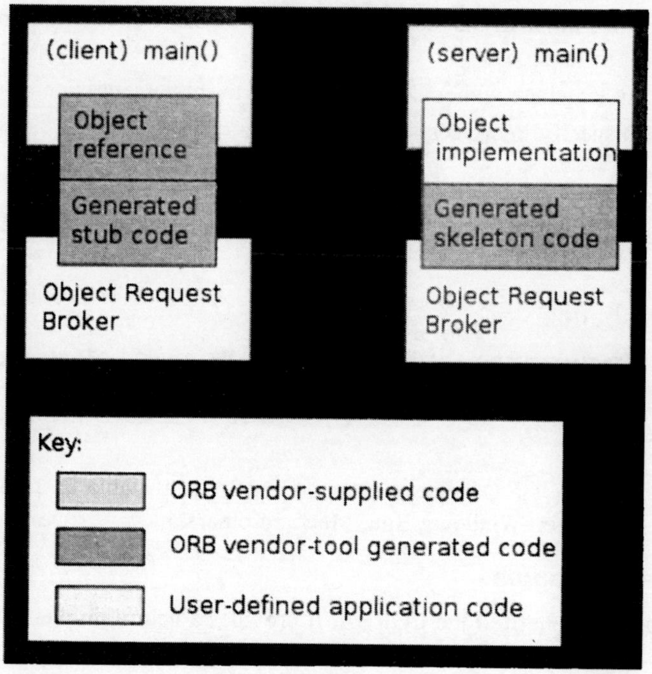

Fig. 9.7.

Architecture of CORBA

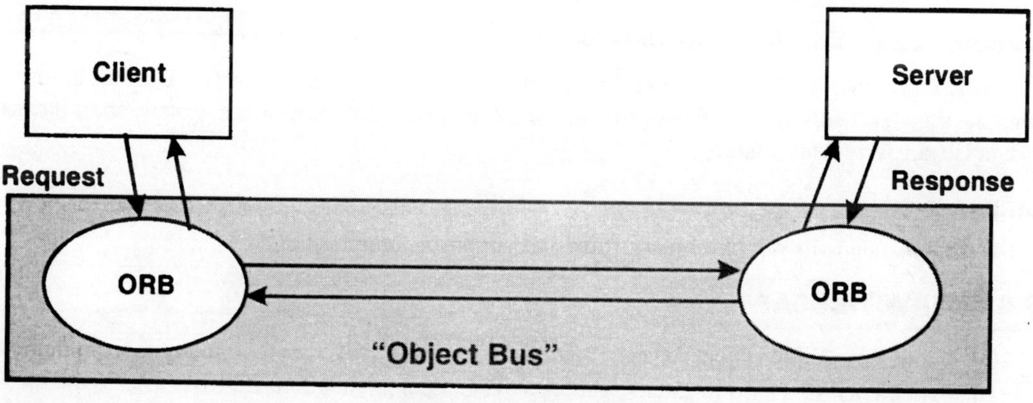

Fig. 9.8.

Steps for implementing CORBA objects

(*i*) Write the interface that specifies how the object works, using IDL

(*ii*) Generate the stub and helper classes using the IDL compiler for the target language.

(*iii*) Add the implementation code for the server objects, using the language of your choice.

(*iv*) Write a server program that creates and registers the server objects using CORBA naming service, a service that is similar to the rmiregistry.

(*v*) Write a client program that look up the server objects and invokes services by them.

(*vi*) Start the naming service and the server program on the server and the client program on the client.

These steps are quite similar to the steps that you use to build distributed applications with RMI. There are two important differences:

- You can use any language with a CORBA binding to implement clients and servers.
- You use IDL to specify interfaces.

Benefits of CORBA

CORBA is a bunch of benefits that is given by no other technology in one package. These benefits are as following:

Language Independence

CORBA supports implementation of an application with a good set of languages, yet there is a standard semantic for that.

OS Independence

CORBA's design supports the OS-independency. CORBA is available in Java (OS-independent), as well as natively for Linux/Unix, Windows, Sun, Mac and others.

Freedom from Technologies

It's the most important benefit of the CORBA. It provides a neutral implementation environment by providing the interfaces between various technologies supported systems.

Strong Data Typing

CORBA is based on IDL, and IDL provides data type mapping with various languages. So that human error is reduced. It will implicitly convert the data types.

Freedom From Data Transfer Details

CORBA provides high level of exception handling during the data transfer. If there is any error during the transfer then it will receive or fire an exception, and if not receiving an exception, it means that the data is transferred successfully.

Compression

CORBA marshals its data in a binary form and supports compression.

9.6. RMI WITH SOAP

SOAP stands for Simple Object Access Protocol. SOAP is a XML based protocol for communication between applications. It is used in Internet and is a W3C recommendation. CORBA uses SOAP protocol to communicate between objects. SOAP is defined by WSDL (Web Services Description Language). SOAP provides a way to communicate between applications running on different operating systems with different technologies and programming languages.

A SOAP message is an XML document having elements

(1) An envelope element that identifies the XML document as a SOAP message.

(2) A header element that contains header information.

(3) A body element contains call and response information.

(4) A fault element contains error and status information.

The Syntax of SOAP message is as follows:

```
<?xml version="1.0"?>
```

```
<soap:Envelope
xmlns:soap="http://www.w3.org/2001/12/soap-envelope"
soap:encodingStyle="http://www.w3.org/2001/12/soap-encoding">
<soap:Header>
  ...
</soap:Header>
<soap:Body>
  ...
<soap:Fault>
    ...
</soap:Fault>
</soap:Body>
</soap:Envelope>
```

The current W3C specification and implementation of the SOAP API from Apache present only HTTP binding for SOAP. Due to this, the web services have to be deployed in a web server container. The possibility of opening SOAP based interface to the existing RMI services may be limited by the following factor: if we convert the existing RMI services as SOAP based and deploy them in web server container, the existing clients to those services have to be rewritten or reconfigured. We may also maintain two interfaces each for RMI and SOAP for the same set of services which is not desirable.

REVIEW QUESTIONS

Q. 1. Discuss the following concepts:

(a) JAVA IDL

(b) CORBA **[MDU, May 2011]**

Ans. (a) *Refer Section 9.4.*

(b) *Refer Section 9.5.*

Q. 2. What is RMI? How do you setup RMI for a web server? **[MDU, May 2011]**

Ans. *Refer Section 9.2.*

Q. 3. What is RMI? Differentiate between stub and skeletons with their usage.

[MDU, May 2007, May 2008, Dec. 2009]

Ans. Sometimes it is necessary to invoke a method of an object that is begin executed and a remote machine as part of distributed computing. This shares computation between 2 machines. Fortunately java has provided this mechanism. This mechanism is called RMI. (*Refer 9.2*)

Q. 4. What is distributed programming? How is it accomplished in java? Explain with suitable example.

[MDU, May 2007, May 2008, Dec. 2008]

Ans. The goals for supporting distributed objects in the Java programming language are:

- Support seamless remote invocation on objects in different virtual machines
- Support callbacks from servers to applets
- Integrate the distributed object model into the Java programming language in a natural way while retaining most of the Java programming language's object semantics
- Make differences between the distributed object model and local Java platform's object model apparent
- Make writing reliable distributed applications as simple as possible
- Preserve the type-safety provided by the Java platform's run-time environment
- Support various reference semantics for remote objects; for example live (non-persistent) references, persistent references, and lazy activation

- Maintain the safe environment of the Java platform provided by security managers and class loaders (*Refer Introduction*)

Q. 5. What is CORBA ? Explain architecture of CORBA. Explain the steps to implement CORBA objects. **[MDU, Dec. 2007]**

Ans. Steps for implementing CORBA objects

- Write the interface that specifies how the object works, using IDL.
- Generate the stub and helper classes using the IDL compiler for the target language.
- Add the implementation code for the server objects, using the language of your choice.
- Write a server program that creates and registers the server objects using CORBA naming service, a service that is similar to the rmiregistry.
- Write a client program that look up the server objects and invokes services by them.
- Start the naming service and the server program on the server and the client program on the client.

  These steps are quite similar to the steps that you use to build distributed applications with RMI. There are two important differences:
- You can use any language with a CORBA binding to implement clients and servers.
- You use IDL to specify interfaces.

Q. 6. Explain Java IDL. **[MDU, May 2007, Dec. 2009]**

Ans. Java IDL (Java Interface Definition Language) is a technology design for distributed objects; those are communicating with each other and running on different platform across the network. It is similar to RMI but with one advantage that interaction among the object written in any programming language like C, C++ etc., including Java.

Q. 7. Explain setup for Remote Method Invocation.

Ans. In section 9.2.1 Setup for Remote Method Invocation is explained. In the following order explain the given points

- How does RMI implemented and works?
- Creating Distributed Applications by Using RMI
- **First.** Implementation of RemoteInterface
- **Second.** Implementation of RemoteClass
- **Third.** Generation of Stub class
- **Fourth.** Implementing Server Class and Binding RemoteClass object
- **Fifth.** Implementing Client Class and Locating RemoteClass object
- **Sixth.** Deployment
- **Seventh.** Start RMI Registry
- Eighth. Run Server Program
- Ninth. Run Client Program

Q. 8. Explain parameter passing in Remote method Invocation.

Ans. The details is given in section 9.2.2 Parameter Passing in Remote Methods :

- Primitive Parameters
- Object Parameters
- Remote Parameters

Q. 9. Explain method call with SOAP? **[MDU, May 2007, May 2008, Dec. 2008, May 2009]**

Ans. Simple Object Access Protocol (SOAP) also allows the same mechanism as above, but it uses XML for transmission of request and response.

❑❑❑

10
CHAPTER

Swings

INSIDE THIS CHAPTER

Swing is a set of packages that are built on top of AWT which provides a large number of inbuilt predefined classes. Swing is a part of Java Foundation Class (JFC). Swing class provides a large set of components such as toolbars, progressbar, text fields and timers. Swings provide a large set of built-in controls such as trees, image buttons, sliders, tabbed panes, toolbars, tables, color choosers etc. Swing is customizable that helps in changing certain properties such as text alignment, adding images, changing borders or separating the internal appearance from visual appearance. Swing provides several UI components of which, few are listed below:

- **JApplet:** It is an extended version of *java.applet*.
- **JButton:** A push or command button.
- **JCheckBox:** A checkbox can be selected or deselected displaying its state.
- **JColorChooser:** It allows a user to select a color.
- **JComboBox:** It is a combination of text field and dropdown list.
- **JComponent:** It is the base class for Swing components.
- **JDialog:** It is the base class for creating a dialog window.
- **JEditorPane:** It is a text component that allows the user to edit various kinds of content.
- **JFrame:** It is an extension to *java.awt.Frame*
- **JLabel:** It is a display area for a short text string.
- **JList:** It is a component that allows a user to select one or more objects from a list.
- **JMenu:** A pop-up menu containing *JMenuItem* object's that are displayed when the user selects it in the *JMenuBar* component.
- **JOptionPane:** Makes it easy to pop up a standard dialog box.
- **JPanel:** A generic lightweight container.
- **JPasswordField:** It allows the editing of a line of text where the user views some special characters in place of original characters.
- **JPopupMenu:** It provides a pop-up menu.
- **JProgressBar:** It is a component that displays an integer value within an interval.

- **JRadioButton**: A radio button can be selected or deselected displaying its state.
- **JScrollBar:** It gives an implementation of the scrollbar.
- **JSeparator:** It provides a menu separator.
- **JSlider:** It is a component that lets the user select a value by sliding a knob.
- **JTable:** It presents the data in a two-dimensional format.
- **JTextArea:** It provides a multiline area that displays plain text.
- **JTextField**: It allows the editing of a single line of text.
- **JTextPane**: It is a text component that can be marked up with attributes.
- **JToolBar:** It displays the toolbar that is used for displaying commonly used controls.
- **Tree:** It displays a set of hierarchical data as an outline.
- **JViewport:** It provides a viewport through which the user can see the information.
- **JWindow:** It provides a window that can be displayed anywhere on the desktop.

Besides the above listed UI components, Swings in Java provide other features also:

1. **Borders:** The users can draw borders in various different styles and patterns around the components using *setBorder* method.
2. **Graphics Debugging:** The users can use the *setDebuggingGraphicsOptions* method to set up graphics debugging.
3. **Mouseless Operations:** Swings make it much easier to connect keystrokes to components.
4. **Tooltips:** The users can use the *setToolTipText* method of *JComponent.*
5. **Scrolling:** Swings overcome an important drawback of AWT, that is, it allows scrolling to various components.
6. **Pluggable look:** The users can set the appearance of applets and applications to any of the standard looks.
7. **New layout managers:** Swing introduces new layout managers like *BoxLayout* and *OverlayLayout.*

10.1. LISTS

Lists are supported in Swings by the *JList* class. It is a very popular control as it allows the user to present a list of items in an easier and efficient manner. It also allows hiding the long list of items by making the lists scrollable. *JList* has been inherited from a long list of classes. The inheritance diagram for *JList* is as follows:

java.lang.Object

↓

java.awt.Component

↓

java.awt.Container

↓

javax.swing.JComponent

↓

javax.swing.JList

Fig. 10.1. Inheritance Diagram for JList

Table 10.1 below shows the constructors of the *JList* class:

Table 10.1. Constructors of JList

Constructor	Function
JList()	Constructs a JList object
JList(ListModel dataModel)	Constructs a JList object that displays the elements in the indicated data model
JList(Object[ ] listData)	Constructs a JList object that displays the elements in the indicated array
JList(Vector listData	Constructs a JList object that displays the elements in the indicated vector

The simple example for creating a list using Swings that displays 15 items and reports which ones the user clicks is illustrated below.

```
import java.awt.*;
import javax.swing.*;
import javax.swing.event.*;
/*<APPLET code=list1.class WIDTH=300 HEIGHT=200> </APPLET>*/
public class list1 extends JApplet implements ListSelectionListener
{
    JList jlist;
    public void init()
   {
     Container c= getContentPane();
     String[] str=new String[15];
     for(int i=0;i<15;i++)
     {
       str[i]="Item_Selection" +(i+1);
     }
     jlist= new JList(str);
     JScrollPane scp= new JScrollPane(jlist);
     jlist.setVisibleRowCount(5);
    jlist.setSelectionMode(ListSelectionModel.SINGLE_SELECTION);
   jlist.addListSelectionListener(this);
   c.setLayout(new FlowLayout());
   c.add(scp);
  }
 public void valueChanged(ListSelectionEvent lse)
 {
   String st="Choose Item";
   st+=jlist.getSelectedIndex();
   showStatus(st);
  }
}
```

The output for the same would look like:

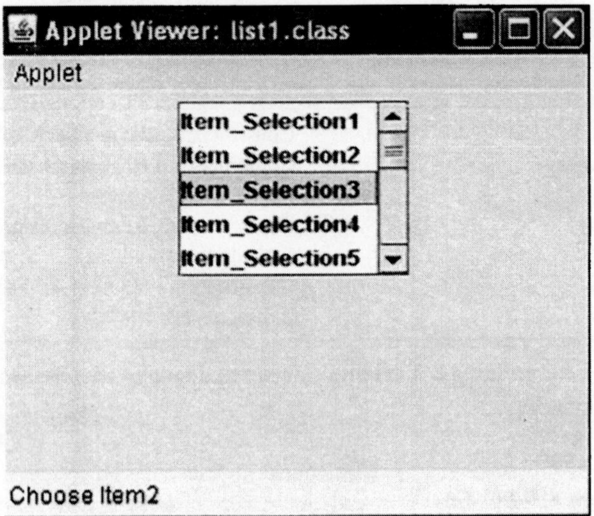

10.2. TREES

The *JTree* class defines a component that displays data organized in a tree-like structure. The elements of the tree are referred to as nodes. The root node is also called the parent node through which other nodes are attached, and are called the child nodes. The nodes that do not have any children are called leaf nodes. The table below illustrates the interfaces of JTree class:

Table 10.2. Constructors of JTree

Constructor	Function
MutableTreeNode	Defines the requirements for a tree node object that can change by adding or removing child nodes, or by changing the contents of a user object stored in the node
RowMapper	Defines the requirements for an object that translates paths in the tree into display rows
TreeCellEditor	Adds to CellEditor the extensions necessary to configure an editor in a tree
TreeCellRenderer	Defines the requirements for an object that displays a tree node

The example below shows a simple application for creating a tree. Program is an example of JTree that is showing the Table of Content in Tree format.

```
import java.awt.*;
import javax.swing.*;
import javax.swing.tree.*;
import java.awt.event.*;
public class SimpleTree extends JFrame {
  public static void main(String[] args) {
```

```
    new SimpleTree();
  }
  public SimpleTree() {
    super("Creating a Simple JTree");
    WindowUtilities.setNativeLookAndFeel();
    addWindowListener(new ExitListener());
    Container content = getContentPane();
    Object[] hierarchy =
      {"1. CHAPTER 1",
    new Object[] {"1.1 INTRODUCTION TO JAVA",
    "1.1.1 History of JAVA",
    "1.1.2 JAVA and Internet",
    "1.1.3 JAVA Virtual Machine",
    "1.1.4 JAVA's Magic: The Byte Code",
    "1.1.5 Features of JAVA",
    "1.1.6 Java Environment"},
new Object[] {"1.2    DATA TYPES & OPERATORS"},
new Object[] {"1.3    ARRAYS",
    "1.3.1 Declaration of Array Variables",
    "1.3.2 Array Construction",
    "1.3.3 Initialization of Array",
    "1.3.4 Array Access",
    "1.3.5 Arrays of Characters"},
new Object[]{"1.4     CONTROL STATEMENTS"},
new Object[]{"1.5     A SAMPLE PROGRAM IN JAVA"},
new Object[]{"1.6     CLASSES & METHODS",
    "1.6.1 Constructor",
    "1.6.2 Garbage Collection"},
new Object[]{"1.7     INHERITANCE",
    "1.7.1 Single Inheritance",
    "1.7.2 Multiple Inheritance",
    "1.7.3 Multilevel Inheritance",
    "1.7.4 Hybrid Inheritance"},
new Object[]{"1.8     INTERFACE"},
new Object[]{"1.9     PACKAGES"},
new Object[]{"1.10    EXCEPTION HANDLING",
    "1.10.1      try and catch with multiple catch clause",
    "1.10.2      throw and throws",
    "1.10.3      finally"},
new Object[]{"1.11    MULTITHREADING",
    "1.11.1      States of a Thread"},
new Object[]{"1.12    COLLECTIONS"},
```

```
new Object[]{"1.13   I/O STREAMS",
   "1.13.1      Working with InputStream and OutputStream",
   "1.13.2      Working with Reader and Writer"},
new Object[]{"1.14   AWT "},
new Object[]{"1.15   APPLET PROGRAMMING"}};
   DefaultMutableTreeNode root = processHierarchy(hierarchy);
   JTree tree = new JTree(root);
   content.add(new JScrollPane(tree), BorderLayout.CENTER);
   setSize(275, 300);
   setVisible(true);
  }
  private DefaultMutableTreeNode processHierarchy(Object[] hierarchy)
{
   DefaultMutableTreeNode node =  new
                             DefaultMutableTreeNode(hierarchy[0]);
   DefaultMutableTreeNode child;
   for(int i=1; i<hierarchy.length; i++) {
     Object nodeSpecifier = hierarchy[i];
     if (nodeSpecifier instanceof Object[])  // Ie node with children
       child = processHierarchy((Object[])nodeSpecifier);
     else
       child = new DefaultMutableTreeNode(nodeSpecifier); // Ie Leaf
     node.add(child);
   }
   return(node);
  } }
import javax.swing.*;
import java.awt.*;
public class WindowUtilities {
  public static void setNativeLookAndFeel() {
    try {
   UIManager.setLookAndFeel(UIManager.getSystemLookAndFeelClassName());
    } catch(Exception e) {
      System.out.println("Error setting native LAF: " + e);
    }
  }
  public static void setJavaLookAndFeel() {
    try {
   UIManager.setLookAndFeel(UIManager.getCrossPlatformLookAndFeelClassName());
    } catch(Exception e) {
      System.out.println("Error setting Java LAF: " + e);
```

```
    }
  }
   public static void setMotifLookAndFeel() {
    try {
UIManager.setLookAndFeel("com.sun.java.swing.plaf.motif.MotifLookAndFeel");
    } catch(Exception e) {
      System.out.println("Error setting Motif LAF: " + e);
    }
  }
  public static JFrame openInJFrame(Container content,
                                    int width,
                                    int height,
                                    String title,
                                    Color bgColor) {
    JFrame frame = new JFrame(title);
    frame.setBackground(bgColor);
    content.setBackground(bgColor);
    frame.setSize(width, height);
    frame.setContentPane(content);
    frame.addWindowListener(new ExitListener());
    frame.setVisible(true);
    return(frame);
  }
  public static JFrame openInJFrame(Container content, int width, int height,
                                                          String title)
{
    return(openInJFrame(content, width, height, title, Color.white));
  }
  public static JFrame openInJFrame(Container content, int width, int height)
{
    return(openInJFrame(content, width, height,
                        content.getClass().getName(),
                        Color.white));
  }
}
import java.awt.*;
import java.awt.event.*;
public class ExitListener extends WindowAdapter {
  public void windowClosing(WindowEvent event) {
    System.exit(0);
  }
}
```

The output interface would look like:

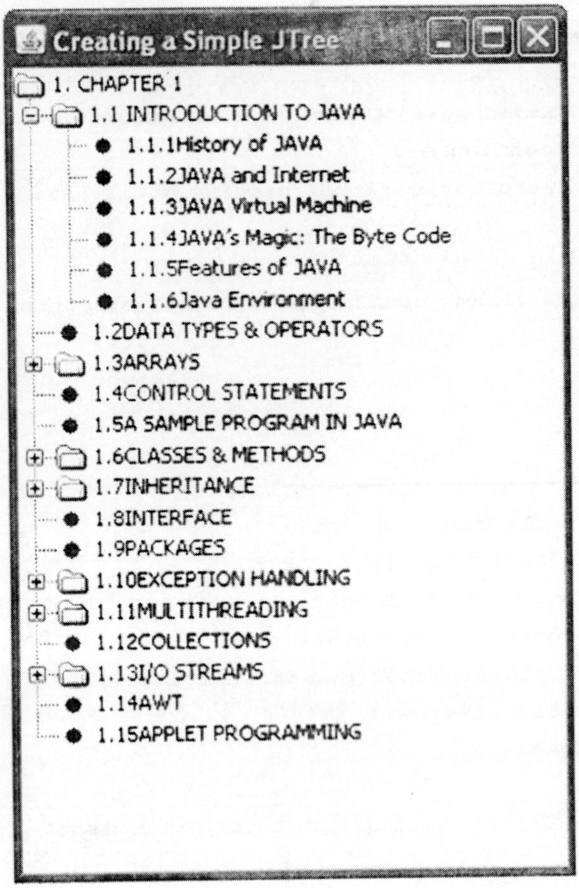

10.3. TABLES

The tables are very powerful components in Swing. They were created to provide an interface to JDBC that makes programming quite flexible for the user. *JTable* component allows the user to show and edit tabular data. *JTable* components display a rectangular array of data on the screen. The items in the table can have different types of data. The interface for java.swing.JTable are listed in the table below:

Table 10.3. Interfaces of JTable

Interface	Function
TableCellEditor	This defines the method any object that would like to be an editor of values for components such as JListBox, JComboBox, JTree, JTable needs to implement
TableCellRenderer	This defines the method required by any object that would like to be a renderer for cells in a JTable
TableColumnModel	Defines the requirements for a table column model object suitable for use with JTable
TableModel	This specifies the methods the JTable will use to interrogate a tabular data model.

We can create a *JTable* object directly from such an object by passing a reference of type *TableModel* to a *JTable* constructor as:

```
JTable table = new JTable(model);
```

model is a variable of type TableModel that stores a reference to an object that encapsulates the data to be displayed.

The rows and columns can also be added at runtime after creating TableModel interface. The example below shows a simple application for creating a table. The application is written in Swing with JDBC. The motive to design a search form for a AddBook table. That will search on three fields as following:

- Book Title - Search by Book Name
- Publisher - Search by Publisher Name
- Keywords - Search by main topic covered by the book, and this details is already stored in the database.

It will apply some conditon as AND or OR, so that the user can apply more than one searching criteria to search the data from AddBook table.

The structure of the table is given below.

AddBooks

Field Name	Data Type	Description
bookno	Text	
accno	Text	
name	Text	
author	Text	
publisher	Text	
issued	Text	
topic	Text	

The code for the application is:

```
import java.awt.*;
import java.awt.event.*;
import javax.swing.*;
import java.sql.*;
import java.io.*;
import java.util.*;
import javax.swing.table.*;
class Search extends JFrame implements ActionListener{
JCheckBox cbgbooktitle, cbgpublisher, cbgkeyword;
JTextField txtbooktitle,txtpublisher,txtkeyword;
JComboBox cb,cb1;
Container cp;
JButton btsearch,refresh;
ScrollPane sp;
JPanel p;
JTable table;
String escapeStr="%";
Connection con=null;
Vector columnNames;
```

```
Vector data;
DefaultTableModel model ;
public Search()
{
model = new DefaultTableModel(new Object[]{"bookno","accno","name",
   "author","publisher","topic","issued"}, 0);
try{
Class.forName("sun.jdbc.odbc.JdbcOdbcDriver");
con=DriverManager.getConnection("jdbc:odbc:MYDSN");
}
catch(SQLException e){System.out.println(e);}
catch(ClassNotFoundException cl){System.out.println(cl);}
cp=getContentPane();
p=new JPanel();
cp.add(p);
p.setLayout(null);
columnNames = new Vector();
data = new Vector(10);
columnNames.addElement("bookno");
columnNames.addElement("accno");
columnNames.addElement("name");
columnNames.addElement("author");
columnNames.addElement("publisher");
columnNames.addElement("topic");
columnNames.addElement("issued");
table = new JTable(model);
JScrollPane scrollPane = new JScrollPane( table );
p.add(scrollPane);
scrollPane.setBounds(20,250,750,200);
cbgbooktitle=new JCheckBox("Book Title");
cbgpublisher=new JCheckBox("Publisher");
cbgkeyword=new JCheckBox("Keyword");
txtbooktitle=new JTextField(30);
txtpublisher=new JTextField(30);
txtkeyword=new JTextField(30);
cb=new JComboBox();
cb.addItem("Or");
cb.addItem("And");
cb1=new JComboBox();
cb1.addItem("Or");
cb1.addItem("And");
btsearch=new JButton("Search");
p.add(cbgbooktitle);
p.add(cbgkeyword);
p.add(cbgpublisher);
p.add(txtkeyword);
```

```
p.add(txtpublisher);
p.add(txtbooktitle);
p.add(btsearch);
p.add(cb);
p.add(cb1);
p.setBounds(0,0,270,250);
cbgbooktitle.setBounds(20,20,100,20);
txtbooktitle.setBounds(120,20,250,20);
cb.setBounds(270,60,100,20);
cbgpublisher.setBounds(20,100,100,20);
txtpublisher.setBounds(120,100,250,20);
cb1.setBounds(270,140,100,20);
cbgkeyword.setBounds(20,180,100,20);
txtkeyword.setBounds(120,180,250,20);
btsearch.setBounds(160,210,100,20);
btsearch.addActionListener(this);
}
public void actionPerformed(ActionEvent ae){
int a[]=new int[3];
String escapeStr="%";
if(ae.getSource()==btsearch){
int i=0;
for (;i<model.getRowCount(); )
{
model.removeRow(i);
}
String qstr="Select bookno,accno,name,author,publisher,topic,issued
    from AddBooks where ";
if(cbgbooktitle.isSelected() && !cbgpublisher.isSelected() &&
    !cbgkeyword.isSelected())
{
qstr=qstr+" name like ?";
a[0]=1;
}
else if(cbgbooktitle.isSelected() && cbgpublisher.isSelected() &&
    !cbgkeyword.isSelected())
{
qstr=qstr+" name like ? "+cb.getSelectedItem()+" publisher like ?";
a[0]=1;
a[1]=2;
}
else if(cbgbooktitle.isSelected() && cbgpublisher.isSelected() &&
    cbgkeyword.isSelected())
{
qstr=qstr+" name like ? "+cb.getSelectedItem()+" publisher like ?
    "+cb1.getSelectedItem()+" topic like ? ";
a[0]=1;
```

```
a[1]=2;
a[2]=3;
}
else if(!cbgbooktitle.isSelected() && cbgpublisher.isSelected()  &&
   !cbgkeyword.isSelected()){
qstr=qstr+" publisher like ? ";
a[1]=1;
}
else if(!cbgbooktitle.isSelected() && cbgpublisher.isSelected()  &&
   cbgkeyword.isSelected()){
qstr=qstr+" publisher like ? "+cb1.getSelectedItem()+" topic like ? ";
a[1]=1;
a[2]=2;
}
else if(cbgbooktitle.isSelected() && !cbgpublisher.isSelected()  &&
   cbgkeyword.isSelected()){
qstr=qstr+" name like ? "+cb.getSelectedItem()+" topic like ? ";
a[0]=1;
a[2]=2;
}
else if(cbgkeyword.isSelected()){
qstr=qstr+"topic like ? ";
a[2]=1;
}
try{
PreparedStatement p=con.prepareStatement(qstr);
if(a[0]>0)
p.setString(a[0],(escapeStr+txtbooktitle.getText()+escapeStr));
if(a[1]>0)
p.setString(a[1],(escapeStr+txtpublisher.getText()+escapeStr));
if(a[2]>0)
p.setString(a[2],(escapeStr+txtkeyword.getText()+escapeStr));
ResultSet rs=p.executeQuery();
String stdata[]=new String[7];
while(rs.next()){
stdata[0]=rs.getString("bookno");
stdata[1]=rs.getString("accno");
stdata[2]=rs.getString("name");
stdata[3]=rs.getString("author");
stdata[4]=rs.getString("publisher");
stdata[5]=rs.getString("topic");
stdata[6]=rs.getString("issued");
model.addRow(stdata);
}}catch(Exception e){e.printStackTrace();}
}}}
public class BookSearch{
public static void main(String args[]){
JFrame jfr=new Search();
jfr.setSize(800,500);
jfr.setVisible(true);
```

```
jfr.setResizable(false);
jfr.setDefaultCloseOperation(WindowConstants.DISPOSE_ON_CLOSE);
jfr.addWindowListener(new WindowAdapter(){
public void WindowClosing(WindowEvent we){
System.exit(0);
}}}}
```

The output interface would look like:

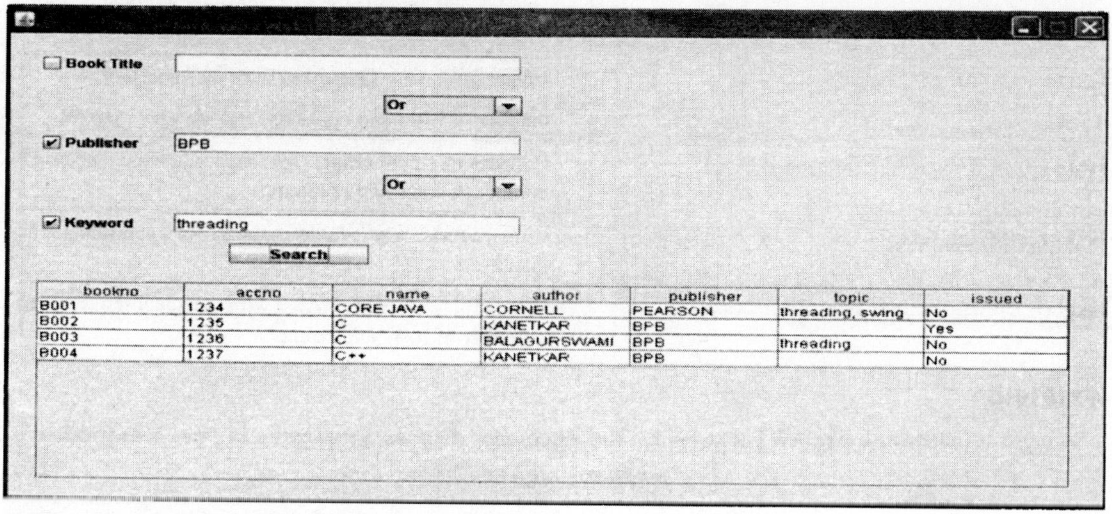

10.4. STYLED TEXT COMPONENTS

This section includes Text Area, Text Field and Editor Pane components of Swing class.

JTextArea

It is the subclass of JTextComponent that allows the editing of multiline text. Scrollbars can easily be added by placing JTextArea components in a JScrollPane container. The inheritance diagram for *JTextArea* is as follows:

java.lang.Object
↓
java.awt.Component
↓
java.awt.Container
↓
javax.swing.JComponent
↓
javax.swing.text.JTextComponent
↓
javax.swing.JTextArea

Fig. 10.2. Inheritence Diagram for JTextArea

Table 10.4 below shows the constructors of TextArea class:

Table 10.4. Constructors of TextArea class

Constructors	Function
JTextArea()	Constructs a new TextArea
JTextArea(Document doc)	Constructs a new JTextArea with the given document model, and defaults for all of the other arguments (null, 0, 0).
JTextArea(Document doc, String text, int rows, int columns)	Constructs a new JTextArea with the specified number of rows and columns, and the given model.
JTextArea(int rows, int columns)	Constructs a new empty TextArea with the specified number of rows and columns.
JTextArea(String text)	Constructs a new TextArea with the specified text displayed.
JTextArea(String text, int rows, int columns)	Constructs a new TextArea with the specified text and number of rows and columns.

JTextField

Swing text fields work like AWT text fields. The inheritance diagram for JTextField class is as follows:

java.lang.Object
↓
java.awt.Component
↓
java.awt.Container
↓
javax.swing.JComponent
↓
javax.swing.text.JTextComponent
↓
javax.swing.JTextField

Fig. 10.3. Inheritance Diagram for JTextField

The Table 10.5 below shows the constructors of TextField class

Table 10.5. Constructors of TextField Class

Constructors	Function
JTextField()	Constructs a new TextField.
JTextField(Document doc, String text, int columns)	Constructs a new JTextField that uses the given text storage model and the given number of columns.
JTextField(int columns)	Constructs a new empty TextField with the specified number of columns.
JTextField(String text)	Constructs a new TextField initialized with the specified text.
JTextField(String text, int columns)	Constructs a new TextField initialized with the specified text and columns.

JEditorPane

It enables the users to implement sophisticated editing facilities quite easily. The table below shows the constructors of JEditorPane class:

Table 10.6. Constructors of JEditorPane Class

Constructors	Function
JEditorPane()	Creates a new JEditorPane.
JEditorPane(String url)	Creates a JEditorPane based on a string containing a URL specification.
JEditorPane(String type, String text)	Creates a JEditorPane that has been initialized to the given text.
JEditorPane(URL initialPage)	Creates a JEditorPane based on a specified URL for input.

10.5. PROGRESS INDICATORS

Progress indicators or progress bars are the new controls that give the users some indications of the progress of an operation.

POINTS TO REMEMBER

They were originally developed to check the progress of installation processes but are now frequently used for all time-consuming operations.

When we call JProgressBar, we can set the maximum and minimum value for the progress bar. The inheritance diagram for JProgressBar class is as follows:

java.lang.Object
↓
java.awt.Component
↓
java.awt.Container
↓
javax.swing.JComponent
↓
javax.swing.JProgressBar

Fig. 10.4. Inheritance Diagram for JProgressBar

The Table 10.7 below shows the fields of JProgressBar class

Table 10.7. Fields of JProgressBar class

Field	Function
changeEvent	Only one ChangeEvent is needed per instance since the event's only interesting property is the immutable source, which is the progress bar.
changeListener	Listens for change events sent by the progress bar's model, redispatching them to change-event listeners registered upon this progress bar.

Field	Function
model	The object that holds the data for the progress bar.
orientation	Whether the progress bar is horizontal or vertical.
paintBorder	Whether to display a border around the progress bar.
paintString	Whether to textually display a string on the progress bar.
progressString	An optional string that can be displayed on the progress bar.

The Table 10.8 shows the Constructors of JProgressBar class

TABLE 10.8. Constructors of JProgressBar class

Constructor	Function
JProgressBar()	Creates a horizontal progress bar that displays a border but no progress string.
JProgressBar(BoundedRangeModel newModel)	Creates a horizontal progress bar that uses the specified model to hold the progress bar's data.
JProgressBar(int orient)	Creates a progress bar with the specified orientation, which can be either JProgressBar.VERTICAL or JProgressBar.HORIZONTAL.
JProgressBar(int min, int max)	Creates a horizontal progress bar with the specified minimum and maximum.
JProgressBar(int orient, int min, int max)	Creates a progress bar using the specified orientation, minimum, and maximum.

A JProgressBar is a Swing component that indicates progress. A ProgressMonitor is a dialog box that contains a progress bar. A ProgressMonitorInputStream displays a progress monitor dialog box while the stream is read. A progress bar is a simple component just a rectangle that is partially filled with color to indicate the progress of an operation. By default, progress is indicated by a string "n%". You can see a progress bar in the bottom right of Fig below

You construct a progress bar much as you construct a slider, by supplying the minimum and maximum value and an optional orientation:

```
progressBar = new JProgressBar(0, 1000);
progressBar = new JProgressBar(SwingConstants.VERTICAL, 0, 1000);
```

A progress bar is a simple component that can be placed inside a window. In contrast, a ProgressMonitor is a complete dialog box that contains a progress bar (see Fig below). The dialog box contains a Cancel button. If you click it, the monitor dialog box is closed.

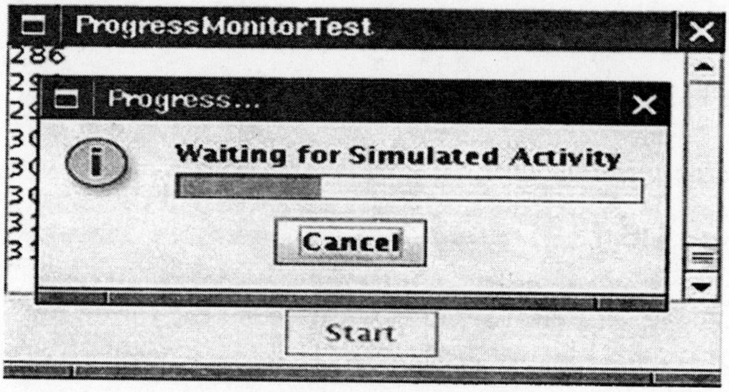

You construct a progress monitor by supplying the following:

- The parent component over which the dialog box should pop up;
- An object (which should be a string, icon, or component) that is displayed on the dialog box;
- An optional note to display below the object;
- The minimum and maximum values.

There are two conditions for termination. The activity might have completed, or the user might have canceled it. In each of these cases, we close down:

- the timer that monitored the activity;
- the progress dialog box;
- the activity itself (by interrupting the thread).

The example below shows a simple application for creating a progress bar.

```
class ProgressBar extends JFrame
{
  JProgressBar current;
   JTextArea ta;
   int num = 0;
  public ProgressBar()
  {
       Container pane=getContentPane();
        ta=new JTextArea("");
        pane.setLayout(new GridLayout());
        current = new JProgressBar(0, 100);
        current.setValue(0);
        current.setStringPainted(true);
        pane.add(current);
        pane.add(ta);
}
public void iterate()
{
   while (num <= 100)
   {
```

```
                current.setValue(num);
                 try
                  {
              Thread.sleep(500);
            }
             catch (InterruptedException e) { }
            num += 10;
              If(num>100)break;
            else
             ta.setText(num+"%completed");
          } }
      public static void main(String[] arguments) {
      ProgressBar pb = new ProgressBar();
       pb.pack();
      pb.setVisible(true);
      pb.iterate( );
    } }
```

10.6. COMPONENT ORGANIZERS

In this section, we will discuss about *JSplitPane, JTabbedPane & Desktope panes.* Swing support split panes to let programs present two components side by side. The inheritance diagram for JSplitPane class is as follows:

java.lang.Object
↓
java.awt.Component
↓
java.awt.Container
↓
javax.swing.JComponent
↓
javax.swing.JSplitPane

Fig. 10.5. Inheritance Diagram for Component Organizers

Split panes split a component into two parts, with an adjustable boundary in between. Fig below shows a frame with two split panes.

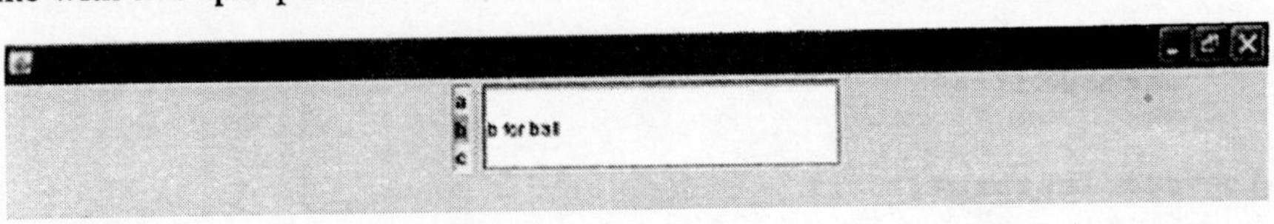

You construct a split pane by specifying the orientation, one of JSplitPane.HORIZONTAL_SPLIT or JSplitPane.VERTICAL_SPLIT, followed by the two components.

For example,

```
JSplitPane innerPane = new JSplitPane(JSplitPane.HORIZONTAL_SPLIT,
   planetList,planetImage);
```

The example below shows a simple application for creating a split pane.

```
public class splitpanecmd extends JFrame implements ListSelectionListener
{
  String ab[]={"hi a","hi b","hi c","hi d"};
  JList l1=new JList();
  JTextArea l1Description=new JTextArea("hello");
  DefaultListModel model;
  Container c=getContentPane();
    public splitpanecmd()
    {
        String[] b={"a","b","c","d"};
        model=new DefaultListModel();
        for(int i=0;i<b.length;i++)
        model.add(i, b[i]);
        l1.setModel(model);
        JSplitPane innerPane
        = new JSplitPane(JSplitPane.VERTICAL_SPLIT, l1,l1Description);
        c.add(innerPane);
        l1.addListSelectionListener(this);    }
public void valueChanged(ListSelectionEvent e)
    {
        int g=l1.getSelectedIndex();
         l1Description.setText(ab[g]);
    }
    public static void main(String[] a)
    {
        splitpanecmd s=new splitpanecmd();
        s.setVisible(true);
    }
```

Tabbed Panes

Tabbed panes are a familiar user interface device to break up a complex dialog box into subsets of related options. You can also use tabs to let a user flip through a set of documents or images.

To create a tabbed pane, you first construct a JTabbedPane object, then you add tabs to it.

```
JTabbedPane tabbedPane = new JTabbedPane();
tabbedPane.addTab(title, icon, component);
```

The example below shows a simple application for creating a tabbed pane.

```
public class tabbedpanedemo extends JApplet
{
public void init()
{
    JTabbedPane jtp = new JTabbedPane();
    jtp.addTab("Cities", new CitiesPanel());
    jtp.addTab("Colors", new ColorsPanel());
    jtp.addTab("Flavors", new FlavorsPanel());
    getContentPane().add(jtp);
}
}
class CitiesPanel extends JPanel
{
public CitiesPanel()
{
    JButton b1 = new JButton("New York");
    add(b1);
    JButton b2 = new JButton("London");
    add(b2);
    JButton b3 = new JButton("Hong Kong");
    add(b3);
    JButton b4 = new JButton("Tokyo");
    add(b4);
}
}
class ColorsPanel extends JPanel
{
public ColorsPanel()
{    JCheckBox cb1 = new JCheckBox("Red");
    add(cb1);
    JCheckBox cb2 = new JCheckBox("Green");
    add(cb2);
    JCheckBox cb3 = new JCheckBox("Blue");
    add(cb3);
}}
class FlavorsPanel extends JPanel {
public FlavorsPanel() {
JComboBox jcb = new JComboBox();
jcb.addItem("Vanilla");
```

```
jcb.addItem("Chocolate");
jcb.addItem("Strawberry");
add(jcb);
}}
```

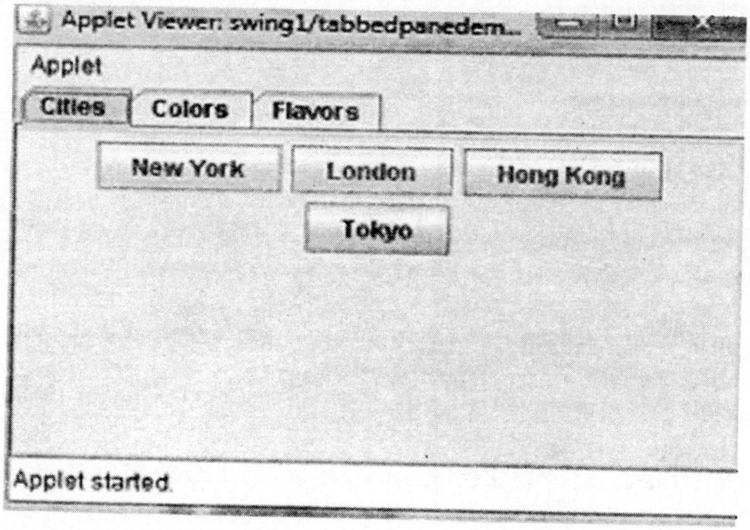

Desktop Panes and Internal Frames

Many applications present information in multiple windows that are all contained inside a large frame. If you minimize the application frame, then all of its windows are hidden at the same time. In the Windows environment, this user interface is sometimes called the multiple document interface or MDI.

REVIEW QUESTIONS

Q. 1. What are Java Swings? How to implement a tree using Java Swings? **[MDU, May 2011]**

Ans. *Refer Section 10.2.*

Q. 2. Write an application to implements trees and styled text components.

[MDU, May 2007, May 2009]

Ans. Every computer user who uses a hierarchical file system has encountered *tree* displays directories and files in forms of treelike organizations. *(Refer Section 10.2, 10.4)*

Q. 3. Write an application to implements tables and progress indicators. **[MDU, Dec. 2007, Dec. 2009]**

Ans. The tables are very powerful components in Swing. They were created to provide an interface to JDBC that makes programming quite flexible for the user. Progress indicators or progress bars are the new controls that give the users some indications of the progress of an operation. They were originally developed to check the progress of installation processes but are now frequently used for all time-consuming operations. When we call JProgressBar, we can set the maximum and minimum value for the progress bar. *(Refer Sections 10.3, 10.5)*

Q. 4. Write short note on lists, Trees, Styled Text Components. **[MDU, May 2008, Dec. 2008]**

Ans. List is a swing component used to show group of elements with scrollbar. Every computer user who uses a hierarchical file system has encountered *tree* displays directories and files in forms of treelike organizations. *(Refer Sections 10.1, 10.2, 10.4)*

Q. 5. What are component organizers? How the Swing components differ from AWT components?

[MDU, May 2012, Dec. 2008]

Ans. AWT components are heavy-weight, whereas Swing components are lightweight. Hence Swing works faster than AWT. Heavy weight components depend on the local windowing toolkit. For example, java.awt.Button is a heavy weight component. Pluggable look and feel possible using java Swing. Also, we can switch from one look and feel to another at runtime in swing which is not possible in AWT. *(Refer 10.6)*

Q. 6. Name the containers which use Border Layout as their default layout?

Ans. Window, Frame and Dialog classes.

Q. 7. Name Container classes.

Ans. Window, Frame, Dialog, FileDialog, Panel, Applet, or ScrollPane

Q. 8. How are the elements of different layouts organized?

Ans. A layout manager is an object that is used to organize components in a container. The different layouts available are FlowLayout, BorderLayout, CardLayout, GridLayout and GridBagLayout.

FlowLayout: The elements of a FlowLayout are organized in a top to bottom, left to right fashion.

BorderLayout: The elements of a BorderLayout are organized at the borders (North, South, East and West) and the center of a container.

CardLayout: The elements of a CardLayout are stacked, on top of the other, like a deck of cards.

GridLayout: The elements of a GridLayout are of equal size and are laid out using the square of a grid.

GridBagLayout: The elements of a GridBagLayout are organized according to a grid. However, the elements may be different sizes and may occupy more than one row or column of the grid. In addition, the rows and columns may have different sizes.

Q. 9. What advantage do Java's layout managers provide over traditional windowing systems?

Ans. Java uses layout managers to lay out components in a consistent manner across all windowing platforms. Since Java's layout managers aren't tied to absolute sizing and positioning, they are able to accommodate platform-specific differences among windowing systems.

Q. 10. What are component organizers? Write a program in Java to implement Trees and Tables? **[MDU, May 2012]**

Ans. Refer 10.1, 10.2 and 10.6.

Q. 11. What is the JFC?

Ans. JFC stands for "Java Foundation Classes". This is the term for several capabilities included in the java 2 Platform. (Some pieces are available for JDK 1.1)

- Swing components
- Accessibility
- Pluggable look-and-feel.
- Java 2D (JDK 1.2 only)
- Drag and drop (JDK 1.2 only)

Q. 12. How do you put my information in a table?

Ans. Either subclass *AbstractTableModel* and implement the required three methods, or use a *DefaultTableModel* (which has a built-in *Vector*). The method required by *AbstractTableModel* are very simple to implement, so subclassing is the preferred method. Using *DefaultTableModel* requires the expense of copying your data from its native from into *DefaultTableModel.*

Q. 13. How do you add and delete rows in a JTable?

Ans. JTable is only the view onto the real table. To change the table, you need to work with the table model. Each model will have its own method to do this. If you've subclass *AbstractTableModel,* you'll have to create your own add/delete methods. *DefaultTableModel* contains methods to add/delete/records rows.

Q. 14. How do you hide a column in a table?

Ans. Sometimes it's done by setting all the column widths (minimum, preferred and maximum) to 0 pixels, but this leaves the column in the table. A better approach is to use the *removeColumn()* method:

```
JTable table = ......
```

TableColumn deleteColumn = table.getColumn("Column Title");

Table.removeColumn(deletedColumn);

It can later be added back into the table by *addColumn()*. This only affects what column is shownby the JTable; the model is unaffected.

Q. 15. How can you remove a column from JTable?

Ans. You have to remove it from the ColumnModel.

```
TableColumnModel model = table.getColumnModel();
Model.removeColumn(model.getColumn(index));
```

Q. 16. How can you add a column to a JTable?

Ans. Use *DefaultTableModel's* addColumn method.

Create a new JTable model.

Update your own JTable's TableModel (Since your Table Model is a subclass of Abstract Table Model(there isn't a strong reason for it not to be) you just need to call fire Table Structure Changed() when the update is complete).

❑❑❑

11

CHAPTER

Abstract Window Toolkit (AWT)

The Java programming language class library provides a user interface toolkit called the Abstract Windowing Toolkit, or the AWT. AWT is both powerful and flexible. Newcomers, however, often find that its power is implied. The class and method descriptions found in the distributed documentation provide little guidance for the new programmer. Furthermore, the available examples often leave many important questions unanswered. Of course, newcomers should expect some difficulty. Effective graphical user interfaces are inherently challenging to design and implement, and sometimes complicated interactions between classes in the AWT only make this task more complex. However, with proper guidance, the creation of a graphical user interface using the AWT is not only possible, but relatively straightforward.

The AWT contains numerous classes and methods that allow us to create and manage windows. The main purpose of the AWT is to support applet windows. It can also be used to create stand-alone windows that run in a GUI environment, such as Windows.

INSIDE THIS CHAPTER

11.1. WINDOW FUNDAMENTALS

The AWT defines windows according to a class hierarchy that adds functionality to each level. The two most common windows are those derived from **Panel**, which is used by applets, and those derived from **Frame**, which creates a standard window. Much of the functionality of these windows is derived from their parent classes. The classes are described in the section below.

Component

Component class is at the top of AWT hierarchy. **Component** is an abstract class that encapsulates all of the attributes of a visual component. All user interface elements that are displayed on the screen and that interact with the user are subclasses of **Component**. It defines over a hundred public methods that are responsible for managing events, such as mouse and keyboard input, positioning and sizing the window, and repainting. A **Component** object is responsible for remembering the current foreground and background colors and the currently selected text font.

Container

The **Container** class is a subclass of **Component**. It has additional methods that allow other **Component** objects to be nested within it. **Container** objects can be stored inside of a **Container.** A container is responsible for positioning any components that it contains. It does this through the use of various layout managers.

Panel

The **Panel** class is a concrete subclass of **Container**. It doesn't add any new methods but simply implements **Container**. **Panel** is the superclass for **Applet**. When screen output is directed to an applet, it is drawn on the surface of a **Panel** object. A **Panel** is a window that does not contain a title bar, menu bar, or border. When an applet is run using an applet viewer, the applet viewer provides the title and border. Other components can be added to a **Panel** object by its **add( )** method. Once these components have been added, we can position and resize them manually using the **setLocation( )**, **setSize( )**, or **setBounds( )** methods defined by **Component**.

Window

The **Window** class creates a top-level window. A *top-level window* is not contained within any other object; it sits directly on the desktop. We do not create **Window** objects directly. Rather, we will use a subclass of **Window** called **Frame**.

Frame

Frame is a subclass of **Window** and has a title bar, menu bar, borders, and resizing corners. If we create a **Frame** object from within an applet, it will contain a warning message to the user that an applet window has been created. This message warns users that the window they see was started by an applet and not by any other software running on their computer. When a **Frame** window is created by a program, a normal window is created.

11.2. THE RENDERING PIPELINE

Graphics is an object given by JDK in java.awt package that has properties and provides methods to work on drawing operations. The properties are coordinate transformation, color, paint, etc. How the figures and images are rendered that depends on the properties of the Graphic object. JDK provides graphics libraries to support this. But when the object has more complex properties as arbitrary affine transforms, rendering hints, transparency, these are not supported by the native graphics libraries.

POINTS TO REMEMBER

The Graphics2D API are providing rendering pipeline. Graphics2D methods and properties are supported by some native graphics libraries.

The initial state for graphics object of most raster displays would call for two levels of indirection:

Graphics2D object → IntegerGraphicsState → DirectRasterGraphics

POINTS TO REMEMBER

The *Graphics2D object* is an instance of a concrete implementation subclass of java.awt.Graphics2D.

This object manages a state object that represents the current pipeline configuration. The *Graphics2D object* forwards most of the requests to the state object. The *Graphics2D object* itself only administers properties that are not specific for a certain state. It stores properties such as foreground and background color so that the state object is not bothered with get-requests for these properties.

You select color and paint mode, and call methods of the Graphics class such as drawRect or fillOval.

The Java 2D API supports following options:

- You can easily create a ample range of *shapes.*
- You will have the control on the *stroke,* the pen that traces shape boundaries.
- Shapes can be filled with solid colors, You can *fill* shapes with solid colors, varying shade, and repeating patterns.
- *Transformation* is used to move, scale, rotate, or stretch shapes.
- *Clipping* can be possible.
- By select *composition rule,* you can describe how to combine the pixels of a new shape with existing pixels.
- To make trade-offs between speed and drawing quality, you can give *rendering hints.*

To draw a shape, you follow the following steps:

(1) Obtain an object of the *Graphics2D* class. This class is a subclass of the Graphics class.

(2) Use the *setRenderingHints* method to set rendering hints

(3) Use the setStroke method to set the *stroke.*

(4) Use the setPaint method to set the *paint.*

(5) Use the setClip method to set the clipping region.

(6) Use the setTransform method to set a *transformation* from user space to device space.

(7) Use the setComposite method to set a *composition rule* that describes how to combine the new pixels with the existing pixels.

(8) Create a shape. The Java 2D API supplies many shape objects and methods to combine shapes.

Draw or fill the shape. If you draw the shape, its outline is stroked. If you fill the shape, the interior is painted.

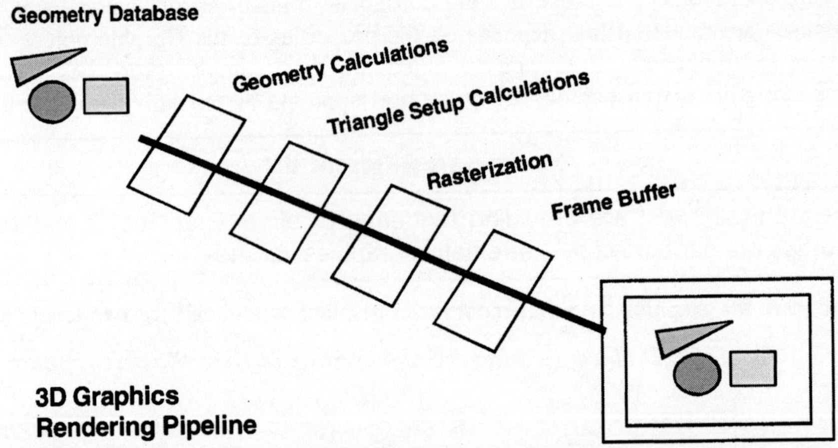

Fig. 11.1.

To render a shape the following steps have to be taken:

1. Set the path of the shape is stroked.
2. Transformed the shape.
3. Clip the shape, if there is no intersection between the shape and the clipping area, then stops the process.
4. The remain portion of the shape after clipping is filled.
5. The pixels of the filled shape are composed with the existing pixels.

An application can be used to draw onto components, for such purposes the Graphic class is given by Java for all graphics contexts. It's an abstract class. Graphics object contains all the information that is required to perform rendering operation of a graphics.

POINTS TO REMEMBER

The Graphics class is an abstract base class for all graphics contexts that allow an application to draw onto components that are realized on various devices, as well as onto off-screen images.

This information includes the following properties:

- The Component object on which to draw.
- A translation origin for rendering and clipping coordinates.
- The current clip.
- The current color.
- The current font.
- The current logical pixel operation function (XOR or Paint).
- The current XOR alternation color.

The Graphics2D class extends the Graphics to provide more properties for sophisticated control.

About Rendering Process

The Rendering Process is divided in four phases. Cashing the results for future calls is the way to optimize these steps. The steps in the rendering process are:

1. Decide what to render.
2. Constrain the rendering operation to the current Clip.
3. Decide what colors to render.
4. Apply the colors to the destination drawing surface.

The three types of rendering operations, along with details of each of their particular rendering processes are:

1. Shape operations

1. Draw (Shape) operation.
2. Transforming the Shape.
3. Extracted outline of the Shape.
4. Get the Path Iterator to compress the Shape.
5. Acquire the PaintContext for color.

2. Text operations

1. Set of glyphs required to render the indicated String.

2. Queried the Font to outlines for the indicated glyphs.
3. Outlines of the character are filled.
4. Queries for a PaintContext, which specifies the colors to render.

3. Image Operations

1. Defining the region of interest by bounding box.This bounding box is specified in
2. AffineTransform is used to transform the bounding box
3. Perform the curret Transformation of bounding box.
4. Determines what colors to render by Image object.

All graphics states are derived from a class AbstractGraphicsState:

```
public abstract class AbstractGraphicsState implements Cloneable {
Graphics2DImpl frontend;
    public void setFrontend(Graphics2DImpl frontend);
    public void dispose();
    // Graphics methods:
    public abstract void setColor(Color color);
    public abstract void setPaintMode();
    public abstract void setXORMode(Color altColor);
    public abstract void setFont(Font font);
    public abstract FontMetrics getFontMetrics(Font font);
    public abstract void setClip(Shape clip);
    public abstract Shape getClip();
    public abstract Rectangle getClipBounds();
    public abstract void copyArea(int x, int y, int width, int height,
        int dx, int dy);
    public abstract void drawLine(int x1, int y1, int x2, int y2);
    public abstract void fillRect(int x, int y, int width, int height);
    public abstract void clearRect(int x, int y, int width, int height);
    public abstract void drawRoundRect(int x, int y, int width, int
        height, int arcWidth, int arcHeight);
    public abstract void fillRoundRect(int x, int y, int width, int
        height int arcWidth, int arcHeight);
    public abstract void drawOval(int x, int y, int width, int height);
    public abstract void fillOval(int x, int y,  int width, int height);
    public abstract void drawArc(int x, int y, int width, int height, int
        startAngle, int arcAngle);
    public abstract void fillArc(int x, int y, int width, int height, int
        startAngle, int arcAngle);
    public abstract void drawPolyline(int[] xPoints, int[] yPoints, int
        nPoints);
    public abstract void drawPolygon(int[] xPoints, int[] yPoints, int
        nPoints);
    public abstract void fillPolygon(int[] xPoints, int[] yPoints, int
        nPoints);
    public abstract boolean drawImage(Image image, int x, int y,
        ImageObserver observer);
```

```
        // Graphics2D methods:
        public abstract void draw(Shape shape);
        public abstract void fill(Shape shape);
        public abstract boolean hit(Rectangle rect, Shape text, boolean
          onStroke);
        public abstract void drawString(String text, int x, int y);
        public abstract void drawString(String text, float x, float y);
        public abstract void translate(int x, int y);
        public abstract void translate(double tx, double ty);
        public abstract void rotate(double theta);
        public abstract void rotate(double theta, double x, double y);
        public abstract void scale(double scaleX, double scaleY);
        public abstract void shear(double shearX, double shearY);
    }
```

POINTS TO REMEMBER

DirectRasterGraphics is an interface that describes a common denominator between most raster based graphics devices.

11.3. SHAPES

The Shape is an interface that provides description for an object that represents a geometric shape. The Shape is described by a PathIterator object, which express the outline of the Shape that gives details about the interior and exterior point of the 2D plane.

The point lies inside the shape is call *insideness* of point. A point is considered to lie inside a Shape if and only if:

- it lies completely inside *theShape* boundary *or*
- it lies exactly on the Shape boundary *and* the space immediately adjacent to the point in the increasing X direction is entirely inside the boundary *or*
- it lies exactly on a horizontal boundary segment **and** the space immediately adjacent to the point in the increasing Y direction is inside the boundary.

Here are some of the methods in the Graphics class to draw shapes:

- drawLine
- drawRectangle
- drawRoundRect
- draw3DRect
- drawPolygon

Instead of methods, there are classes:

- Line2D
- Rectangle2D
- RoundRectangle2D
- Ellipse2D
- Arc2D
- QuadCurve2D
- CubicCurve2D

General Path

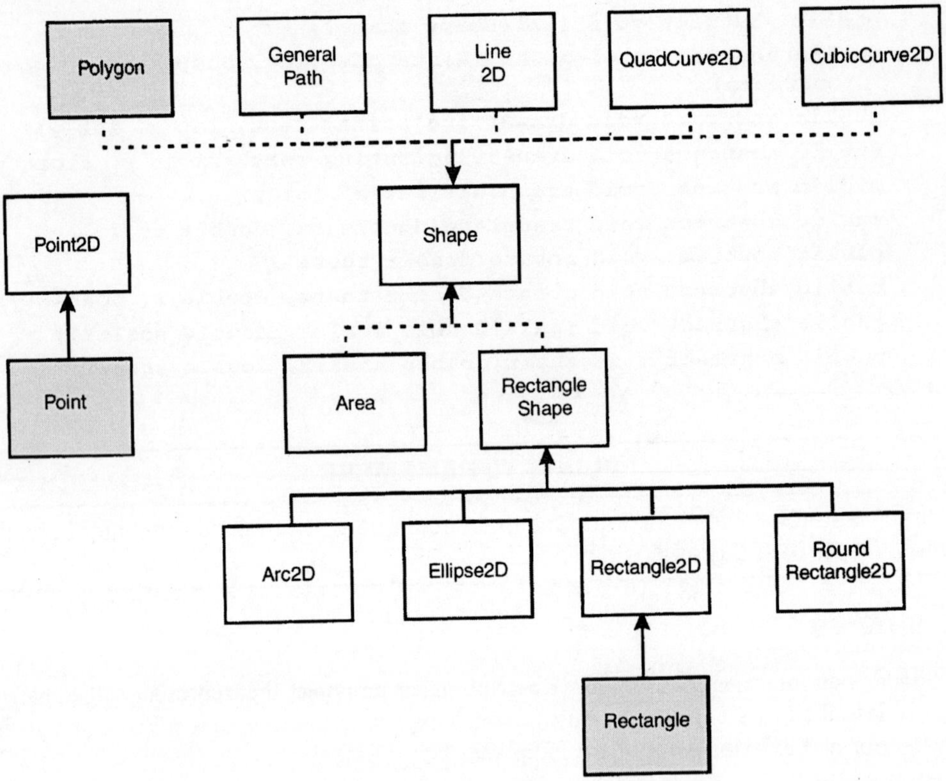

Fig. 11.2. Relationships between the shape classes

Example of DrawLine

```
public class LineDemo2D extends JApplet {
   public void init() {
      setBackground(Color.white);
      setForeground(Color.white);
   }
   public void paint(Graphics g) {
      Graphics2D g2 = (Graphics2D) g;
      g2.setRenderingHint(RenderingHints.KEY_ANTIALIASING,
          RenderingHints.VALUE_ANTIALIAS_ON);
      g2.setPaint(Color.gray);
      int x = 5;
      int y = 7;
      g2.draw(new Line2D.Double(x, y, 200, 200));
      g2.drawString("Line2D", x, 250);
   }
   public static void main(String s[]) {
      JFrame f = new JFrame("ShapesDemo2D");
      f.addWindowListener(new WindowAdapter() {
         public void windowClosing(WindowEvent e) {
            System.exit(0);
         }
```

```
        });
        JApplet applet = new LineDemo2D();
        f.getContentPane().add("Center", applet);
        applet.init();
        f.pack();
        f.setSize(new Dimension(300, 300));
        f.show();
    }
}
```

Fig. 11.3.

Example of DrawRectangle

```
import java.awt.Graphics;

import javax.swing.JComponent;
import javax.swing.JFrame;

class MyCanvas extends JComponent {

  public void paint(Graphics g) {
    g.drawRect (10, 10, 200, 200);
  }
}

public class DrawRect {
  public static void main(String[] a) {
    JFrame window = new JFrame();
    window.setDefaultCloseOperation(JFrame.EXIT_ON_CLOSE);
    window.setBounds(30, 30, 300, 300);
    window.getContentPane().add(new MyCanvas());
    window.setVisible(true);
  }
}
```

11.4. AREAS

For manipulation of resolution-independent description is stored in an Area object of 2-dimenstional space. Area object is responsible to perform Constructive Area Geometry (CAG) and transformation. The CAG operations are ADDITION, SUBTRACTION, INTERSECTION, and EXCLUSIVE OR.

The Area class implements the Shape interface, and provides methods to support for all of its hit-testing and path iteration facilities, but an Area is more specific than a generalized path in the following ways:

- Only closed paths and sub-paths are stored.
- The interiors of the individual stored sub-paths.
- The geometry of the path describing the outline.

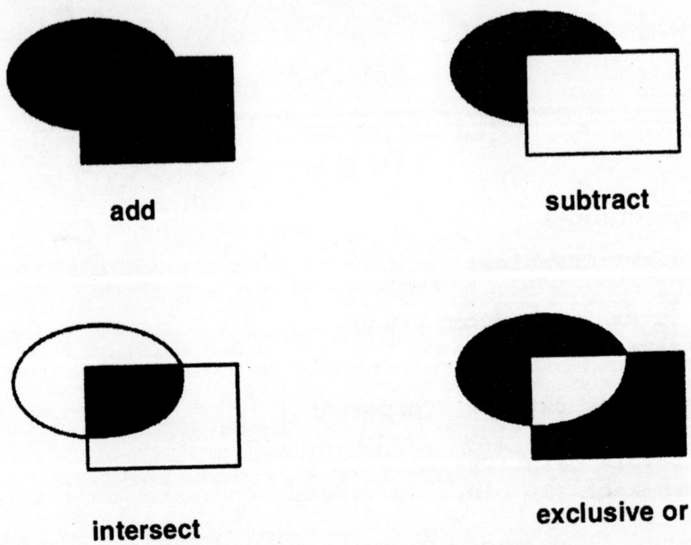

Fig. 11.4. Shows these operations.

Example of add()— First and second area points are combined.

```
public class AreaAdding extends JPanel {
      DrawingCanvas canvas= new DrawingCanvas();

      public AreaAdding() {
        add(canvas);
```

```
        canvas.area1 = new Area(canvas.gp);
        canvas.area1.add(canvas.area2);
        canvas.repaint();
    }
    public static void main(String arg[]) {
        JFrame frame = new JFrame();
        frame.setDefaultCloseOperation(JFrame.EXIT_ON_CLOSE);
        frame.getContentPane().add("Center", new AreaAdding());
        frame.pack();
        frame.setSize(new Dimension(400, 300));
        frame.setVisible(true);
    }
    class DrawingCanvas extends Canvas {
        GeneralPath gp;
        Ellipse2D ellipse;
        Area area1, area2;

        public DrawingCanvas() {
            setBackground(Color.white);
            setSize(350, 250);
            int w = getWidth();
            int h = getHeight();

            gp = new GeneralPath();
            gp.moveTo(w / 8, h / 2);
            gp.lineTo(w / 2, h / 4);
            gp.lineTo(7 * w / 8, h / 2);
            gp.lineTo(w / 2, 3 * h / 4);
            gp.closePath();
            area1 = new Area(gp);
            ellipse = new Ellipse2D.Double(w / 4, h / 4, w / 2, h / 2);
            area2 = new Area(ellipse);
        }

        public void paint(Graphics g) {
            Graphics2D g2D = (Graphics2D) g;
            g2D.setStroke(new BasicStroke(2.0f));

            g2D.draw(area1);
            g2D.draw(area2);
        }

    }
}
```

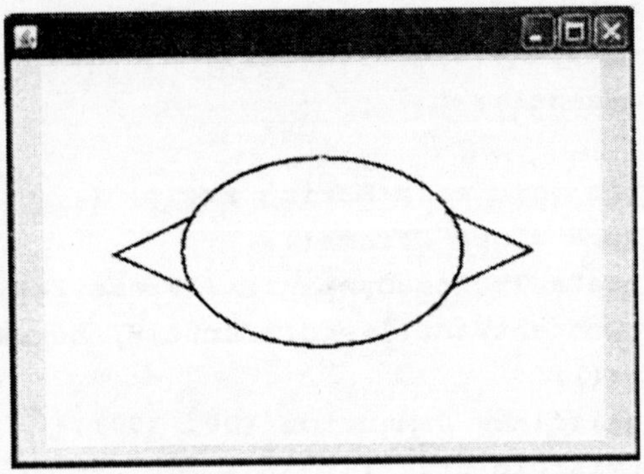

Example of intersect()

The combined area contains all points that are in the first and the second area.

```
public class AreaIntersecting extends JPanel {
    DrawingCanvas canvas;

    public AreaIntersecting() {
      canvas = new DrawingCanvas();
      add(canvas);

      canvas.area1 = new Area(canvas.gp);
      canvas.area1.intersect(canvas.area2);
      canvas.repaint();
    }
    public static void main(String arg[]) {
      JFrame frame = new JFrame();
      frame.setDefaultCloseOperation(JFrame.EXIT_ON_CLOSE);
      frame.getContentPane().add("Center", new AreaIntersecting());
      frame.pack();
      frame.setSize(new Dimension(400, 300));
      frame.setVisible(true);
    }
    class DrawingCanvas extends Canvas {
      GeneralPath gp;
      Ellipse2D ellipse;
      Area area1, area2;

      public DrawingCanvas() {
        setBackground(Color.white);
        setSize(350, 250); // width and height of canvas
        int w = getWidth();
        int h = getHeight();

        gp = new GeneralPath();
        gp.moveTo(w / 8, h / 2);
        gp.lineTo(w / 2, h / 4);
        gp.lineTo(7 * w / 8, h / 2);
        gp.lineTo(w / 2, 3 * h / 4);
        gp.closePath();
        area1 = new Area(gp); // General path area object
```

```
        ellipse = new Ellipse2D.Double(w / 4, h / 4, w / 2, h / 2);
        area2 = new Area(ellipse); // Ellipse area object
      }

      public void paint(Graphics g) {
        Graphics2D g2D = (Graphics2D) g;
        g2D.setStroke(new BasicStroke(2.0f));

        g2D.draw(area1);
        g2D.draw(area2);
      }

    }
  }
```

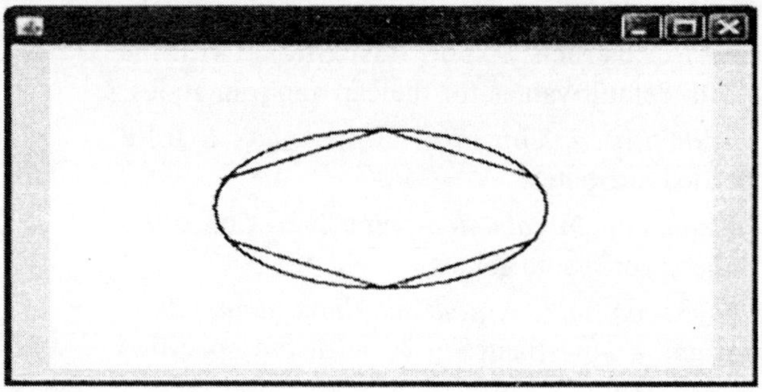

11.5. STROKES

The Stroke is an interface that allows a Graphics2D object to get a Shape, which has bedecked outline, and having stylish representation of outline. Stroking is a Shape that is created like a marking pen. It's of different shapes and size. Where the pen leaves the mark, the area is enclosed by the outline Shape.

POINTS TO REMEMBER

The object of the classes that implements the Stroke must be read-only. It can't be cloned, because Graphic2D object is itself cloned.

The Graphics2d class draws the boundary of a shape of given or selected *stroke* type. By default stroke is a solid line, of one pixel wide. The setStroke() method is used to select any type of stroke, by passing the class object that implements the Stroke interface. The Java 2D API defines only one such class, called BasicStroke.

```
public class BasicStroke extends Object implements Stroke {
...
...
}
```

POINTS TO REMEMBER

The *BasicStorke* class provides method to render the graphics with its attributes for the outline, which are rendered with a Graphics2D object. The *BasicStroke* describes the shape mark.

The *BasicStorke* class provides method to render the graphics with its attributes for the outline. which are rendered with a Graphics2D object. The *BasicStroke* describes the shape mark. These rendering attributes include:

- *width* : the pen width,
- *end caps* : the decoration applied to the ends of unclosed subpaths and dash segments. The three different decorations are: CAP_BUTT, CAP_ROUND, and CAP_SQUARE.
- *line joins* : intersection of two path segments and at the intersection of the endpoints of a subpath. The three different decorations are: JOIN_BEVEL, JOIN_MITER, and JOIN_ROUND.
- *miter limit* : the limit to trim a line
- *dash attributes* : how to make a dash pattern by alternating between opaque and transparent sections.

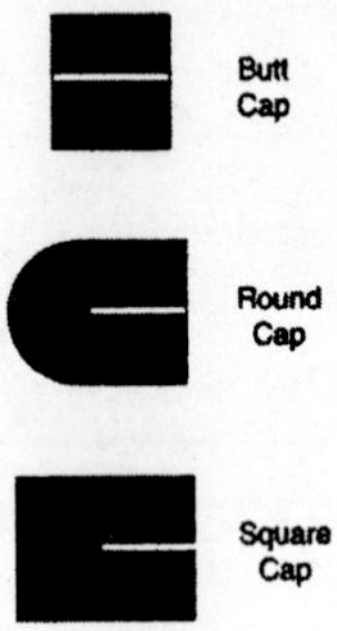

Fig. 11.5.

BasicStroke() Constructs a new BasicStroke with defaults for all attributes.

BasicStroke(float width) Constructs a solid BasicStroke with the specified line width and with default values for the cap and join styles.

BasicStroke(float width, int cap, int join) Constructs a solid BasicStroke with the specified attributes.

BasicStroke(float width, int cap, int join, float miterlimit) Constructs a solid BasicStroke with the specified attributes.

BasicStroke(float width, int cap, int join, float miterlimit, float[] dash, float dash_phase) Constructs a new BasicStroke with the specified attributes.

When a stroke is more than a pixel thick, then the *end* of the stroke can have different styles.

- A *butt cap* simply ends the stroke at its end point.
- A *round cap* adds a half-circle to the end of the stroke.
- A *square cap* adds a half-square to the end of the stroke.

11.6. PAINT

The Paint is an interface that defines how color patterns are generated for the graphics. Graphics2D is supported for all graphics related operations. The Paint interface is added with Graphics2D context to provide the color pattern by draw and fill methods.

POINTS TO REMEMBER

Instances of classes implementing Paint must be read-only because the Graphics2D does not clone these objects when they are set as an attribute with the setPaint method or when the Graphics2D object is itself cloned.

Color Class is used to define the color in which the graphics will be rendered.

Constructors

Color(ColorSpace cspace, float[] components, float alpha)

Creates a color in the specified ColorSpace with the color components specified in the float array and the specified alpha.

Color(float r, float g, float b)

Creates an opaque sRGB color with the specified red, green, and blue values in the range (0.0 - 1.0).

Color(float r, float g, float b, float a)

Creates an sRGB color with the specified red, green, blue, and alpha values in the range (0.0 - 1.0).

Color(int rgb)

Creates an opaque sRGB color with the specified combined RGB value consisting of the red component in bits 16-23, the green component in bits 8-15, and the blue component in bits 0-7.

Color(int rgba, boolean hasalpha)

Creates an sRGB color with the specified combined RGBA value consisting of the alpha component in bits 24-31, the red component in bits 16-23, the green component in bits 8-15, and the blue componem in bits 0-7.

The GradientPaint class varies colors by interpolating between two given color values (see Figure).

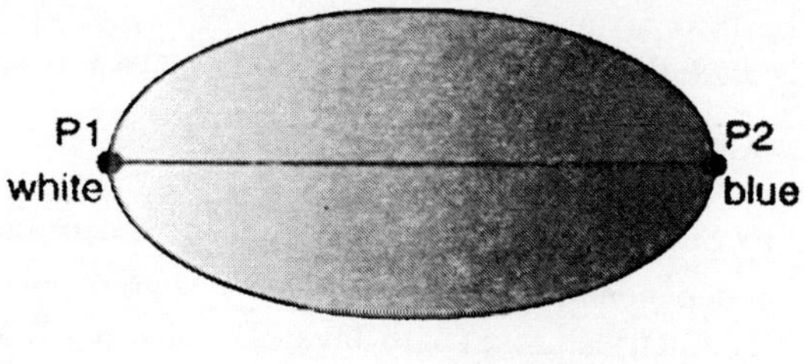

Fig. 11.6. Gradient paint

If Point P1 with Color C1 and Point P2 with Color C2 are specified in user space, the Color on the P1, P2 connecting line is proportionally changed from C1 to C2. Any point P not on the extended P1, P2 connecting line has the color of the point P' that is the perpendicular projection of P on the extended P1, P2 connecting line. Points on the extended line outside of the P1, P2 segment can be colored in one of two ways.

> **DO YOU KNOW**
>
> Tho GradientPaint class offer a way to fill a Shape with a line color.

- If the gradient is cyclic then the points on the extended P1, P2 connecting line cycle back and forth between the colors C1 and C2.
- If the gradient is acyclic then points on the P1 side of the segment have the constant Color C1 while points on the P2 side have the constant Color C2.

Constructors

GradientPaint(float x1, float y1, Color color1, float x2, float y2, Color color2) Constructs a simple acyclic GradientPaint object.

GradientPaint(float x1, float y1, Color color1, float x2, float y2, Color color2, boolean cyclic) Constructs either a cyclic or acyclic GradientPaint object depending on the boolean parameter.

GradientPaint(Point2D pt1, Color color1, Point2D pt2, Color color2) Constructs a simple acyclic GradientPaint object.

GradientPaint(Point2D pt1, Color color1, Point2D pt2, Color color2, boolean cyclic) Constructs either a cyclic or acyclic GradientPaint object depending on the boolean parameter.

- **The TexturePaint class fills an area with repetitions of an image (see Figure).**

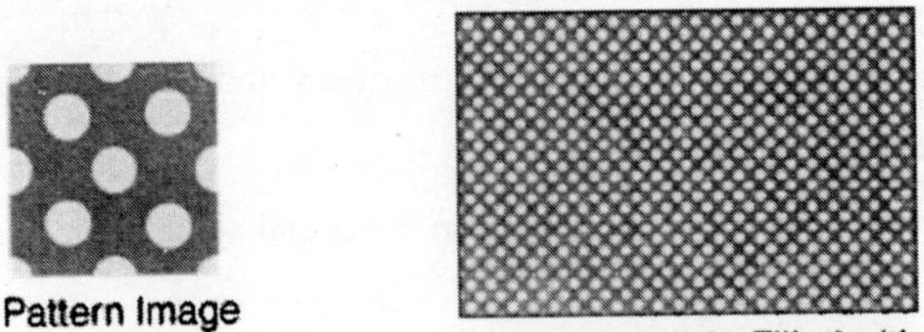

Fig. 11.7. Texture paint

The TextturePaint is a class that provides a method to fill the Shape with a texture that is specified by BufferedImage. It is necessary that the size of the BufferedImage must be small so that it can be easily copied to the TexturePaint object. The texture is replicated infinitely in all the direction in the user space.

Constructor

TexturePaint(BufferedImage txtr, Rectangle2D anchor) Constructs a TexturePaint object.

11.7. COORDINATE TRANSFORMATIONS

As defined in Java tutorial by Sun the Java 2D API maintains two coordinate systems:

- *User space* is a device-independent, logical coordinate system. Applications use this coordinate system exclusively; all geometries passed into Java 2D rendering routines are specified in user space.
- *Device space* is a device-dependent coordinate system that varies according to the target rendering device. In a multi-screen environment with a virtual desktop where a window can span more than one physical screen device, the device coordinate system that's used is the coordinate system of the virtual desktop that encompasses all of the screens.

POINTS TO REMEMBER

Java 2D allows you to easily translate, rotate, scale, or shear the coordinate system. This features is very easy in moving the coordinate system.

The easiest way to picture what is happening in a transformation is to imagine that the person doing the drawing has a picture frame that he lays down on top of a sheet of paper. The drawer always sits at the bottom of the frame. To apply a translation, you move the frame (also moving the drawer) and do the drawing in the new location. You then move the frame back to its original location, and what you now see is the final result. Similarly, for a rotation, you spin the frame (and the drawer), draw, then spin back to see the result. Similarly for scaling and shears: modify the frame without touching the underlying sheet of paper, draw then reverse the process to see the final result.

You can also perform complex transformations by directly manipulating the underlying arrays that control the transformations. This type of manipulation is a bit more complicated to envision than the basic translation, rotation, scaling, and shear transformations. The idea is that a new point (x_2, y_2) can be derived from an original point (x_1, y_1) as follows:

$$\begin{bmatrix} x_2 \\ y_2 \\ 1 \end{bmatrix} = \begin{bmatrix} m_{00} & m_{01} & m_{02} \\ m_{10} & m_{11} & m_{12} \\ 0 & 0 & 1 \end{bmatrix} = \begin{bmatrix} m_{00}x_1 + m_{01}y_1 + m_{02} \\ m_{10}x_1 + m_{11}y_1 + m_{12} \\ 1 \end{bmatrix}$$

> **NOTE**
>
> Note that you can only supply six of the nine values in the transformation array (the m_{xx} values). The coefficients m_{02} and m_{12} provide x and y *translation* of the coordinate system. The other four transformation coefficients (m_{00}, m_{01}, m_{10}, m_{11}) provide *rotation* of the system. For the transformation to preserve orthogonality ("straightness" and "parallelness" of lines), the Jacobian (determinant) of the transformation matrix must equal 1. The bottom row is fixed at [0 0 1] to guarantee that the transformations does not rotate the shape out of the *x-y* plane (produce components along the *z*-axis). There are several ways to supply this array to the AffineTransform constructor; see the AffineTransform API for details.

There are four fundamental transformations:

- Scaling: blowing up, or shrinking, all distances from a fixed point;
- Rotation: rotating all points around a fixed center;
- Translation: moving all points by a fixed amount;
- Shear: leaving one line fixed and "sliding" the lines parallel to it by an amount that is proportional to the distance from the fixed line.

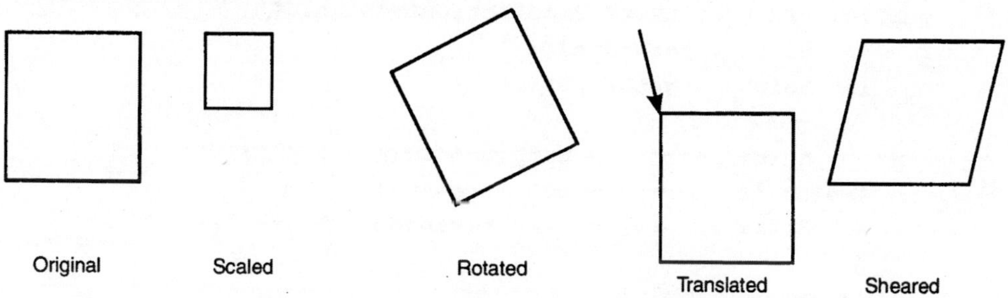

Fig. 11.8. The fundamental transformations

11.7.1. Clipping

Any portion of the drawing is restricting to be rendered on the plan, is called the clipping. Its path is a part of the Graphics2D. You ca shrink the clipping path by making the call to the clip method and passing in another Shape.

Example:

```
public class Clipping extends JPanel implements ActionListener {
    private int posx = 8;
    private int posy = 8;
    private int radius = 90;
    Timer timer;
    Image image;
    private double delta_size[] = { 3, 3 };
    public Clipping() {
        image = new ImageIcon("images.jpg").getImage();
        timer = new Timer(30, this);
        timer.start();
    }
    public void paint(Graphics g) {
        super.paint(g);
```

```
       Graphics2D g2d = (Graphics2D) g;
      RenderingHints rh = new RenderingHints(
  RenderingHints.KEY_ANTIALIASING, RenderingHints.VALUE_ANTIALIAS_ON);
    rh.put(RenderingHints.KEY_RENDERING,
             RenderingHints.VALUE_RENDER_QUALITY);
       g2d.setClip(new Ellipse2D.Double(posx, posy, radius, radius));
       g2d.drawImage(image,  5, 5, null);
   }
   public static void main(String[] args) {
       JFrame frame = new JFrame("Clipping Demo");
       frame.add(new Clipping());
       frame.setDefaultCloseOperation(JFrame.EXIT_ON_CLOSE);
       frame.setSize(400, 300);
       frame.setLocationRelativeTo(null);
       frame.setVisible(true);
   }
   public void actionPerformed(ActionEvent e) {
       int width = getWidth();
       int height = getHeight();
       if (posx < 0) {
           delta_size[0] = Math.random() % 8 + 9;
       } else if (posx > width - radius) {
           delta_size[0] = -(Math.random() % 8 + 9);
       }
       if (posy < 0 ) {
           delta_size[1] = Math.random() % 8 + 9;
       } else if (posy > height - radius) {
           delta_size[1] = -(Math.random() % 8 + 9);
       }
       posx += delta_size[0];
       posy += delta_size[1];
       repaint();
   } }
```

11.7.2. Transparency

Java2D allows you to give transparency value to drawing operations, so that the graphic to be drawn, partially shows that underlying images. AlphaComposite object has to be created by Alphacomposite.getInstance to provide and set the transparency. The object of AlphaComposite will be passed to setComposite method of Graphics2D object. There are 8 built-in mixing rules, but the one normally used for drawing with transparency settings is AlphaComposite.SRC_OVER. Alpha values range from 0.0F (completely transparent) to 1.0F (completely opaque).

Example:

```
public class TransparentRectangles extends JPanel {
    public void paint(Graphics g) {
        super.paint(g);
        Graphics2D g2d = (Graphics2D)g;
        g2d.setColor(Color.BLUE);
        for (int i = 1; i <= 10; i++) {
                g2d.setComposite(AlphaComposite.getInstance(
                                    AlphaComposite.SRC_OVER,i * 0.1f));
            g2d.fillRect(50 * i, 20, 40, 40);
        }
    }
    public static void main(String[] args) {
        JFrame frame = new JFrame("Transparent Rectangles");
        frame.add(new TransparentRectangles());
        frame.setDefaultCloseOperation(JFrame.EXIT_ON_CLOSE);
        frame.setSize(590, 120);
        frame.setLocationRelativeTo(null);
        frame.setVisible(true);
    }
}
```

11.7.3. Composition

Composing transformations

$$\begin{bmatrix} x_{new} \\ y_{new} \\ 1 \end{bmatrix} = \begin{bmatrix} a & c & e \\ b & d & f \\ 0 & 0 & 1 \end{bmatrix} . \begin{bmatrix} x \\ y \\ 1 \end{bmatrix}$$

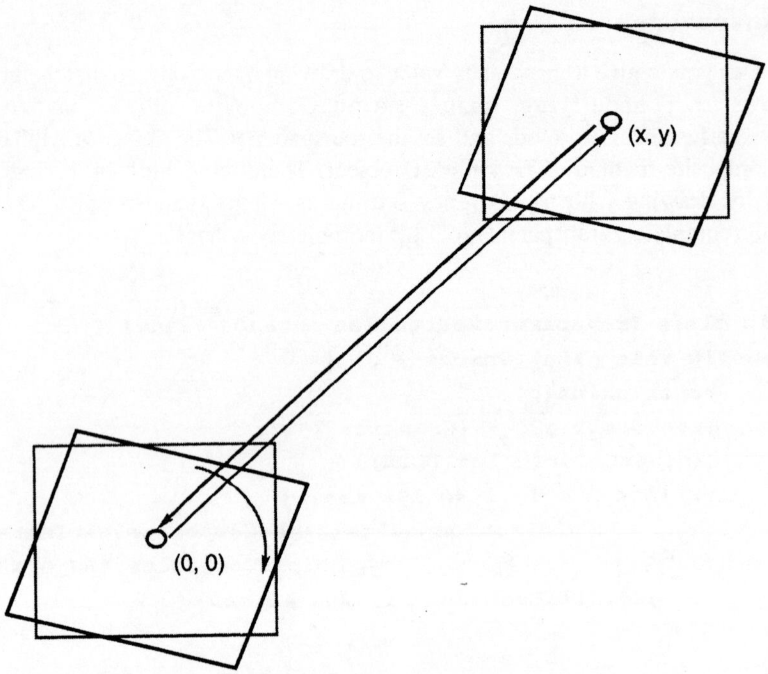

Fig. 11.9. Composing transformations

11.8. RENDERING HINTS

The class java.awt.RenderingHints is a another graphics attribute used by Graphics2D. It's a combination of rendering hint name and rendering hint value, so that it is mapped to java.util.Map.

Another graphics attribute used by the Graphics2D class is java.awt.RenderingHints. This class is a mapping (a java.util.Map) from a set of rendering hint names to a set of rendering hint values. Unlike with other attributes, Graphics2D defines more than one method to set the rendering hints attribute. setRenderingHints() specifies a new set of hints that replaces the old set of hints, while addRenderingHints() adds a set of hints to the existing set and setRenderingHint() sets the value for a single hint in the current set of hints.

Rendering hints are suggestions to Java 2D about how it should perform its rendering. The RenderingHints class defines a number of constants whose names begin with KEY_. These constants represent the kind of hints you can give. The class also defines a number of constants whose names begin with VALUE_. These are the legal values for the various hints. The names of the VALUE constants make it clear with which hint KEY constant each value is associated.

The purpose of the hints is to allow you to request that Java 2D turn a particular feature, such as antialiasing, on or off. In addition, the hints allow you to suggest what kind of speed versus quality trade-offs Java 2D should make. Remember that these are hints and suggestions to Java 2D, not commands. Not all Java 2D implementations support all the hints, and different implementations have different default values for the hints. Furthermore, the meanings of the hints are not precisely defined, so different implementations may interpret the hints differently.

```
static class RenderingHints.Key
```

Defines the base type of all keys used to control various aspects of the rendering and imaging pipelines.

Constructor

RenderingHints(Map<RenderingHints.Key,?> init)

RenderingHints(Map<RenderingHints.Key,?> init) Constructs a new object with keys and values initialized from the specified Map object (which may be null). **RenderingHints**(RenderingHints.Key key, Object value) Constructs a new object with the specified key/value pair.

11.9. READERS AND WRITERS FOR IMAGES

To load an image, the static read method of the ImageIO class is used:

```
File f = . . .;
BufferedImage image = ImageIO.read(f);
```

The ImageIO class picks an appropriate reader, based on the file type. It may consult the file extension and the "magic number" at the beginning of the file for that purpose. If no suitable reader can be found or the reader can't decode the file contents, then the read method returns null. Writing an image to a file is just as simple:

```
File f = . . .;
String format = . . .;
ImageIO.write(image, format, f);
```

Here the format string is a string identifying the image format, such as "JPEG" or "PNG". The ImageIO class picks an appropriate writer and saves the file. For more advanced image reading and writing operations that go beyond the static read and write methods of the ImageIO class, you first need to get the appropriate ImageReader and ImageWriter objects. The ImageIO class enumerates readers and writers that match one of the following:

An image format (such as "JPEG")

A file suffix (such as "jpg")

A MIME type (such as "image/jpeg")

11.10. IMAGE MANIPULATION

If we have an image and we want to improve its appearance. We then need to access the individual pixels of the image and replace them with other pixels. Or perhaps we want to compute the pixels of an image from scratch, for example, to show the result of physical measurements or a mathematical computation. The BufferedImage class gives the control over the pixels in an image, and classes that implement the BufferedImageOp interface let us transform images.

This is a major change from the image support in JDK 1.0. At that time, the image classes were optimized to support *incremental rendering*. The original purpose of the classes was to render GIF and JPEG images that are downloaded from the Web, a scan line at a time, as soon as partial image data is available. In fact, scan lines can be *interlaced,* with all even scan lines coming first, followed by the odd scan lines. This mechanism lets a browser quickly display an approximation of the image while fetching the remainder of the image data. The ImageProducer, ImageFilter, and ImageConsumer interfaces in JDK 1.0 expose all the complexities of incremental rendering. Writing an image manipulation that fit well into that framework was quite complex. Fortunately, the need for using these classes has completely gone away. JDK 1.2 replaced the "push model" of JDK 1.0 with a "direct" model that lets us access pixels directly and conveniently. The only disadvantage of the direct model is that it requires all image pixels to be in memory. Future versions of the Java platform may support a "pull" model with which a processing pipeline can reduce memory consumption and increase speed by fetching and processing pixels only when they are actually needed.

Most of the images that we manipulate are simply read in from an image file. They were either produced by a device such as a digital camera or scanner, or constructed by a drawing program. To create an image, construct a BufferedImage object in the usual way.

```
image = new BufferedImage(width, height, BufferedImage.TYPE_INT_ARGB);
```

Now, call the getraster method to obtain an object of type WritableRaster. You use this object to access and modify the pixels of the image.

```
WritableRaster raster = image.getRaster();
```

The setPixel method lets us set an individual pixel. The complexity here is that we can't simply set the pixel to a Color value. We must know how the buffered image specifies color values. That depends on the *type* of the image. If the image has a type of TYPE_INT_ARGB, then each pixel is described by four values, for red, green, blue, and alpha, each of which is between 0 and 255. We supply them in an array of four integers.

```
int[] black = { 0, 0, 0, 255 };
raster.setPixel(i, j, black);
```

In the lingo of the Java 2D API, these values are called the *sample values* of the pixel. Consider the example below:

```
import java.awt.*;
import java.applet.*;
public class abc extends Applet
{
  Image img;
  public void init( )
  {
    URL base;
   Base=getDocumentBase();
   img=getImage(base,"c:\abc.gif");
  }
 public void paint(Graphics g)
 {
  WritableRaster raster=img.getRaster();
  Raster.setPixel(20,20,Black);
  g.drawImage(img,50,70,this);
  }
}
```

11.11. PRINTING IN AWT

print class of awt package is used for printing in awt. Consider the example below for printing in AWT:

```
import java.awt.*;
import java.awt.print.*;
class MyPage implements Printable
{
 public void paint (Graphics g)
```

```
  {
   Graphics2D g2d = (Graphics2D) g;
   g2d.setFont(new Font("Arial", Font.BOLD, 30));
   g2d.drawString ("Hello World of Paper!", 200, 150);
  }
  public int print (Graphics g, PageFormat f, int i)
  {
   if (i != 0)
    return NO_SUCH_PAGE;
   paint (g);
   return PAGE_EXISTS;
  }
 }
class MyPrinter
{
 public static void main (String[] args)
 {
  PrinterJob job = PrinterJob.getPrinterJob();
  System.out.println ("Staring printer...");
  job.setPrintable(new MyPage());
  if (job.printDialog())
  {
   try
   {
    job.print();
   }
   catch (PrinterException e)
   {
     e.PrintStackTrace();
   }
 }
```

11.12. THE CLIPBOARD

In Java, there are actually two kinds of Clipboard - system and local. The system clipboard corresponds to the usual idea of sharing data between otherwise independent applications running on the same computer. Local Clipboards are visible only within a single Java application. They are created by simply passing a name to the Clipboard constructor. A clipboard acts as a temporary staging area where application users store cut or copied information, including text, graphics, objects, and other data types. It is a heavily exploited data sharing facility. Therefore, when developing new applications, it makes sense to expend extra effort to support the clipboard. Neglecting clipboard functionality can result in an application that gets complaints, requires new development, or—worst of all—never gets used.

11.13. DRAG & DROP

Java introduced the ability to transfer data using Drag & Drop metaphor. Java 2 developed many new classes to develop D&D metaphor in the package java.awt.dnd. The figure below shows a GUI component representing the data source & destination of D&D operation that maintains an association with a java.awt.dnd.DropSource object and java.awt.dnd.DropTarget object. It depicts a java.awt.datatransfer.Transferable object that encapsulates the data transfer between DragSource and DragTarget objects.

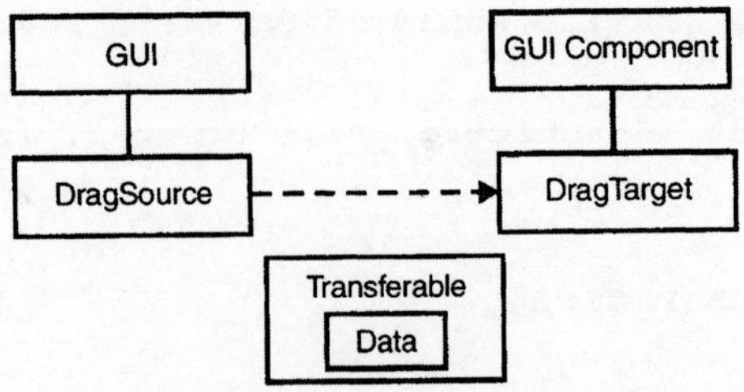

Fig. 11.10. Drag & Drop

When the Transferable object encapsulates data, it makes the data available to DragTarget. For a local transfer within the same JVM, Transferable provides an object reference. When invoking a Drag & Drop operation, various drag and drop operations are requested such as:

```
ACTION_NONE     no action taken
ACTION_COPY     the DragSource leaves the data intact
ACTION_MOVE     the DragSource deletes the data upon successful
                     completion of the drop
```

REVIEW QUESTIONS

Q. 1. Explain the following:

(*a*) Clipping

(*b*) Rendering **[MDU, May 2011]**

Ans. *Refer Sections 11.7, 11.8.*

Q. 2. What are AWT components ?How image manipulation is accomplished using AWT components*?* **[MDU, May 2007, Dec. 2007, Dec. 2008]**

Ans. AWT stands for Abstract windows Toolkit. AWT is used for graphics and windows programming. The different AWT components are button, checkbox, radiobutton, textbolx, combo box etc.

Coordinate transformation changes the objects position, objects size and structures. There are four fundamental transformations.

Scaling: blowing up, or shrinking, all distances from a fixed point

Rotation: rotating all points around a fixed center

Translation: moving all points by a fixed amount

Shear: leaving one line fixed and "sliding" the lines parallel to it by an amount that is proportional to the distance from the fixed line.

Diagram for fundamental transformations:

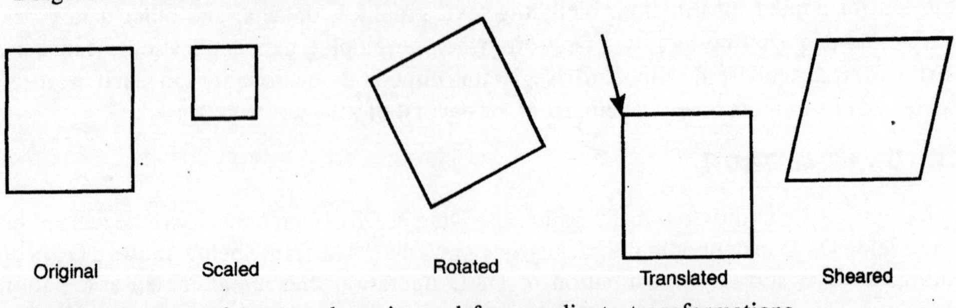

Graphics2D class of awt package is used for coordinate transformations.

Q. 3. How printing is done using AWT components in Java? Explain with the help of an example. **[MDU, May 2007, Dec. 2008]**

Ans. print class of awt package is used for printing in AWT.

Q. 4. Discuss Readers and Writers for images. What is image manipulation? **[MDU, May 2008, May 2009]**

Ans. To load an image, use the static read method of the ImageIO class:

```
File f = . . .;
BufferedImage image = ImageIO.read(f);
```

The ImageIO class picks an appropriate reader, based on the file type. It may consult the file extension and the "magic number" at the beginning of the file for that purpose. If no suitable reader can be found or the reader can't decode the file contents, then the read method returns null.

Writing an image to a file is just as simple:

```
File f = . . .;
String format = . . .;
ImageIO.write(image, format, f);
```

Here the format string is a string identifying the image format, such as "JPEG" or "PNG". The ImageIO class picks an appropriate writer and saves the file.

Image Manipulation: Image Manipulation is the process of modifying the image.

Suppose you have an image and you would like to improve its appearance. You then need to access the individual pixels of the image and replace them with other pixels. Or perhaps you want to compute the pixels of an image from scratch, for example, to show the result of physical measurements or a mathematical computation. The **BufferedImage** class gives you control over the pixels in an image.

Accessing Image Data: To create an image, construct a BufferedImage object in the usual way.

```
image = new BufferedImage(width, height, BufferedImage.TYPE_INT_ARGB);
```

Now, call the getraster method to obtain an object of type WritableRaster. You use this object to access and modify the pixels of the image.

```
WritableRaster raster = image.getRaster();
```

The setPixel method lets you set an individual pixel. The complexity here is that you can't simply set the pixel to a Color value. You must know how the buffered image specifies color values. That depends on the *type* of the image. If your image has a type of TYPE_INT_ARGB, then each pixel is described by four values, for red, green, blue, and alpha, each of which is between 0 and 255. You supply them in an array of four integers.

```
int[] black = { 0, 0, 0, 255 };
raster.setPixel(i, j, black);
```

In the lingo of the Java 2D API, these values are called the *sample values* of the pixel.

Q. 5. How Coordinate Transformation and Image Manipulation is done in AWT? **[MDU, May 2012]**

Ans. *Refer 11.7, 11.10.*

Q. 6. Write a program to implement Drag & Drop in AWT. **[MDU, May 2012]**

Ans. *Refer 11.13.*

Q. 7. Explain Transparency, composition and clipboard. **[MDU, Dec. 2009]**

Ans. Java2D allows you to give transparency value to drawing operations, so that the graphic to be drawn, partially shows that underlying images.

Q. 8. What is rendering Hints. Explain with example.

Ans. Section 11.8 gives explanation about the rendering hints. An example is associated with it.

❑❑❑

12
CHAPTER

Java Beans Components

Java Bean is a software component written in Java that provides portability, platform-independent component model, and reusability. You can add these components into applets, applications, or with other bean components. JavaBeans components are known as *beans*. By nature, beans are dynamic means like other software components; these can be changed or customized.

JavaBeans is a portable, platform-independent component model written in Java programming language. JavaBeans architecture was built through a collaborative industry effort and enables developers to write reusable components in the Java programming language.

INSIDE THIS CHAPTER

12.1. BEANS

With JavaBeans API you can create reuseable, platform-independent components. Using JavaBeans-compliant application builder tools, you can combine these components into applets, applications, or composite components.

There are three types of beans:

- ***Visible Bean*** : A bean component that is visible to an end user, such as a button, or any graphical component, that is use with GUI.
- ***Invisible Bean*** : A bean, that added with any application but not visible to the end user on GUI, for example checking the spelling of a document.
- ***Container Bean*** : A bean can contain another bean component, is called the container bean, for example a tree bean can contain another component like images, labels, etc.

The bean is created by combing the following components.

- ***Properties*** define the characteristics (appearance or behavior) of the bean. It can be changed at design time or run time if it is not read only and some mutator metods are given in its implementation.
- ***Events*** to communicate with other beans.
- ***Methods*** can be called from other beans or from the environment or application in that it is added.

12.2. ADVANTAGES OF JAVA BEANS

Bean development is a standard mechanism to develop the component for the software. There are following advantages of Java Bean components.

- It supports or provide "write-once, run-anywhere" paradigm.
- Three component of the bean are exposed to an application builder tool.
- It has to run in different environment, so that, it is designed correctly.
- Its configuration setting is stored and helps to persist its state and can be used later.
- A Bean may receive events and can communicate with others by generating the events, if it is registered with events.

12.3. STATES OF BEAN

Bean is a reusable software component. So as software has its development life cycle similarly bean component has its own states. These states define its lifecycle, and are as following:

- *Construction State:* In this state, the bean is under coding phase. The developer is busy writing the code for it.
- *Build State:* Bean is in build state when after construction it is added with any application.
- *Run State:* As the container application is runs then bean is in run state.

12.4. FUNCTIONAL CHARACTERISTICS OF BEAN

12.4.1. Communication

One bean can communicate with other bean and application in that it is build. By message passing to method, it does the communication.

12.4.2. Introspection

It is an automatic process to analyze the properties, events and methods.

12.4.3. Persistence

A bean persists by storing and retrieving the properties from storage. The state of components is to be stored in a non-volatile place for later retrieval.

12.4.4. Customizers

Modifying the appearance and behavior at the time of building, and run time.

12.5. USING THE BEAN DEVELOPER KIT (BDK)

The Bean Development Kit (BDK) is provided by the JavaSoft. It is a tool to create it and configure.

Installing the BDK

Be ensure before installing BDK on your machine, SDK must be installed on your machine. Download the BDK from http://java.sun.com. It is packaged as one file that is a self-extracting archive. Follow the instructions to install it on your machine.

Starting the BDK

To start the BDK, follow these steps:

1. Change to the directory **c:\bdk\beanbox**.
2. Execute the batch file called **run.bat**.

This causes the BDK to display the three windows shown in figure below. ToolBox lists all of the different Beans. BeanBox provides an area to layout to draw the beans from ToolBox. Properties provide the ability to configure a selected Bean.

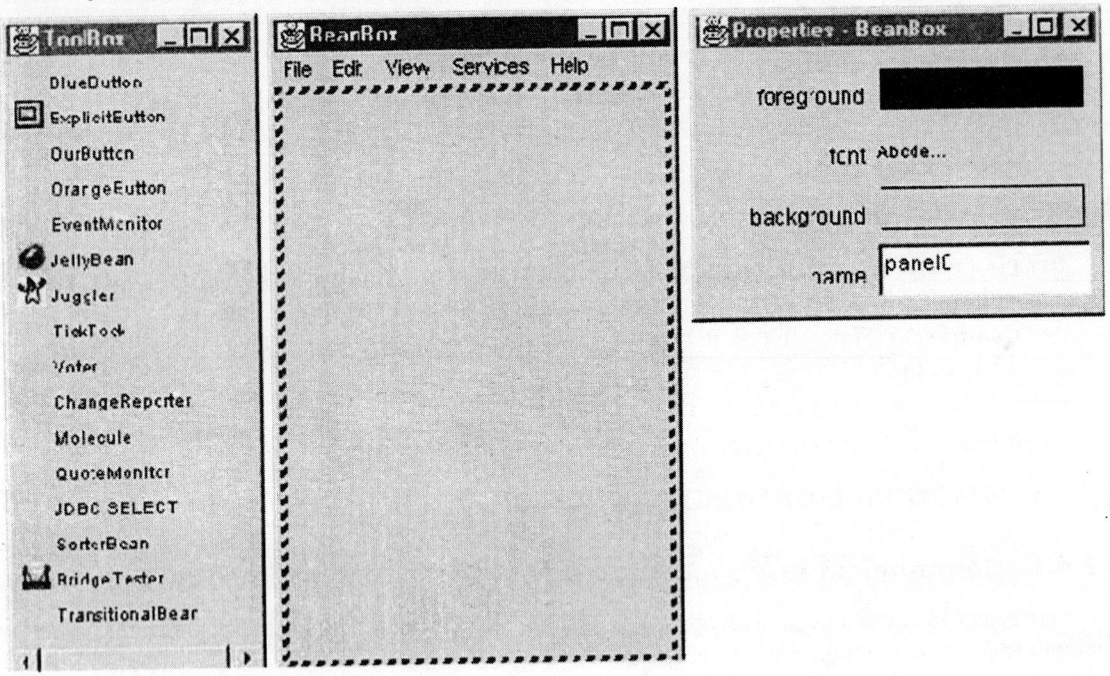

12.6. BEAN-WRITING PROCESS

Beans writing is a process to defe the properties, registering with events and implementing the methods. Bean writing is the construction state of the bean. In this section we'll study how to define the properties, methods and events

Here are the steps that you must follow to create a new Bean:

1. Create a directory for the new Bean.
2. Create the Java source file(s).
3. Compile the source file(s).
4. Create a manifest file.
5. Generate a JAR file.
6. Start the BDK.
7. Test.

12.7. USING BEANS TO BUILD AN APPLICATION

Following components are required to build an application using Java Beans.

12.7.1. Properties and Properties Design Patterns

There are two types of methods for the properties. One is to the set the value of the property, called the ***mutator*** or setter method, and other is to get or retrieve the value of the property, called ***accessor*** or getter method. But, there is a naming convention for the method names. For the setter method, 'set' is used as prefix to identifier, similarly for getter method 'get' is used as prefix to identifier, and the name of property will be added

For Example:

Property name is 'height'. The property name follows the prefix, and its first character will be in uppercase.

```
void setHeight(int height){}
int getHeight(){}
```

Set method will receive the value and assign to property, while get method returns the

Simple Properties: A property has only value, and independent to change. If any property value gets change, then there is no affect on other.

To add simple properties to a bean, add appropriate get*XXX* and set*XXX* methods (or is*XXX* and set*XXX* methods for a boolean property).

The names of these methods follow specific rules called *design patterns*. These design pattern-based method names allow builder tools such as the NetBeans GUI Builder, to provide the following features:

- Discover a bean's properties
- Determine the properties' read/write attributes
- Determine the properties' types
- Locate the appropriate property editor for each property type
- Display the properties (usually in the Properties window)
- Alter the properties (at design time)

Example:

```
public class MyFirstBean {
/*Property caption.*/
private String caption;
/* Constructor of MyFirsBean */
public MyFirstBean() {
}
/*Getter for property caption.
and return Value of property caption.
*/
public String getCaption() {
return this.caption;
}
/*Setter for property caption.
and parameter caption New value of property caption.
*/
public void setTitle(String caption) {
this.title = title;
}
}
```

Bound Property: As the value of the property is changed then its notification will be send to some other bean.

Yes accessor methods are provided in the same manner. But here, you've to register the method for listener PropertyChangeListener class, and fire a PropertyChangeEvent, the propertyChange() method is called.

PropertyChangeSupport class offers your bean to implements these methods. Either you can inherit it or use it as inner class. PropertyChangeEvent class includeds property change details, and is sent from the property change event source to the object of property change listener with the propertyChange method.

Implementing Bound Property Support Within a Bean

To implement a bound property in your application, follow these steps:

1. To get the access to the PropertyChangeSupport class import the java.beans package.
2. Instantiate a PropertyChangeSupport object.
3. Implement methods to maintain the property change listener list.
4. Modify a property's set method to fire a property change event when the property is changed.

Example:

```
public class MyFirstBean extends JComponent implements Serializable
{
    private String caption;
    private final PropertyChangeSupport ob = new PropertyChangeSupport(
        this );
    public String getCaption()
    {
        return this.caption;
    }
    public void setCaption( String caption)
    {
        String oldcaption = this.caption;
        this.caption = caption;
        this.ob.firePropertyChange( "caption", oldcaption, caption );
    }
    public void addPropertyChangeListener( PropertyChangeListener listener
    )
    {
        this.ob.addPropertyChangeListener( listener );
    }
    public void removePropertyChangeListener( PropertyChangeListener
        listener )
    {
        this.ob.removePropertyChangeListener( listener );
    }
    protected void paintComponent( Graphics g )
    {
        g.setColor( getForeground() );
        int height = g.getFontMetrics().getHeight();
        paintString( g, this.caption, height );
    }
    private void paintString( Graphics g, String str, int height )
    {
        if ( str != null )
```

```
            g.drawString( str, 0, height );
        }
    }
```

Index properties: A bean property with more than one value. An indexed property is an array of properties or objects that supports a range of values.

```
Indexed properties are specified by the following methods:
    //Methods to access individual values
    public PropertyElement getPropertyName(int index)
    public void setPropertyName(int index, PropertyElement element)
and
    //Methods to access the entire indexed property array
    public PropertyElement[] getPropertyName()
    public void setPropertyName(PropertyElement element[])
```

Note that the distinction between the get and set methods for indexed properties is subtle. The getter method either has an argument that is the array index of the property, or returns an array. The setter method either has two arguments, namely an integer array index and the property element object that is being set, or has the entire array as an argument.

Example:

```
public class MyRectamgleBean {
private double coordinate[ ];
public double getCoordinate(int index) {
return cordinate[index];
}
public void setCoordinate(int index, double value) {
coordinate[index] = value;
}
public double[ ] getCoordinate( ) {
return data;
}
public void setData(double[ ] values) {
coordinate = new double[values.length];
System.arraycopy(values, 0, data, 0, values.length);
}
}
```

Constrained Properties: A bean property value will change on the validation of the property. The other bean may refuse to modify if it is not suitable. A bean property is *constrained* if the bean supports the VetoableChangeListener and PropertyChangeEvent classes, and if the set method for this property throws a PropertyVetoException. Constrained properties are more complicated than bound properties.

The following operations in the set*XXX()* method must be implemented in following order:

1. Save the old value in case the change is vetoed.
2. Notify listeners of the new proposed value, allowing them to veto the change.
3. If no listener vetoes the change, set the property to the new value.

The accessor methods for a constrained property are defined in the same way as simple properties, with the addition that the mutator method throws a PropertyVetoException exception. The syntax is as follows:

```
public void setPropertyName(PropertyType pt) throws PropertyVetoException
{
      . . .
      }
```

Example:

```
public class MyFirstBean extends JComponent implements Serializable
{
    private String caption;
        private final PropertyChangeSupport ob1 = new
                  PropertyChangeSupport( this );
      private final VetoableChangeSupport ob2 = new
              VetoableChangeSupport( this );
    public String getCaption()
    {
        return this.caption;
    }
        public void setCaption( String caption) throws
                                      PropertyVetoException

    {
            String old = this.title;
            this.ob2.fireVetoableChange( "caption", old, caption );
            this.caption = caption;
            this.ob1.firePropertyChange( "caption", old, caption );

    }
    public void addPropertyChangeListener( PropertyChangeListener listener )
     {
        this.ob1.addPropertyChangeListener( listener );
     }
     public void removePropertyChangeListener( PropertyChangeListener
listener )
     {
        this.ob1.removePropertyChangeListener( listener );
     }
   public void addVetoableChangeListener( VetoableChangeListener listener )
     {
        this.ob2.addVetoableChangeListener( listener );
     }
     public void removeVetoableChangeListener( VetoableChangeListener
        listener )
     {
        this.ob2.removeVetoableChangeListener( listener );
     }

     protected void paintComponent( Graphics g )
     {
```

```
            g.setColor( getForeground() );

            int height = g.getFontMetrics().getHeight();
            paintString( g, this.caption, height );
        }
        private void paintString( Graphics g, String str, int height )
        {
            if ( str != null )
                g.drawString( str, 0, height );
        }
    }
```

12.7.2. Events Design/Naming Patterns

Event *means the mechanism by which components communicate with each other. As any event fires, the corresponding component broadcasts the events. The notified components perform some action.*

To build a bean with event handling you must know the event, listener and source. The source of an event is the object that originates or fires the event. The source must define the events it will fire, as well as the methods for registering listeners of these events. A listener is an object that indicates that it is to be notified of events of a particular type. Listeners register for events using the methods defined by the sources of those events.

Creating a Bean Manifest File

A manifest file is an important file for beans creation. The file contains some information about the bean component. Bean component is a collection of classes, interfaces, images, etc. So, their paths, which class file is bean class, all these related information is stored in the manifest file. There are some rules to create it. The extension of the manifest file is ".mft".

Example:

```
Name: mybean/image1.gif
Name: mybean/image2.gif
Name: mybean/image3.gif
Name: mybean/image4.gif
Name: mybean/MyFirstBean.class
Java-Bean: True
```

Creating jar file

The jar file can be created using JAR utility. A utility is used to generate a JAR file. Its syntax is shown here:

```
jar options files
```

Option	Description
c	A new archive is to be created.
C	Change directories during command execution.
f	The first element in the file list is the name of the archive that is to be created or accessed.
i	Index information should be provided.
m	The second element in the file list is the name of the external manifest file.

M	Manifest file not created.
t	The archive contents should be tabulated.
u	Update existing JAR file.
v	Verbose output should be provided by the utility as it executes.
x	Files are to be extracted from the archive.
0	Do not use compression.

Example: The following command creates a JAR file named **XYZ.jar** that contains all of the **.class** and **.gif** files in the current directory:

```
jar cf XYZ.jar MyMFT.mft  *.class *.gif
```

Tabulating the Contents of a JAR File

The following command lists the contents of **XYZ.jar**:

```
jar tf Xyz.jar
```

Extracting Files from a JAR FileYZ

The following command extracts the contents of **XYZ.jar** and places those files in the current directory:

```
jar xf Xyz.jar
```

Updating an Existing JAR File

The following command adds the file **file1.class** to **XYZ.jar**:

```
jar -uf Xyz.jar file1.class
```

Recursing Directories

The following command adds all files below **directoryX** to **X.jar**:

```
jar -uf Xyz.jar -C directoryX *
```

12.8. BEANINFO CLASSES

Design patterns are in use to determine the information that is given to the bean user. But BeanInfo class gives control on this process as follows:

This interface defines several methods, including these:

```
PropertyDescriptor[ ] getPropertyDescriptors( )
EventSetDescriptor[ ] getEventSetDescriptors( )
MethodDescriptor[ ] getMethodDescriptors( )
```

Each corresponding methods returns the information about the properties, events, and methods of a given bean.

SimpleBeanInfo is a class that provides default implementations of the *BeanInfo* interface, including the three methods just shown. You may extend this class and override one or more of them. The class has to overrides *getPropertyDescriptors( )* in order to designate which properties are presented to a Bean user. This method creates a *PropertyDescriptor* object for the given property. The *PropertyDescriptor* constructor that is used is shown here:

```
PropertyDescriptor(String propertyName, Class beanClass) throws
   IntrospectionException
```

Example:

```
import java.beans.*;
public class MyFirstBeanInfo extends SimpleBeanInfo {
public PropertyDescriptor[] getPropertyDescriptors() {
try {
PropertyDescriptor captionDesc = new PropertyDescriptor("caption",
   MyFirstBean.class);
PropertyDescriptor pdesc[] = {captionDesc};
return pdesc;
}
catch(Exception e) {
}
return null;
}
}
```

You must compile this file MyFirstBean.java, and MyFirstBeanInfo.java. After this file is successfully compiled, the **myMFT.mft** file can be have these contents:

```
Name: mybean/MyFirstBeanInfo.class
Name: mybean/MyFirstBean.class
Java-Bean: True
Use the JAR utility to create jar file.
```

12.9. PROPERTY EDITORS

A property editor is a tool for customizing a particular property type. Property editors are activated in the Properties window. This window determines a property's type, searches for a relevant property editor, and displays the property's current value in a relevant way. Property editors must implement the PropertyEditor interface, which provides methods to specify how a property should be displayed in a property sheet. The following figure represents the Properties window containing myBean1 properties:

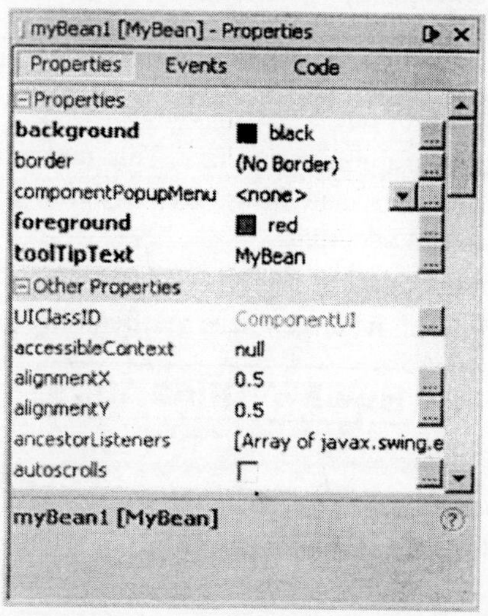

We begin the process of editing these properties by clicking the property entry. Clicking most of these entries will bring up separate panels. The support class PropertyEditorSupport provides a default implementation of the PropertyEditor interface. By subclassing property editor from PropertyEditorSupport, we can simply override the methods you need.

To support the custom property editor, override two more methods. Override supports CustomEditor to return true, and then override getCustomEditor to return a custom editor instance. ColorEditor.getCustomEditor returns this. In addition, the PropertyEditorSupport class maintains a PropertyChangeListener list, and fires property change event notifications to those listeners when a bound property is changed.

Property editors are discovered and associated with a given property in the following ways:

- Explicit association by way of a BeanInfo object. The editor of the title's property is set with the following line of code: pd.setPropertyEditorClass(TitleEditor.class);
- Explicit registration by way of the java.beans.PropertyEditorManager.registerEditor method. This method takes two arguments: the bean class type, and the editor class to be associated with that type.
- Name search. If a class has no explicitly associated property editor, then the PropertyEditorManager searchs for that class's property editor in the following ways:
  (*a*) Appending "Editor" to the fully qualified class name. For example, for the my.package.ComplexNumber class, the property editor manager would search for the my.package.ComplexNumberEditor class.
  (*b*) Appending "Editor" to the class name and searching a class path.

12.10. CUSTOMIZERS

Customization *provides a means for modifying the appearance and behavior of a bean within an application builder so it meets your specific needs. There are several levels of customization available for a bean developer to allow other developers to get maximum benefit from a bean's potential functionality. The JavaBeans specification provides for user-defined customizers, through which we can define a higher level of customization for bean properties than is available with property editors.*

When we use a bean *Customizer*, we have complete control over how to configure or edit a bean. A Customizer is an application that specifically targets a bean's customization. Sometimes properties are insufficient for representing a bean's configurable attributes. Customizers are used where sophisticated instructions would be needed to change a bean, and where property editors are too primitive to achieve bean customization.

All customizers must:

- Extend java.awt.Component or one of its subclasses.
- Implement the java.beans.Customizer interface This means implementing methods to register PropertyChangeListener objects, and firing property change events at those listeners when a change to the target bean has occurred.
- Implement a default constructor.
- Associate the customizer with its target class via BeanInfo.getBeanDescriptor.

REVIEW QUESTIONS

Q. 1. Explain the following:
(*a*) Bean Writing Process
(*b*) Naming Patterns for Bean Components **[MDU, May 2012]**

Ans. Refer 12.6 and 12.7(*b*)

Q. 2. What are Java Beans? Explain how Java Beans can be used for reusability of code? Explain with example. **[MDU, May 2011]**

Ans. *(Refer 12.1)*

Q. 3. What are beans? Explain the bean writing process in details with the help of an example. **[MDU, May 2007, Dec. 2007, May 2008, Dec. 2008, May 2009, Dec. 2009]**

Ans. Java Bean is a software component written in Java, provides portability, platform-independent component model, and reusability. You can add these components into applets, applications, or with other bean components. JavaBeans components are known as *beans*. By nature, beans are dynamic like other software components; these can be changed or customized. JavaBeans is a portable, platform-independent component model written in the Java programming language. The JavaBeans architecture was built through a collaborative industry effort and enables developers to write reusable components in the Java programming language.

Q. 4. Explain how to build an application using Beans.

Ans. The creation steps are given in summary as well and explained in different section. Take the steps and get the explanation from the section.

Q. 5. Difference between java bean & java class.

Ans. Java class may contain main method and can be directly executable, but java bean classes will never have main method. They are dependent to run on another application. It has be build with some another bean, and finally an application to run.

❑❑❑

13
CHAPTER

Security

JAVA applications are made of components that can be deployed into different containers. These components build a multilevel enterprise application. The goal of the JAVA security architecture is to achieve end-to-end security by securing each level. The levels can contain both protected and unprotected resources.

INSIDE THIS CHAPTER

POINTS TO REMEMBER

Authorization provides controlled access to protected resources which is based on identification and authentication. *Identification* is a process that enables recognition of an entity by a system, and *authentication* is a process that verifies the identity of a user, device, or other entity in a computer system.

Authorization provides controlled access to protected resources which is based on identification and authentication. *Identification* is a process that enables recognition of an entity by a system, and *authentication* is a process that verifies the identity of a user, device, or other entity in a computer system.

Authorization is not required to access unprotected resources. Since authorization is built upon authentication, authentication is also not needed to access unprotected resources.

DO YOU KNOW

Accessing a resource without authentication is referred to as *unauthenticated* or *anonymous* access.

Three mechanisms in Java technology help ensure safety:

1. Language design features that perform bound checking on arrays, legal type conversions, no pointer arithmetic.
2. An access control mechanism that controls what the code can do such as file access, network access.
3. Code signing, where code authors can use standard cryptographic algorithms to authenticate Java programming language code. JVM checks bad pointers and invalid array offsets.

The Fig. 13.1 below shows the core Java security architecture.

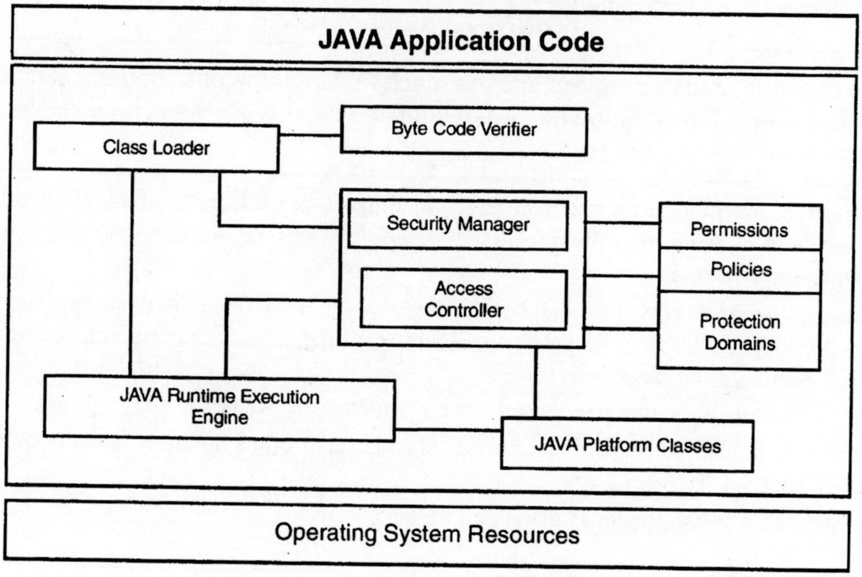

Fig. 13.1. Core Java Security Architecture

The security architecture for core Java consists of various components such as bytecode verifier, class loader, security manager, access controller, permissions, policies, and protection domains. The functions and working of these components are discussed in the proceeding sections.

When java first came into existence the excitement was not about a well-crafted programming language but also about the possibility of safely executing applets that are delivered over internet. Obviously, delivering executable is applets is practical only when the recipients are sure that the code can't wreak havoc on their machines. For this reason security was the major concern of both designers and the users of java technology. This means in java security mechanisms are implemented as an integral part of java technology unlike the case with other languages and systems where security was implemented as an afterthought or a reaction to break-ins.

Three mechanisms help ensure safety:

- Language design features (bounds checking on arrays, legal type conversions only, no pointer arithmetic).
- An access control mechanism that controls what the code can do (such as file access, network access).
- Code signing, whereby code authors can use standard cryptographic algorithms to authenticate java code. Then, the users of the code can determine exactly who created the code & whether the code has been altered after it was signed.

When class files are loaded into the virtual machine, they are checked for integrity. For maximum security, both the default mechanism for loading a class and a custom class need to work with a *security manager* class that controls what actions code can perform.

13.1. CLASS LOADERS

These class files need to be interpreted by a program that can translate the instruction set of the virtual machine into the machine language of the target machine. The virtual machine interpreter loads only those class files that are needed for the execution of a program. For example, suppose program execution starts with *NewProgram.class*. The steps that the virtual machine carries out are:

POINTS TO REMEMBER

JVM converts source program into the machine language of a theoretical machine, called the *virtual machine*. The virtual machine code is stored in a class file with a *.class* extension.

1. The virtual machine has a mechanism for loading class files, for example, by reading the files from disk or by requesting them from the Web. It uses this mechanism to load the contents of the *NewProgram* class file.
2. If the *NewProgram* class has instance variables or superclasses of another class type, these class files are loaded as well. The process of loading all the classes that a given class depends on is called *resolving* the class.
3. The virtual machine then executes the main() method in *NewProgram*.
4. If the main() method requires additional classes, then other methods that the main() calls are loaded next.

The architecture of class loader is discussed below:

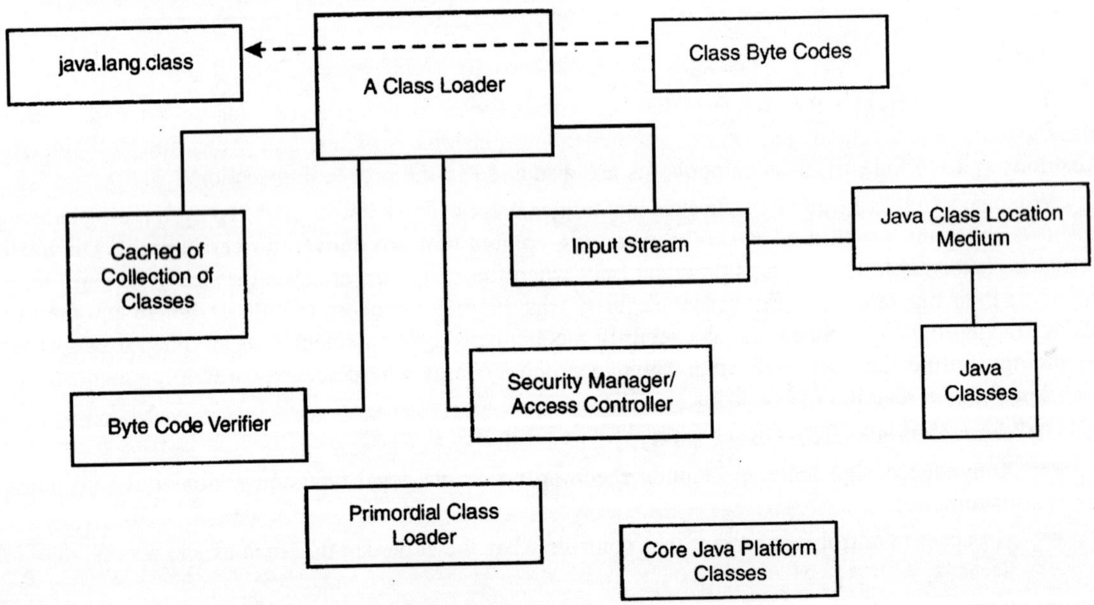

Fig. 13.2. Class Loader Architecture

The above figure shows the basic class loader architecture assumed by Java security model. When a class is loaded from an input stream, the class loader first checks its cached collection of classes to check if the class has already been loaded. If it is already loaded, the class loader can return without further processing. The security manager and access controller can then be consulted to verify if the access to that particular class was permitted. Then, the primordial class loader determines whether it can load that class. If the class is not loaded by the primordial class loader, the class is then read by the input stream. The bytes then read from the input stream are then passed to the *bytecode* verifier. If successful *bytecode* verification is done, then, a *java.lang.Class* object is created that describes the class to the virtual machine. Again the security manager is consulted if the class is permitted. Finally, if the class is loaded, the class is then cached with its collection of other loaded classes.

The class loading mechanism uses at least three class loaders:

- **The bootstrap class loader:** The bootstrap class loader loads the system classes. It is an integral part of the virtual machine and is usually implemented in C. There is no *ClassLoader* object corresponding to the bootstrap class loader. For example, *String.class.getClassLoader()* returns null.
- **The extension class loader:** The extension class loader loads a standard extension from the jre/lib/ext directory. The extension class loader is implemented in Java.
- **The system class loader:** The system class loader loads the application classes. It locates classes in the directories. This class loader is also implemented in Java.

Class loaders have a *parent/child* relationship. Every class loader except for the bootstrap class loader has a parent class loader. Given a class loader, you load a new class as follows:

```
ClassLoader loader = new MyClassLoader();
String className = ". . .";
Class cl = loader.loadClass(className);
```

POINTS TO REMEMBER

To write your own class loader, you simply extend the *ClassLoader* class and override the method. To implement this method, we need to load the bytecodes for the class from the local file system or from some other source. Then, call the *defineClass* method of the *ClassLoader* superclass to present the bytecodes to the virtual machine.

13.2. BYTECODE VERIFICATION

When a class loader presents the *bytecodes* of a newly loaded Java platform class to the virtual machine, the *bytecodes* are first inspected by a *verifier.* The *verifier* checks that the instructions cannot perform actions that are damaging. All classes except system classes are verified. We can deactivate verification with the undocumented—*noverify* option. For example,

DO YOU KNOW

The *verifier* checks that the instructions cannot perform actions that are damaging.

```
java -noverify NewProgram
```

POINTS TO REMEMBER

The *bytecode verifier* traverses the *bytecodes*, constructs the type state information, and verifies the types of the parameters to all the *bytecode* instructions.

The figure below shows the flow of data and control from the source code to the class loader and *bytecode* verifier and finally on to the Java virtual machine, which contains the interpreter and runtime system. Java class loader and the *bytecode* verifier make no assumptions about the primary source of the bytecode stream. The *bytecode verifier* acts gatekeeper as it ensures that code passed to the Java interpreter is in a good state to be executed and can run without the fear of breaking the Java interpreter. Imported code is not allowed to execute by any means until after it has passed the verifier's tests.

Once the verifier is done, a number of important properties are known. Knowing these properties makes the Java interpreter much faster, because it doesn't have to check anything. There are no operand type checks and no stack overflow checks. The interpreter can thus function at full speed without compromising reliability. The properties are:

- There are no operand stack overflows or underflows
- The types of the parameters of all bytecode instructions are known to always be correct
- Object field accesses are known to be legal—private, public, or protected

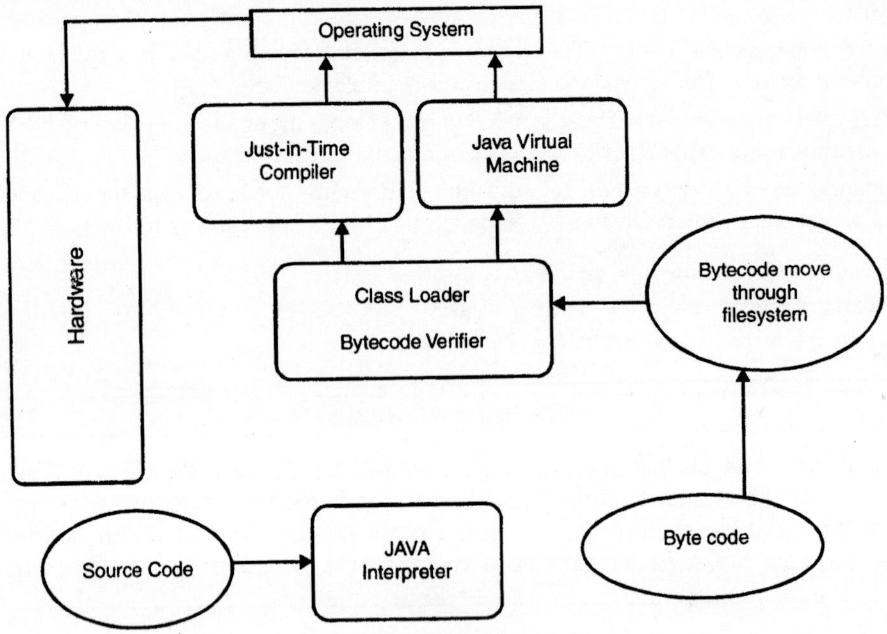

Fig. 13.3. Flow of Data and Control

POINTS TO REMEMBER

The verifier permits only some bytecode sequences in valid programs and the verifier ensures that any given instruction operates on a fixed stack location that allows the JIT compiler to transform stack accesses into fixed register accesses.

Due to this, the JVM does not imply a speed penalty for emulation on register-based architectures when using a JIT compiler. In the face of the code-verified JVM architecture, it makes no difference to a JIT compiler whether it gets named imaginary registers or imaginary stack positions that need to be allocated to the target architecture's registers.

POINTS TO REMEMBER

Code verification makes the JVM different from a classic stack architecture whose efficient emulation with a JIT compiler is more complicated and is carried out by a slow interpreter.

The following verification checks are performed by the bytecode verifier. It verifies that

- the byte stream is neither truncated nor padded with extra bytes.
- the class bytes adhere to the correct format for the class itself, methods, fields and other components.
- there are no overflows of operands on the stack.
- there are no underflows of operands on the stack.
- the proper access specifiers of methods and field variables is defined.

- final classes are not subclassed.
- final methods are not overridden.
- final variables are assigned an initial value and not modified.
- methods are invoked with the correct number and types of arguments.
- field variables of the class are assigned with values of the correct type.
- no local variable is accessed before it has an assigned value.
- the class has one superclass.

13.3. SECURITY MANAGERS AND PERMISSIONS

POINTS TO REMEMBER

Once a class has been loaded into the virtual machine by a class loader or by the default class loading mechanism and checked by the verifier, the *security manager* starts its function.

A security manager is a class that checks if a specific operation is permitted. Various operations checked by a security manager include:

- if the current thread can create a new class loader
- if the current thread can create a subprocess
- if the current thread can halt the virtual machine
- if the current thread can load a dynamic link library
- if a class can access a member of another class
- if the current thread can access a specified package
- if the current thread can define classes in a specified package
- if the current thread can access or modify system properties
- if the current thread can read from or write to a specified file
- if the current thread can delete a specified file
- if the current thread can accept a socket connection from a specified host and port number
- if the current thread can open a socket connection to the specified host and port number
- if the current thread can wait for a connection request on a specified local port number
- if the current thread can use IP multicast
- if the current thread can invoke a *stop, suspend, resume, destroy, setPriority/setMaxPriority, setName, or setDaemon* method of a given thread or thread group
- if the current thread can set a socket or stream handler factory
- if a class can start a print job
- if a class can access the system clipboard
- if a class can access the AWT event queue
- if current thread is trusted to bring up a top-level window.

Once a class has been loaded into virtual machine by a class loader or by the default class loading mechanism and checked by the verifier, the third security mechanism of the Java platform springs into action: the **security manager**. **A security manager is a class that controls whether a specific operation is permitted**. Other operations checked by a security manager include the following:

- Whether the current thread can create a new class loader

- Whether the current thread can halt the virtual machine
- Whether the current thread can open a socket connection to an external host
- Whether a class can start a print job
- Whether a class can access the system clipboard.
- Whether a class can access the AWT event queue
- Whether the current thread is trusted to bring up a top-level window.

For example, applets are not allowed to exit the virtual machine. If they try calling the exit method, then a security exception is thrown. Here is the code of the exit method:

```
public void exit(int status)
{
  SecurityManager security = System.getSecurityManager( );
  if (security != null)
  security.checkExit(status);
  exitInternal (status);
}
```

If the security manager agrees with the exit request, then the checkExit method simply returns and normal processing continues. However, if the security manager doesn't want to grant the request, the checkExit method throws a SecurityException.

When you run a Java application, the default is that no security manager is running. Your program can install a specific security manager by a call to the static setSecurityManager method in the System class. Once your program installs a security manager, any attempt to install a second security manager succeeds only if the first security manager agrees to be replaced .Thus although it is possible to have multiple class loaders, a program in the Java programming language can be governed by only one security manager.

Local classes had full permissions, and remote classes were confined to the *sandbox*. Just like a child that can only play in a sandbox, remote was only allowed to paint on the screen and interact with the user. The applet security manager denied all access to a local resource. A security policy maps code sources to permission *sets*. (See Fig. 13.4 below)

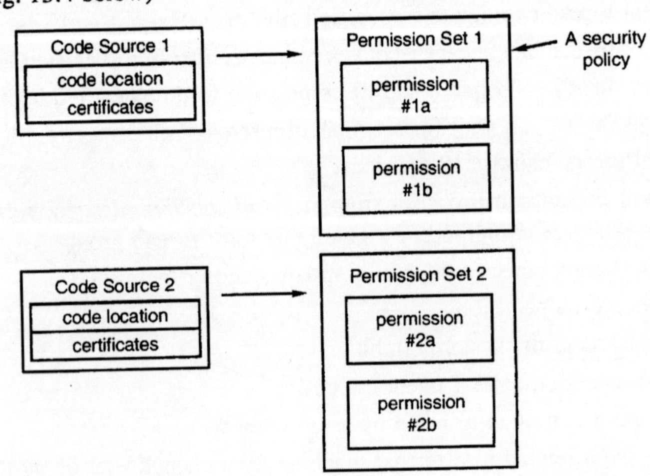

Fig. 13.4.

. A source code has two properties: the **code location** (for eg, an HTTP URL for remote code, or a file URL for local in a JAR file) & **certificates**. Certificates implies that how code can be certified by trusted parties.

A permission is any property that is checked by a security manager. For example,

```
Filepermission p = new FilePermission("/tmp/* ", "read, write");
```

The above instance of the **FilePermission** class states that it is ok to read and write any file in the /tmp directory. Fig below shows the hierarchy of permission classes that are supplied with jdk 1.2

No security manager is installed by default. The security manager checks if the *exit* request came from the browser or an individual applet. If the security manager agrees with the *exit* request, then the *checkExit* method simply returns, and normal processing continues. If the security manager doesn't want to grant the request, the *checkExit* method throws a *SecurityException.*

The *exit* method continues only if no exception occurred. Any code that attempts to exit the virtual machine must go through the *exit* method and through the *checkExit* security check without triggering a security exception. The default security manager of the Java allows both programmers and system administrators fine-grained control over individual security permissions. When the SecurityManager needs to check a permission, it looks at the classes of all methods currently on the call. It then gets the protection domains of all classes and asks each protection domain if its permission collection allows the operation that is currently being checked. If all domains agree, then the check passes. Otherwise, a SecurityException is thrown. Figure 13.5 depicts the permission hierarchy in JDK 1.2.

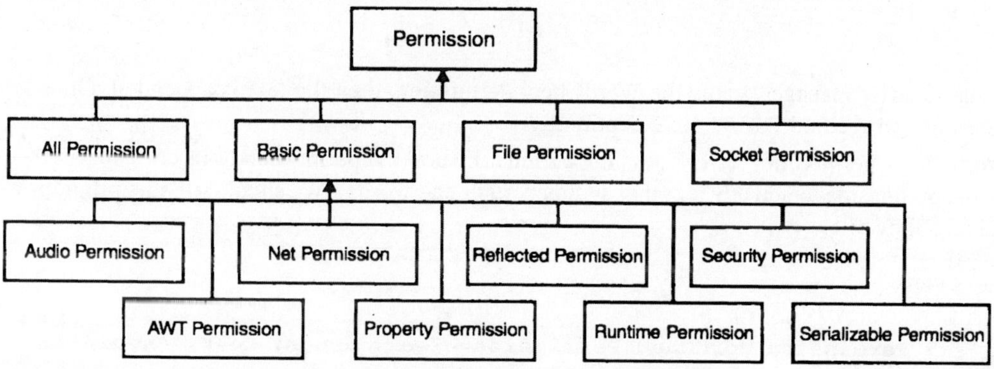

Fig. 13.5. Permission hierarchy in JDK 1.2

Custom Permissions

In this section, we discuss how we can supply our own permission class that users can refer to in their policy files. To implement your permission class, you extend the Permission class and supply the following methods:

- A constructor with two String parameters, for the target and the action list .
- String getActions()
- boolean equals()
- int hashCode()
- boolean implies(Permission other)

Permissions have an ordering, in which more general permissions reply more specific ones. Consider the file permission

```
p1=new FilePermission("/tmp/-","read/write");
```

This permission allows reading & writing of any file in the /tmp directory & any of its subdirectories.

A file permission p1 implies another file permission p2 if

(1) The target file set of p1 contains the target file set of p2.

(2) The action set of p1 contains the action set of p2.

Implementation of a Permission Class

Here we implement a new permission for monitoring the insertion of text into a text area. The program ensures that you cannot add "bad words" such as sex, drugs, and C++ into a text area. We use a custom permission class so that the list of bad words can be supplied in a policy file.

The following subclass of JTextArea asks the security manager whether it is okay to add new text.

```
Class WordCheckTextArea extends JTextArea
{
    Public void append(String text)
    {
        WordCheckPermission p= new WordCheckPermission(text, "insert");
        SecurityManager manager=System.getSecurityManager();
        If(manager! =null) manager.checkPermission(p);
        Super.append(text);
    }
}
```

If the security manager grants the WordCheckPermission, then the text is appended. Otherwise, the checkPermission method throws an exception.

Word check permissions have 2 possible actions: **insert** (the permission to insert a specific text) and **avoid** (the permission to add any text that avoids certain bad words). We should run this program with the following policy file:

```
grant
{
    Permission WordCheckPermission= "sex, drugs, C++", "avoid";
}
```

When designing the WordCheckPermission class, we must pay particular attention to the implies method. Here are the rules that control whether permission p1 implies permission p2.

(1) If p1 has action avoid and p2 has action insert, then the target of p2 must avoid all words in p1. for example, the permission

```
WordCheckPermission "sex, drugs, C++", "avoid"
```

Implies the permission

```
WordCheckPermission "Mary had a little lamb ","insert"
```

(2) If p1 and p2 both have action avoid, then the word set of p2 must contain all words in the word set of p1. for example, the permission

```
WordCheckPermission "sex, drugs", "avoid"
```

Implies the permission

```
WordCheckPermission "sex, drugs, C++ ","avoid"
```

(3) If p1 and p2 both have action insert, then the text of p1 must contain the text of p2. for example, the permission

```
WordCheckPermission "Mary had a little lamb ","insert"
```

Implies the permission

```
WordCheckPermission "a little lamb ","insert"
```

A Custom Security Manager

We monitor file access by overriding the checkPermission method of the standard security manager class. If the permission isn't a file read permission, then we simply call super.check-Permission. To check that it is permissible to read from a file, we open the file and scan its contents. We grant access to the file only when it doesn't contain any of the forbidden words.

```
public class WordCheckSecurityManager extends SecurityManager
{
    public void checkPermission(Permission p)
     {
     if (p instanceof FilePermission && p.getActions().equals("read"))
     {
        String fileName = p.getName();
                if (containsBadWords(fileName))
                throw new SecurityException("Bad words in " + fileName);
                 }
                else super.checkPermission(p);
    }
    . . .
}
```

There is just one catch in our file check scenario. Consider one possible flow of events

- A method of some class opens a file.
- Then, the security manager springs into action and uses its checkPermission method.
- The checkPermission method calls the containsBadWords method.

But the containsBadWords must itself read the file to check its contents, which calls the security manager again. This would result in an infinite regression unless the security manager has a way of finding out in which context it was called. The getClassContext method is the way to find out how the method was called. This method returns an array of class objects that gives all the classes whose calls are currently pending.

The class at index 0 gives the currently executing call, but you only get to see the classes, not the names of the pending methods.

13.4. DIGITAL SIGNATURES

Digital signature is an important cryptographic primitives used for authentication, authorization and non-repudiation.

POINTS TO REMEMBER

An asymmetric encryption algorithm such as RSA can be used to create and verify digital signature.

The simplest form of the protocol works as follows:

1. Sender encrypts the document with its private key, thereby singing the document

2. Sender sends the signed document to receiver.
3. Receiver deciphers the document with Sender's public key, thereby verifying the signature.

POINTS TO REMEMBER

The strength of the digital signature lies with the fact that although the public-private key pair for asymmetric encryption is mathematically related, it is computationally infeasible to derive the private key from the corresponding public key.

A digital signature must meet the following two properties:

1. **It must be unforgeable.** If an entity sings a document M with signature S(M), it is not possible for other entity to produce the same pair <M, S(M)>
2. **It must be authentic.** If someone R receives a digital signature from S, R must be able to verify that the signature is really from S.

In reality digital signature creation and verification are performed using the combination of hash function and asymmetric encryption.

POINTS TO REMEMBER

To create a digital signature the sender first computes the message authentication code (MAC) or hash of the original message and append the code with the message.

Then the hash code is encrypted using asymmetric encryption. On the reception end the receiver uses the same hash algorithm to compute the hash code of the message decrypts the encrypted message using the corresponding public key and compares the hash value.

If the message is altered, then the fingerprint of the altered message will not match the fingerprint of the original message. If the message and its fingerprint are delivered separately, then the recipient can check whether the message has been tampered or not. Public key cryptography is based on the notion of a *public* key and *private* key. Even though everyone knows your public key, they can't compute your private key in your lifetime, no matter how many computing resources they have available. It may seem difficult to believe that nobody can compute the private key from the public keys, but nobody has ever found an algorithm to do this for the encryption algorithms. If the keys are sufficiently long, brute force would require more computations. There are two kinds of public/private key pairs: for *encryption* and for *authentication.* If anyone sends you a message that was encrypted with your public encryption key, then you can decrypt it with your private decryption key. The verification passes only for messages that you signed, and it fails if anyone else used his key to sign the message.

Message Digests

A message digest is a digital fingerprint of a block of data. A message digest has two essential properties:

1. If one bit or several bits of the data are changed, then the message digest also changes.
2. A forger who is in possession of a given message cannot construct a fake message that has the same message digest as the original.

A number of algorithms have been designed to compute these message digests: The two best-known are SHA1, the secure hash algorithm developed by the National Institute of Standards and Technology, and

MD5, an algorithm invented by Ronald Rivest of MIT. Both algorithms scramble the bits of a message in ingenious ways.

The Java programming language implements both SHA1 and MD5. The MessageDigest class is a factory for creating objects that encapsulate the fingerprinting algorithms. It has a static method, called getInstance, that returns an object of a class that extends the MessageDigest class.

This means the MessageDigest class serves double duty:

- As a factory class
- As the superclass for all message digest algorithms.

For example, here is how you obtain an object that can compute SHA fingerprints.

```
MessageDigest alg = MessageDigest.getInstance("SHA-1");
```

Message Signing

Digital signatures can authenticate a message. When a message is authenticated, you know:

- The message was not altered.
- The message came from the claimed sender.

How digital signatures work

Public key cryptography is based on the notion of a **public key** and **private key**. The idea is that you tell everyone in the world your public key. However, only you hold the private key, and it is important that you safeguard it and don't release it to anyone else. The keys are matched by mathematical relationships, but it is believed to be practically impossible to compute one from the other. That is, even though everyone knows your public key, they can't compute your private key in your lifetime, no matter how many computing resources they have available.

There are two kinds of public/private key pairs: for **encryption** and for **authentication**.

If anyone sends you a message that was encrypted with your public encryption key, then you can decrypt it with your private decryption key, but nobody else can. Conversely, if you sign a message with your private authentication key, then anyone else can verify the signature by checking with your public key. The verification passes only for messages that you signed, and it fails if anyone else used his or her key to sign the message.

Many cryptographic algorithms, such as RSA and DSA (the Digital Signature Algorithm), use this idea. The exact structure of the keys and what it means for them to match depend on the algorithm. For For example,

Suppose Alice wants to send Bob a message, and Bob wants to know this message came from Alice and not an impostor. Alice writes the message and then *signs* the message digest with her private key. Bob gets a copy of her public key. Bob then applies the public key to *verify* the signature. If the verification passes, then Bob can be assured of two facts:

1. The original message has not been altered.
2. The message was signed by Alice, the holder of the private key that matches the public key that Bob used for verification.

If someone steals Alice's private key or if a government can require her to turn it over, then she is in trouble. The thief or a government agent can impersonate her by sending messages, money transfer instructions, and so on, that others will believe came from Alice.

Let us put the DSA algorithm to work. There are three algorithms:

1. To generate a key pair.
2. To sign a message.
3. To verify a signature.

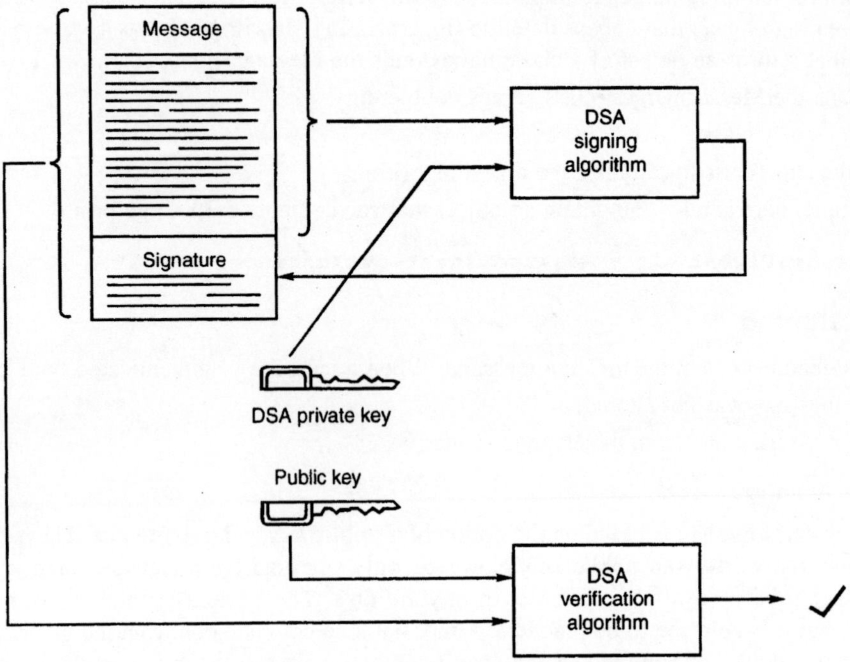

Fig. 13.6. Public key signature exchange with DSA

Of course, you generate a key pair only once and then use it for signing and verifying many messages. To generate a new random key pair, make sure you use truly random numbers.

Message Authentication

Suppose you get a message from a stranger who claims to represent a famous software company, urging you to run the program that is attached to the message.

The stranger even sends you a copy of his public key so you can verify that he authored the message. You check that the signature is valid. This proves that the message was signed with the matching private key and that it has not been corrupted.

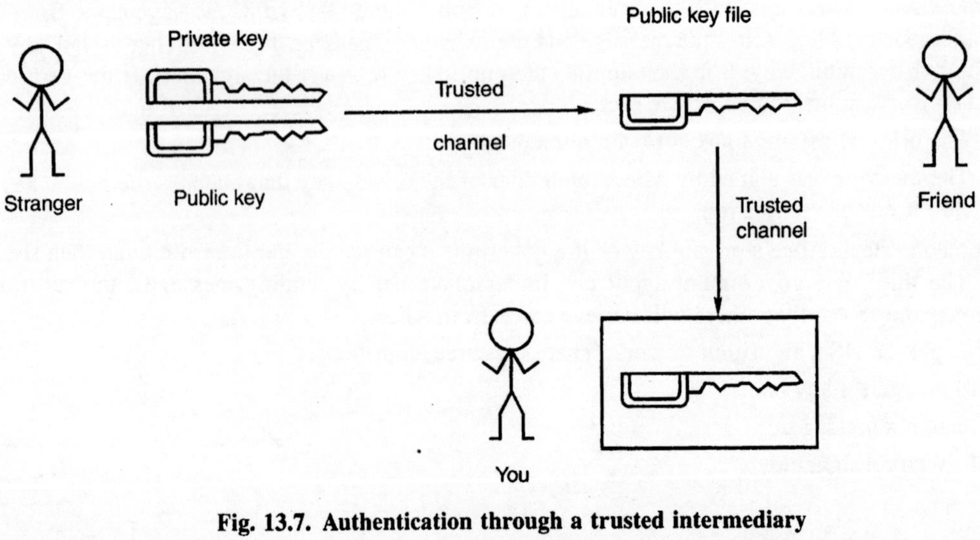

Fig. 13.7. Authentication through a trusted intermediary

Anyone could have generated a pair of public and private keys, signed the message with the private key, and sent the signed message and the public key to you. The problem of determining the identity of the sender is called the **authentication problem.**

The usual way to solve the authentication problem is simple. Suppose the stranger and you have a common acquaintance you both trust. Suppose the stranger meets your acquaintance in person and hands over a disk with the public key. Your acquaintance later meets you, assures you that he met the stranger and that the stranger indeed works for the famous software company, and then gives you the disk. (see fig below) That way, your acquaintance vouches for the authenticity of the stranger.

Infact, your acquaintance does not actually need to meet you. Instead, he can apply his private signature to the stranger's public key file (see fig below)

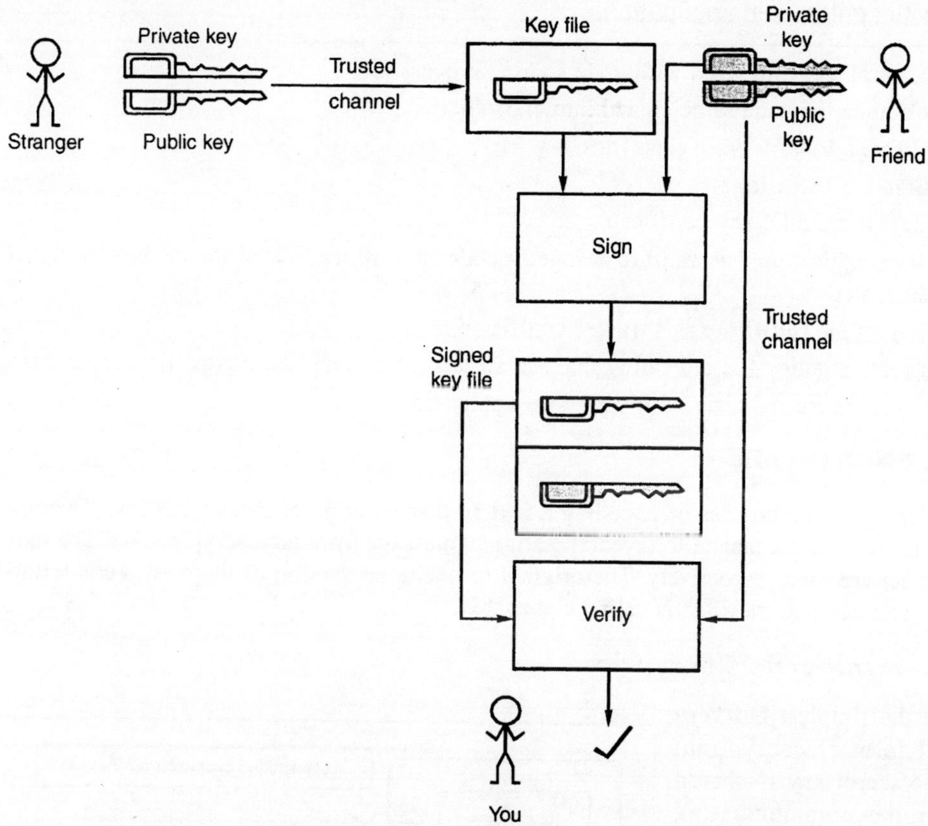

Fig. 13.8. Authentication through a trusted intermediary's signature

When you get the public key file, you verify the signature of your acquaintance, and because you trust him, you are confident that he did check the stranger's credentials before applying his signature.

You will often encounter digital signatures that are signed by one or more entities who will vouch for the authenticity, and you will need to evaluate to what degree you trust the authenticators.

13.5. CODE SIGNING

One of the most important uses of authentication technology is signing executable programs. If a program is downloaded, then we are concerned about damage that a program can do. For example, the program could have been infected by a virus. If we know where the

> **DO YOU KNOW**
>
> The most common use of code signing is to provide security when deploying.

code comes from and that it has not been tampered then we can be comfortable in using the program. For this, we use authentication to verify where the code came from and then run the code with a security policy that enforces the permissions that you want to grant the program, depending on its origin. Almost every code signing implementation will provides some sort of digital signature mechanism to verify the identity of the author or build system, and a checksum to verify that the object has not been modified. Code signing is done on JAR files.

POINTS TO REMEMBER

The Java Archive (JAR) file format enables you to pile multiple files into a single archive file. A JAR file contains the class files and auxiliary resources associated with applets and applications.

The general steps involved with code signing process are:

1. Compile the code to be signed into class files.
2. Create a JAR file from class files.
3. Sign the JAR file.
4. Export a public key certificate.

The Java application that wants to accept the code only from a trusted source that has signed the code follows these steps:

1. Import the certificate as a trusted certificate.
2. Create a policy file indicating that such a principal should be trusted for certain permissions.
3. Load the trusted code with your Java application.

13.6. ENCRYPTION

Encryption is the process of encoding a text so that its original meaning is lost. Decryption is the opposite process, a mechanism to reveal the original message from the encrypted one. The term encipher and decipher are used respectively. The original or unaltered version of the message is termed as plain text and the encrypted message is called cipher text.

13.6.1. Symmetric Encryption

It is the simplest but very efficient form of encryption. Here one secret key is shared between the communicating parties say sender and receiver with encryption and decryption procedures being mirror images of each other. Two parties communicating with symmetric encryption can be explained with the following Fig. 13.9.

The most challenging task in symmetric encryption is to distribute and manage the shared secret Key. DES is an example of symmetric encryption.

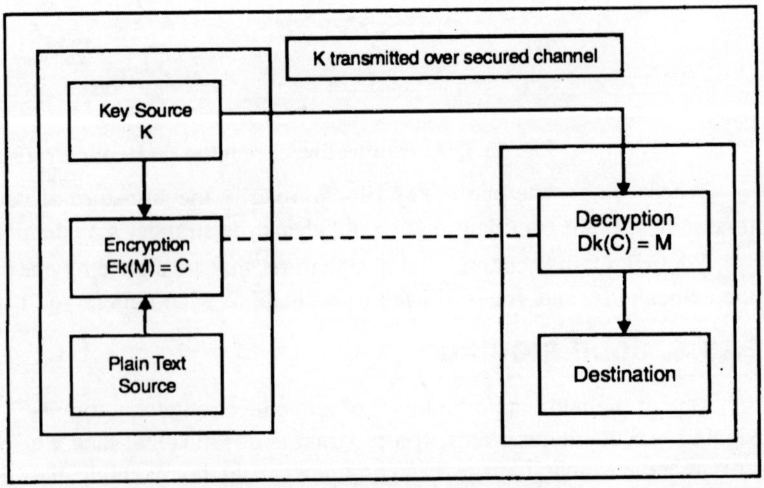

Fig. 13.9. Symmetric Encryption

13.6.2. Asymmetric Encryption/Public Key Encryption

Unlike the symmetric encryption it uses two separate keys for encryption and decryption called private-public key pair. The sender encrypts the message with his private key. Prior to this operation sender must send its corresponding public key to the receiver. On receiving the encrypted text, the public key is used to decipher the original plain text.

POINTS TO REMEMBER

The major advantage in asymmetric encryption lies with the fact that it incurs quite high computational expense for an attacker.

But on the other hand its application is limited where both security and efficiency are desired. RSA is an example of public key encryption.

The problem that lies with asymmetric encryption, is that it demands a huge number of key pairs for a large network.

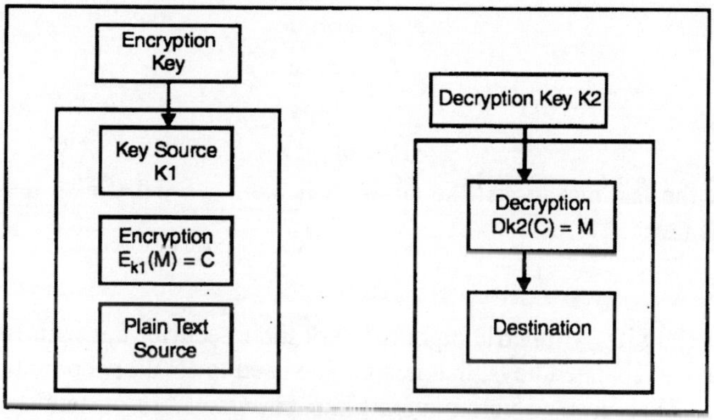

Fig. 13.10. Asymmetric Encryption

13.6.3. Cryptographic Hash Functions

For secure communication it is required that data transmitted is not altered by any entity. Hash functions are the security primitives that ensure data integrity.

POINTS TO REMEMBER

Hash function is often called one-way hash function, because it is quite difficult to compute the inverse function.

For example, consider the cube function $y = x^3$, it is quite easy to compute y given x. But the inverse function, $x = \sqrt[3]{y}$ is much complicated to compute.

The most common use of hash function is as providing digital signatures for entity authentication. With digital signature hash function is applied to the whole message. Then the hashed value is signed. On receiving, the hash value is recomputed and verified that the received signature is unaltered and is from the original source. It saves both time and space as only the hashed value is signed instead of the whole message.

For integrity of data, sender computes the hashed value over the data and sends it along with the original message to the receiver. The destination entity recomputed the hash value from the transmitted message and compares with the hashed value (transmitted). The most common hash functions are MD5 (Message Digest 5) and SHA (Secure Hash Algorithm).

Shift Cipher

Shift cipher encrypts by shifting each plaintext letter by amount (the key) known only by the sender and the authorized recipient. Characters shifted off the end are wrapped around.

For example, shift the plaintext **my dog has fleas** by 2 letters to become the ciphertext as follows :

```
OA FQI  JCU  HNGCU
```

In above example, **y** got pushed off the end and then wrapped back to become **a**.

Shift Cipher Attack

Shift cipher is bad because it has only 25 possible keys. An attacker can easily do the decryption operation with all the 25 possible keys and choose the decryption which makes sense, and since there are 25 possible keys the average steps for the attacker to decrypt a message is 13. For example, if an attacker wants to decrypt the following message **YUNWCHXOBDWBQRWNJUUMJH**. He or she needs to chop off the first few letters of the ciphertext and shift backward until the attacker sees a message that makes sense.

1. yunwch
2. xtmvbg
3. wsluaf
4. vrktze
5. uqjsyd
6. tpirxc
7. sohqwb
8. rngpva
9. qmfouz
10. plenty

Since we see that the last one looks like an english word we will finish decrypting with key = 9. **plenty of sun shine all day**

Public-key Ciphers

The primary problem with symmetric ciphers is not their security but with key exchange. Once the sender and receiver have exchanged keys, that key can be used to securely communicate, but what secure communication channel was used to communicate the key itself? In particular, it would probably be much easier for an attacker to work to intercept the key than it is to try all the keys in the key space. Another problem is the number of keys needed. If there are *n* people who need to communicate, then n(n-1)/2 keys are needed for each pair of people to communicate privately. This may be ok for a small number of people but quickly becomes unwieldly for large groups of people.

Public-key ciphers were invented to avoid the key-exchange problem entirely. A public-key cipher uses a pair of keys for sending messages. The two keys belong to the person receiving the message. One key is a **public key** and may be given to anybody. The other key is a **private key** and is kept secret by the owner. A sender encrypts a message using the public key and once encrypted, only the private key may be used to decrypt it.

This protocol solves the key-exchange problem inherent with symmetric ciphers. There is no need for the sender and receiver to agree upon a key. All that is required is that some time before secret communication the sender gets a copy of the receiver's public key. Furthermore, the one public key can be used by anybody wishing to communicate with the receiver. So only *n* keypairs are needed for *n* people to communicate secretly with one another,

Public-key ciphers are based on one-way trapdoor functions. A one-way function is a function that is easy to compute, but the inverse is hard to compute. For example, it is easy to multiply two prime numbers together to get a composite, but it is difficult to factor a composite into its prime components. A one-way trapdoor function is similar, but it has a trapdoor. That is, if some piece of information is known, it becomes easy to compute the inverse. For example, if you have a number made of two prime factors, then knowing one of the factors makes it easy to compute the second. Given a public-key cipher based on prime factorization, the public key contains a composite number made from two large prime factors, and the encryption algorithm uses that composite to encrypt the message. The algorithm to decrypt the message requires knowing the prime factors, so decryption is easy if you have the private key containing one of the factors but extremely difficult if you do not have it.

As with good symmetric ciphers, with a good public-key cipher all of the security rests with the key. Therefore, key size is a measure of the system's security, but one cannot compare the size of a symmetric cipher key and a public-key cipher key as a measure of their relative security.

REVIEW QUESTIONS

Q. 1. Does Java has security constraints? Explain Class Loaders & Digital Signatures. **[MDU, May 2011]**

Ans. Yes, java has security constraints. The security architecture for core Java consists of various components such as bytecode verifier, class loader, security manager, access controller, permissions, policies, and protection domains.

Q. 2. Write short notes on Class Loaders. **[MDU, May 2007, May 2009]**

Ans. Class loader loads the program into memory. In Java there are three types of class loader

1. bootstrap class loader
2. extension class loader
3. system class loader

Bootstrap class loader loads the system classes. It is an integral part of virtual machine. The extension class loader loads standard extension from the jre/lib/ext directory. The system class loader loads the application classes.

Q. 3. Write short notes on Digital Signature. **[MDU, May 2007, Dec. 2007, May 2009]**

Ans. Digital signature is a signature used in email or in any internet application. It is created by Message digests. A message digest is a digital finger-prints of a block of data. A message digest has two essential properties:

1. If one bit or several bits of the data are changed, then message digest also change.
2. A forger who is possession of a given message cannot construct a fake message that has the same message digest as the original.

Q. 4. Write short notes on Byte code verification. **[MDU, May 2012, May 2007, Dec. 2007, Dec. 2009]**

Ans. When a class loader presents the byte codes of a newly loaded java platform class to the virtual machine, these byte codes are first inspected by a verifier. The verifier checks that the instructions cannot perform actions that are obviously damaging. All classes excepts system classes are verified. We can deactivate verification by—noverify option. For example Java -noverify Hello.

Q. 5. Write short notes on Code Signing and Encryption. **[MDU, May 2012, May 2008, Dec. 2008]**

Ans. Encryption is a process of converting original message ito a coded form called cipher text, that is used to send through communication media and in reverse side decryption process occurs. Encryption is a part of cryptography. It is user to secure the data. Ciphers used in encryption can be divided into two types:

1. Symmetric ciphers
2. Public Key Ciphers.

Q. 6. Write short notes on Security Manager and permission. **[MDU, May 2008, Dec. 2009]**

Ans. A security manager is a class that controls whether a specific operation is permitted. Operations includes the following:

1. Whether a current thread can create a new class loader.
2. Whether a current thread can halt the virtual machine.
3. Whether a class can access a member of another class.

Permissions are classified as follows:

1. Basic permissions
2. File permissions
3. Socket permissions

❑❑❑

APPENDIX

INTERVIEW QUESTIONS

1. What is a transient variable?

Ans. A transient variable is a variable that may not be serialized.

2. Which containers use a border Layout as their default layout?

Ans. The window, Frame and Dialog classes use a border layout as their default layout.

3. Why do threads block on I/O?

Ans. Threads block on i/o (that is enters the waiting state) so that other threads may execute while the i/o Operation is performed.

4. How are Observer and Observable used?

Ans. Objects that subclass the Observable class maintain a list of observers. When an Observable object is updated it invokes the update() method of each of its observers to notify the observers that it has changed state. The Observer interface is implemented by objects that observe Observable objects.

5. What is synchronization and why is it important?

Ans. With respect to multithreading, synchronization is the capability to control the access of multiple threads to shared resources. Without synchronization, it is possible for one thread to modify a shared object while another thread is in the process of using or updating that object's value. This often leads to significant errors.

6. Can a lock be acquired on a class?

Ans. Yes, a lock can be acquired on a class. This lock is acquired on the class's Class object.

7. What's new with the stop(), suspend() and resume() methods in JDK 1.2?

Ans. The stop(), suspend() and resume() methods have been deprecated in JDK 1.2.

8. Is null a keyword?

Ans. The null value is not a keyword.

9. What is the preferred size of a component?

Ans. The preferred size of a component is the minimum component size that will allow the component to display normally.

10. What method is used to specify a container's layout?

Ans. The setLayout() method is used to specify a container's layout.

11. Which containers use a FlowLayout as their default layout?

Ans. The Panel and Applet classes use the FlowLayout as their default layout.

12. What state does a thread enter when it terminates its processing?

Ans. When a thread terminates its processing, it enters the dead state.

13. What is the Collections API?

Ans. The Collections API is a set of classes and interfaces that support operations on collections of objects.

14. Which characters may be used as the second character of an identifier, but not as the first character of an identifier?

Ans. The digits 0 through 9 may not be used as the first character of an identifier but they may be used after the first character of an identifier.

15. What is the List interface?

Ans. The List interface provides support for ordered collections of objects.

16. How does Java handle integer overflows and underflows?

Ans. It uses those low order bytes of the result that can fit into the size of the type allowed by the operation.

17. What is the Vector class?

Ans. The Vector class provides the capability to implement a growable array of objects

18. What modifiers may be used with an inner class that is a member of an outer class?

Ans. A (non-local) inner class may be declared as public, protected, private, static, final, or abstract.

19. What is an Iterator interface?

Ans. The Iterator interface is used to step through the elements of a Collection.

20. What is the difference between the >> and >>> operators?

Ans. The >> operator carries the sign bit when shifting right. The >>> zero-fills bits that have been shifted out.

21. Which method of the Component class is used to set the position andsize of a component?

Ans. setBounds()

22. How many bits are used to represent Unicode, ASCII, UTF-16, and UTF-8 characters?

Ans. Unicode requires 16 bits and ASCII require 7 bits. Although the ASCII character set uses only 7 bits, it is usually represented as 8 bits. UTF-8 represents characters using 8, 16, and 18 bit patterns. UTF-16 uses 16-bit and larger bit patterns.

23. What is the difference between yielding and sleeping?

Ans. When a task invokes its yield() method, it returns to the ready state. When a task invokes its sleep() method, it returns to the waiting state.

24. Which java.util classes and interfaces support event handling?

Ans. The EventObject class and the EventListener interface support event processing.

25. Is sizeof a keyword?

Ans. The sizeof operator is not a keyword.

26. What are wrapped classes?

Ans. Wrapped classes are classes that allow primitive types to be accessed as objects.

27. Does garbage collection guarantee that a program will not run out of memory?

Ans. Garbage collection does not guarantee that a program will not run out of memory. It is possible for programs to use up memory resources faster than they are garbage collected. It is also possible for programs to create objects that are not subject to garbage collection.

28. What restrictions are placed on the location of a package statement within a source code file?

Ans. A package statement must appear as the first line in a source code file (excluding blank lines and comments).

29. Can an object's finalize() method be invoked while it is reachable?

Ans. An object's finalize() method cannot be invoked by the garbage collector while the object is still reachable. However, an object's finalize() method may be invoked by other objects.

30. What is the immediate superclass of the Applet class?

Ans. Panel

31. What is the difference between preemptive scheduling and time slicing?

Ans. Under preemptive scheduling, the highest priority task executes until it enters the waiting or dead states or a higher priority task comes into existence. Under time slicing, a task executes for a predefined slice of time and then reenters the pool of ready tasks. The scheduler then determines which task should execute next, based on priority and other factors.

32. Name three Component subclasses that support painting.

Ans. The Canvas, Frame, Panel, and Applet classes support painting.

33. What value does readLine() return when it has reached the end of a file?

Ans. The readLine() method returns null when it has reached the end of a file.

34. What is the immediate superclass of the Dialog class?

Ans. Window

35. What is clipping?

Ans. Clipping is the process of confining paint operations to a limited area or shape.

36. What is a native method?

Ans. A native method is a method that is implemented in a language other than Java.

37. Can a for statement loop indefinitely?

Ans. Yes, a for statement can loop indefinitely. For example, consider the following: for(;;) ;

38. What are order of precedence and associativity, and how are they used?

Ans. Order of precedence determines the order in which operators are evaluated in expressions. Associatity determines whether an expression is evaluated left-to-right or right-to-left.

39. When a thread blocks on I/O, what state does it enter?

Ans. A thread enters the waiting state when it blocks on I/O.

40. To what value is a variable of the String type automatically initialized?

Ans. The default value of an String type is null.

41. What is the catch or declare rule for method declarations?

Ans. If a checked exception may be thrown within the body of a method, the method must either catch the exception or declare it in its throws clause.

42. What is the difference between a MenuItem and a CheckboxMenuItem?

Ans. The CheckboxMenuItem class extends the MenuItem class to support a menu item that may be checked or unchecked.

43. What is a task's priority and how is it used in scheduling?

Ans. A task's priority is an integer value that identifies the relative order in which it should be executed with respect to other tasks. The scheduler attempts to schedule higher priority tasks before lower priority tasks.

44. What class is the top of the AWT event hierarchy?

Ans. The java.awt.AWTEvent class is the highest-level class in the AWT event-class hierarchy.

45. When a thread is created and started, what is its initial state?

Ans. A thread is in the ready state after it has been created and started.

46. Can an anonymous class be declared as implementing an interface and extending a class?

Ans. An anonymous class may implement an interface or extend a superclass, but may not be declared to do both.

47. What is the range of the short type?

Ans. The range of the short type is -(2^15) to 2^15 - 1.

48. What is the range of the char type?

Ans. The range of the char type is 0 to 2^16 - 1.

49. In which package are most of the AWT events that support the event-delegation model defined?

Ans. Most of the AWT-related events of the event-delegation model are defined in the java.awt.event package. The AWTEvent class is defined in the java.awt package.

50. What is the immediate superclass of Menu?

Ans. MenuItem

51. What is the purpose of finalization?

Ans. The purpose of finalization is to give an unreachable object the opportunity to perform any cleanup processing before the object is garbage collected.

52. Which class is the immediate superclass of the MenuComponent class.

Ans. Object

53. What invokes a thread's run() method?

Ans. After a thread is started, via its start() method or that of the Thread class, the JVM invokes the thread's run() method when the thread is initially executed.

54. What is the difference between the Boolean & operator and the && operator?

Ans. If an expression involving the Boolean & operator is evaluated, both operands are evaluated. Then the & operator is applied to the operand. When an expression involving the && operator is evaluated, the first operand is evaluated. If the first operand returns a value of true then the second operand is evaluated. The && operator is then applied to the first and second operands. If the first operand evaluates to false, the evaluation of the second operand is skipped.

55. Name three subclasses of the Component class.

Ans. Box.Filler, Button, Canvas, Checkbox, Choice, Container, Label, List, Scrollbar, or TextComponent.

56. What is the GregorianCalendar class?

Ans. The GregorianCalendar provides support for traditional Western calendars.

57. Which Container method is used to cause a container to be laid out and redisplayed?

Ans. validate()

58. What is the purpose of the Runtime class?

Ans. The purpose of the Runtime class is to provide access to the Java run-time system.

59. How many times may an object's finalize() method be invoked by the garbage collector?

Ans. An object's finalize() method may only be invoked once by the garbage collector.

60. What is the purpose of the finally clause of a try-catch-finally statement?

Ans. The finally clause is used to provide the capability to execute code no matter whether or not an exception is thrown or caught.

61. What is the argument type of a program's main() method?

Ans. A program's main() method takes an argument of the String[] type.

62. Which Java operator is right associative?

Ans. The = operator is right associative.

63. What is the Locale class?

Ans. The Locale class is used to tailor program output to the conventions of a particular geographic, political, or cultural region.

64. Can a double value be cast to a byte?

Ans. Yes, a double value can be cast to a byte.

65. What is the difference between a break statement and a continue statement?

Ans. A break statement results in the termination of the statement to which it applies (switch, for, do, or while). A continue statement is used to end the current loop iteration and return control to the loop statement.

66. What must a class do to implement an interface?

Ans. It must provide all of the methods in the interface and identify the interface in its implements clause.

67. What method is invoked to cause an object to begin executing as a separate thread?

Ans. The start() method of the Thread class is invoked to cause an object to begin executing as a separate thread.

68. Name two subclasses of the TextComponent class.

Ans. TextField and TextArea

69. What is the advantage of the event-delegation model over the earlier event inheritance model?

Ans. The event-delegation model has two advantages over the event-inheritance model. First, it enables event handling to be handled by objects other than the ones that generate the events (or their containers). This allows a clean separation between a component's design and its use. The other advantage of the event-delegation model is that it performs much better in applications where many events are generated. This performance improvement is due to the fact that the event-delegation model does not have to repeatedly process unhandled events, as is the case of the event-inheritance model.

70. Which containers may have a MenuBar?

Ans. Frame

71. How are commas used in the intialization and iteration parts of a for statement?

Ans. Commas are used to separate multiple statements within the initialization and iteration parts of a for statement.

72. What is the purpose of the wait(), notify(), and notifyAll() methods?

Ans. The wait(),notify(), and notifyAll() methods are used to provide an efficient way for threads to wait for a shared resource. When a thread executes an object's wait() method, it enters the waiting state. It only enters the ready state after another thread invokes the object's notify() or notifyAll() methods.

73. What is an abstract method?

Ans. An abstract method is a method whose implementation is deferred to a subclass.

74. How are Java source code files named?

Ans. A Java source code file takes the name of a public class or interface that is defined within the file. A source code file may contain at most one public class or interface. If a public class or Interface is defined within a source code file, then the source code file must take the name of the public class or interface. If no public class or interface is defined within a source code file, then the file must take on a name that is different than its classes and interfaces. Source code files use the .java extension.

75. What is the relationship between the Canvas class and the Graphics class?

Ans. A Canvas object provides access to a Graphics object via its paint() method.

76. What are the high-level thread states?

Ans. The high-level thread states are ready, running, waiting, and dead.

77. What value does read() return when it has reached the end of a file?

Ans. The read() method returns -1 when it has reached the end of a file.

78. Can a Byte object be cast to a double value?

Ans. No, an object cannot be cast to a primitive value.

79. What is the difference between a static and a non-static inner class?

Ans. A non-static inner class may have object instances that are associated with instances of the class's outer class. A static inner class does not have any object instances.

80. What is the difference between the String and StringBuffer classes?

Ans. String objects are constants. StringBuffer objects are not.

81. If a variable is declared as private, where may the variable be accessed?

Ans. A private variable may only be accessed within the class in which it is declared.

82. What is an object's lock and which object's have locks?

Ans. An object's lock is a mechanism that is used by multiple threads to obtain synchronized access to the object. A thread may execute a synchronized method of an object only after it has acquired the object's lock. All objects and classes have locks. A class's lock is acquired on the class's Class object.

83. What is the Dictionary class?

Ans. The Dictionary class provides the capability to store key-value pairs.

84. How are the elements of a BorderLayout organized?

Ans. The elements of a BorderLayout are organized at the borders (North, South, East, and West) and the center of a container.

85. What is the % operator?

Ans. It is referred to as the modulo or remainder operator. It returns the remainder of dividing the first operand by the second operand.

86. When can an object reference be cast to an interface reference?

Ans. An Object reference be cast to an interface reference when the object implements the referenced interface.

87. What is the difference between a Window and a Frame?

Ans. The Frame class extends Window to define a main application window that can have a menu bar.

88. Which class is extended by all other classes?

Ans. The Object class is extended by all other classes.

89. Can an object be garbage collected while it is still reachable?

Ans. A reachable object cannot be garbage collected. Only unreachable objects may be garbage collected.

90. Is the ternary operator written x : y ? z or x ? y : z ?

Ans. It is written x ? y : z.

91. What is the difference between the Font and FontMetrics classes?

Ans. The FontMetrics class is used to define implementation-specific properties, such as ascent and descent, of a Font object.

92. How is rounding performed under integer division?

Ans. The fractional part of the result is truncated. This is known as rounding toward zero.

93. What happens when a thread cannot acquire a lock on an object?

Ans. If a thread attempts to execute a synchronized method or synchronized statement and is unable to acquire an object's lock, it enters the waiting state until the lock becomes available.

94. What is the difference between the Reader/Writer class hierarchy and the InputStream/OutputStream class hierarchy?

Ans. The Reader/Writer class hierarchy is character-oriented, and the InputStream/OutputStream class hierarchy is byte-oriented.

95. What classes of exceptions may be caught by a catch clause?

Ans. A catch clause can catch any exception that may be assigned to the Throwable type. This includes the Error and Exception types.

96. If a class is declared without any access modifiers, where may the class be accessed?

Ans. A class that is declared without any access modifiers is said to have package access. This means that the class can only be accessed by other classes and interfaces that are defined within the same package.

97. What is the SimpleTimeZone class?

Ans. The SimpleTimeZone class provides support for a Gregorian calendar.

98. What is the Map interface?

Ans. The Map interface replaces the JDK 1.1 Dictionary class and is used associate keys with values.

99. Does a class inherit the constructors of its superclass?

Ans. A class does not inherit constructors from any of its superclasses.

100. For which statements does it make sense to use a label?

Ans. The only statements for which it makes sense to use a label are those statements that can enclose a break or continue statement.

101. What is the purpose of the System class?

Ans. The purpose of the System class is to provide access to system resources.

102. Which TextComponent method is used to set a TextComponent to the read-only state?

Ans. setEditable()

103. How are the elements of a CardLayout organized?

Ans. The elements of a CardLayout are stacked, one on top of the other, like a deck of cards.

104. Is &&= a valid Java operator?

Ans. No, it is not.

105. Name the eight primitive Java types.

Ans. The eight primitive types are byte, char, short, int, long, float, double, and boolean.

106. Which class should you use to obtain design information about an object?

Ans. The Class class is used to obtain information about an object's design.

107. What is the relationship between clipping and repainting?

Ans. When a window is repainted by the AWT painting thread, it sets the clipping regions to the area of the window that requires repainting.

108. Is "abc" a primitive value?

Ans. The String literal "abc" is not a primitive value. It is a String object.

109. What is the relationship between an event-listener interface and an event-adapter class?

Ans. An event-listener interface defines the methods that must be implemented by an event handler for a particular kind of event. An event adapter provides a default implementation of an event-listener interface.

110. What restrictions are placed on the values of each case of a switch statement?

Ans. During compilation, the values of each case of a switch statement must evaluate to a value that can be promoted to an int value.

111. What modifiers may be used with an interface declaration?

Ans. An interface may be declared as public or abstract.

112. Is a class a subclass of itself?

Ans. A class is a subclass of itself.

113. What is the highest-level event class of the event-delegation model?

Ans. The java.util.EventObject class is the highest-level class in the event-delegation class hierarchy.

114. What event results from the clicking of a button?

Ans. The ACtionEvent event is generated as the result of the clicking of a button.

115. How can a GUI component handle its own events?

Ans. A component can handle its own events by implementing the required event-listener interface and adding itself as its own event listener.

116. What is the difference between a while statement and a do statement?

Ans. A while statement checks at the beginning of a loop to see whether the next loop iteration should occur. A do statement checks at the end of a loop to see whether the next iteration of a loop should occur. The do statement will always execute the body of a loop at least once.

117. How are the elements of a GridBagLayout organized?

Ans. The elements of a GridBagLayout are organized according to a grid. However, the elements are of different sizes and may occupy more than one row or column of the grid. In addition, the rows and columns may have different sizes.

118. What advantage do Java's layout managers provide over traditional windowing systems?

Ans. Java uses layout managers to lay out components in a consistent manner across all windowing platforms. Since Java's layout managers aren't tied to absolute sizing and positioning, they are able to accomodate platform-specific differences among windowing systems.

119. What is the Collection interface?

Ans. The Collection interface provides support for the implementation of a mathematical bag - an unordered collection of objects that may contain duplicates.

120. What modifiers can be used with a local inner class?

Ans. A local inner class may be final or abstract.

121. What is the difference between static and non-static variables?

Ans. A static variable is associated with the class as a whole rather than with specific instances of a class. Non-static variables take on unique values with each object instance.

122. What is the difference between the paint() and repaint() methods?

Ans. The paint() method supports painting via a Graphics object. The repaint() method is used to cause paint() to be invoked by the AWT painting thread.

123. What is the purpose of the File class?

Ans. The File class is used to create objects that provide access to the files and directories of a local file system.

124. Can an exception be rethrown?

Ans. Yes, an exception can be rethrown.

125. Which Math method is used to calculate the absolute value of a number?

Ans. The abs() method is used to calculate absolute values.

126. How does multithreading take place on a computer with a single CPU?

Ans. The operating system's task scheduler allocates execution time to multiple tasks. By quickly switching between executing tasks, it creates the impression that tasks execute sequentially.

127. When does the compiler supply a default constructor for a class?

Ans. The compiler supplies a default constructor for a class if no other constructors are provided.

128. When is the finally clause of a try-catch-finally statement executed?

Ans. The finally clause of the try-catch-finally statement is always executed unless the thread of execution terminates or an exception occurs within the execution of the finally clause.

129. Which class is the immediate superclass of the Container class?

Ans. Component

130. If a method is declared as protected, where may the method be accessed?

Ans. A protected method may only be accessed by classes or interfaces of the same package or by subclasses of the class in which it is declared.

131. How can the Checkbox class be used to create a radio button?

Ans. By associating Checkbox objects with a CheckboxGroup.

132. Which non-Unicode letter characters may be used as the first character of an identifier?

Ans. The non-Unicode letter characters $ and _ may appear as the first character of an identifier

133. What restrictions are placed on method overloading?

Ans. Two methods may not have the same name and argument list but different return types.

134. What happens when you invoke a thread's interrupt method while it is sleeping or waiting?

Ans. When a task's interrupt() method is executed, the task enters the ready state. The next time the task enters the running state, an InterruptedException is thrown.

135. What is casting?

Ans. There are two types of casting, casting between primitive numeric types and casting between object references. Casting between numeric types is used to convert larger values, such as double values, to smaller values, such as byte values. Casting between object references is used to refer to an object by a compatible class, interface, or array type reference.

136. What is the return type of a program's main() method?

Ans. A program's main() method has a void return type.

137. Name four Container classes.

Ans. Window, Frame, Dialog, FileDialog, Panel, Applet, or ScrollPane

138. What is the difference between a Choice and a List?

Ans. A Choice is displayed in a compact form that requires you to pull it down to see the list of available choices. Only one item may be selected from a Choice. A List may be displayed in such a way that several List items are visible. A List supports the selection of one or more List items.

139. What class of exceptions are generated by the Java run-time system?

Ans. The Java runtime system generates RuntimeException and Error exceptions.

140. What class allows you to read objects directly from a stream?

Ans. The ObjectInputStream class supports the reading of objects from input streams.

141. What is the difference between a field variable and a local variable?

Ans. A field variable is a variable that is declared as a member of a class. A local variable is a variable that is declared local to a method.

142. Under what conditions is an object's finalize() method invoked by the garbage collector?

Ans. The garbage collector invokes an object's finalize() method when it detects that the object has become unreachable.

143. How are this() and super() used with constructors?

Ans. this() is used to invoke a constructor of the same class. super() is used to invoke a superclass constructor.

144. What is the relationship between a method's throws clause and the exceptions that can be thrown during the method's execution?

Ans. A method's throws clause must declare any checked exceptions that are not caught within the body of the method.

145. What is the difference between the JDK 1.02 event model and the event-delegation model introduced with JDK 1.1?

Ans. The JDK 1.02 event model uses an event inheritance or bubbling approach. In this model, components are required to handle their own events. If they do not handle a particular event, the event is inherited by (or bubbled up to) the component's container. The container then either handles the event or it is bubbled up to its container and so on, until the highest-level container has been tried. In the event-delegation model, specific objects are designated as event handlers for GUI components. These objects implement event-listener interfaces. The event-delegation model is more efficient than the event-inheritance model because it eliminates the processing required to support the bubbling of unhandled events.

146. How is it possible for two String objects with identical values not to be equal under the == operator?

Ans. The == operator compares two objects to determine if they are the same object in memory. It is possible for two String objects to have the same value, but located indifferent areas of memory.

147. Why are the methods of the Math class static?

Ans. So they can be invoked as if they are a mathematical code library.

148. What Checkbox method allows you to tell if a Checkbox is checked?

Ans. getState()

149. What state is a thread in when it is executing?

Ans. An executing thread is in the running state.

150. What are the legal operands of the instanceof operator?

Ans. The left operand is an object reference or null value and the right operand is a class, interface, or array type.

151. How are the elements of a GridLayout organized?

Ans. The elements of a GridBad layout are of equal size and are laid out using the squares of a grid.

152. What an I/O filter?

Ans. An I/O filter is an object that reads from one stream and writes to another, usually altering the data in some way as it is passed from one stream to another.

153. If an object is garbage collected, can it become reachable again?

Ans. Once an object is garbage collected, it ceases to exist. It can no longer become reachable again.

154. What is the Set interface?

Ans. The Set interface provides methods for accessing the elements of a finite mathematical set. Sets do not allow duplicate elements.

155. What classes of exceptions may be thrown by a throw statement?

Ans. A throw statement may throw any expression that may be assigned to the Throwable type.

156. What are E and PI?

Ans. E is the base of the natural logarithm and PI is mathematical value pi.

157. Are true and false keywords?

Ans. The values true and false are not keywords.

158. What is a void return type?

Ans. A void return type indicates that a method does not return a value.

159. What is the purpose of the enableEvents() method?

Ans. The enableEvents() method is used to enable an event for a particular object. Normally, an event is enabled when a listener is added to an object for a particular event. The enableEvents() method is used by objects that handle events by overriding their eventdispatch methods.

160. What is the difference between the File and RandomAccessFile classes?

Ans. The File class encapsulates the files and directories of the local file system. The RandomAccessFile class provides the methods needed to directly access data contained in any part of a file.

161. What happens when you add a double value to a String?

Ans. The result is a String object.

162. What is your platform's default character encoding?

Ans. If you are running Java on English Windows platforms, it is probably Cp1252. If you are running Java on English Solaris platforms, it is most likely 8859_1.

163. Which package is always imported by default?

Ans. The java.lang package is always imported by default.

164. What interface must an object implement before it can be written to a stream as an object?

Ans. An object must implement the Serializable or Externalizable interface before it can be written to a stream as an object.

165. How are this and super used?

Ans. This is used to refer to the current object instance. super is used to refer to the variables and methods of the superclass of the current object instance.

166. What is the purpose of garbage collection?

Ans. The purpose of garbage collection is to identify and discard objects that are no longer needed by a program so that their resources may be reclaimed and reused.

167. What is a compilation unit?

Ans. A compilation unit is a Java source code file.

168. What interface is extended by AWT event listeners?

Ans. All AWT event listeners extend the java.util.EventListener interface.

169. What restrictions are placed on method overriding?

Ans. Overridden methods must have the same name, argument list, and return type. The overriding method may not limit the access of the method it overrides. The overriding method may not throw any exceptions that may not be thrown by the overridden method.

170. How can a dead thread be restarted?

Ans. A dead thread cannot be restarted.

171. What happens if an exception is not caught?

Ans. An uncaught exception results in the uncaughtException() method of the thread's ThreadGroup being invoked, which eventually results in the termination of the program in which it is thrown.

172. What is a layout manager?

Ans. A layout manager is an object that is used to organize components in a container.

173. Which arithmetic operations can result in the throwing of an ArithmeticException?

Ans. Integer / and % can result in the throwing of an ArithmeticException.

174. What are three ways in which a thread can enter the waiting state?

Ans. A thread can enter the waiting state by invoking its sleep() method, by blocking on I/O, by unsuccessfully attempting to acquire an object's lock, or by invoking an object's wait() method. It can also enter the waiting state by invoking its (deprecated) suspend() method.

175. Can an abstract class be final?

Ans. An abstract class may not be declared as final.

176. What is the ResourceBundle class?

Ans. The ResourceBundle class is used to store locale-specific resources that can be loaded by a program to tailor the program's appearance to the particular locale in which it is being run.

177. What happens if a try-catch-finally statement does not have a catch clause to handle an exception that is thrown within the body of the try statement?

Ans. The exception propagates up to the next higher level try-catch statement (if any) or results in the program's termination.

178. What is numeric promotion?

Ans. Numeric promotion is the conversion of a smaller numeric type to a larger numeric type, so that integer and floating-point operations may take place. In numerical promotion, byte, char, and short values are converted to int values. The int values are also converted to long values, if necessary. The long and float values are converted to double values, as required.

179. What is the difference between a Scrollbar and a ScrollPane?

Ans. A Scrollbar is a Component, but not a Container. A ScrollPane is a Container. A ScrollPane handles its own events and performs its own scrolling.

180. What is the difference between a public and a non-public class?

Ans. A public class may be accessed outside of its package. A non-public class may not be accessed outside of its package.

181. To what value is a variable of the boolean type automatically initialized?

Ans. The default value of the boolean type is false.

182. Can try statements be nested?

Ans. Try statements may be tested.

183. What is the difference between the prefix and postfix forms of the ++ operator?

Ans. The prefix form performs the increment operation and returns the value of the increment operation. The postfix form returns the current value all of the expression and then performs the increment operation on that value.

184. What is the purpose of a statement block?

Ans. A statement block is used to organize a sequence of statements as a single statement group.

185. What is a Java package and how is it used?

Ans. A Java package is a naming context for classes and interfaces. A package is used to create a separate name space for groups of classes and interfaces. Packages are also used to organize related classes and interfaces into a single API unit and to control accessibility to these classes and interfaces.

186. What modifiers may be used with a top-level class?

Ans. A top-level class may be public, abstract, or final.

187. What are the Object and Class classes used for?

Ans. The Object class is the highest-level class in the Java class hierarchy. The Class class is used to represent the classes and interfaces that are loaded by a Java program.

188. How does a try statement determine which catch clause should be used to handle an exception?

Ans. When an exception is thrown within the body of a try statement, the catch clauses of the try statement are examined in the order in which they appear. The first catch clause that is capable of handling the exception is executed. The remaining catch clauses are ignored.

189. Can an unreachable object become reachable again?

Ans. An unreachable object may become reachable again. This can happen when the object's finalize() method is invoked and the object performs an operation which causes it to become accessible to reachable objects.

190. When is an object subject to garbage collection?

Ans. An object is subject to garbage collection when it becomes unreachable to the program in which it is used.

191. What method must be implemented by all threads?

Ans. All tasks must implement the run() method, whether they are a subclass of Thread or implement the Runnable interface.

192. What methods are used to get and set the text label displayed by a Button object?

Ans. getLabel() and setLabel()

193. Which Component subclass is used for drawing and painting?

Ans. Canvas

194. What are synchronized methods and synchronized statements?

Ans. Synchronized methods are methods that are used to control access to an object. A thread only executes a synchronized method after it has acquired the lock for the method's object or class. Synchronized statements are similar to synchronized methods. A synchronized statement can only be executed after a thread has acquired the lock for the object or class referenced in the synchronized statement.

195. What are the two basic ways in which classes that can be run as threads may be defined?

Ans. A thread class may be declared as a subclass of Thread, or it may implement the Runnable interface.

196. What are the problems faced by Java programmers who don't use layout managers?

Ans. Without layout managers, Java programmers are faced with determining how their GUI will be displayed across multiple windowing systems and finding a common sizing and positioning that will work within the constraints imposed by each windowing system.

197. What is the difference between an if statement and a switch statement?

Ans. The if statement is used to select among two alternatives. It uses a boolean expression to decide which alternative should be executed. The switch statement is used to select among multiple alternatives. It uses an int expression to determine which alternative should be executed.

198. What happens when you add a double value to a String?

Ans. The result is a String object.

199. What is the List interface?

Ans. The List interface provides support for ordered collections of objects.

❑❑❑

PREVIOUS YEARS EXAMINATION PAPERS

B.E.

8TH SEMESTER EXAMINATION MAY, 2007

(CSE-404-E)

Note : *Attempt any five questions. All questions carry equal marks.*

1. (*a*) What is an Interface? Explain with a suitable example, how multiple inheritance is implemented? What are abstract classes?

(*b*) What are applets? Explain their usage. How are they different from Java programs?

2. With reference to sockets, explain how client/server communication is accomplished in java. Write a program to print the protocol, port, host and file components of a URL.

3. (*a*) Write a function to create a database connection and retrieve all the records from a table that has name="RAHUL".

(*b*) What is the difference between static and dynamic query? Explain in reference to prepared statements.

4. (*a*) What is RMI? Differentiate between stub and skeletons with their usage.

(*b*) What is distributed programming? How is it accomplished in java? Explain with suitable example.

5. (*a*) Explain Java IDL.

(*b*) Explain method call with SOAP?

6. (*a*) What are AWT components? How image manipulation is accomplished using AWT components?

(*b*) How printing is done using AWT components in Java? Explain with the help of an example.

7. (*a*) What are beans? Explain the bean writing process in details with the help of an example.

(*b*) Write an application to implements trees and styled text components

8. Write short notes on:

(*a*) Class Loaders.

(*b*) Digital Signature.

(*c*) Bytecode verification.

B.E.

8TH SEMESTER EXAMINATION DECEMBER, 2007

(CSE-404-E)

Note : *Attempt any five questions. All questions carry equal marks.*

1. Write a program to create a user defined Exception called "No match Exception" that throws an arbitrary message when a string is not equal to "INDIA".

2. (*a*) What is an Interface? Explain with a suitable example, how multiple inheritance is implemented? What are abstract classes?

(*b*) What is multithreading? Explain the life cycle of a thread.

3. With reference to sockets, explain how client/server communication is accomplished in java. Write a program to print the protocol, port, host and file components of a URL.

4. (*a*) Write a function to create a database connection and retrieve all the records from a table that has name="RAHUL".

(*b*) What is JDBC? Explain the different JDBC drivers. Explain the steps of java to database communication in details.

5. What is CORBA? Explain architecture of CORBA. Explain the steps to implement CORBA objects.

6. (*a*) What are AWT components? How image manipulation is accomplished using AWT components?
   (*b*) What are beans? Explain the bean writing process in details with the help of an example.
7. Write an application to implements tables and progress indicators.
8. Write short notes on :
   (*a*) Digital Signature.
   (*b*) Bytecode verification.

B.E.
8TH SEMESTER EXAMINATION MAY, 2008
(CSE-404-E)

Note : *Attempt any five questions. All questions carry equal marks.*

1. (*a*) How the platform independent feature of Java implemented? Explain the role of main() method in the execution of Java program.
   (*b*) What exceptional errors during the execution of a program? Discuss the usage of "finally" block?
2. Discuss with example how a client application read a file from server through URL Connection.
3. (*a*) Explain various JDBC programming concepts.
   (*b*) Write a program to connect to a database query and display the results.
4. (*a*) What is RMI? Differentiate between stub and skeletons with their usage.
   (*b*) What is distributed programming? How is it accomplished in java? Explain with suitable example.
5. (*a*) Explain method call with SOAP?
   (*b*) Discuss Readers and Writers for images. What is image manipulation?
6. What are beans? Explain the bean writing process in details with the help of an example.
7. Write short note on lists, Trees, Styled Text Components.
8. Write short notes on:
   (*a*) Encryption.
   (*b*) Security Manager and permission.

B.E.
8TH SEMESTER EXAMINATION DECEMBER, 2008
(CSE-404-E)

Note : *Attempt any five questions. All questions carry equal marks.*

1. (*a*) What do you mean by bytecode? Explain its working.
   (*b*) Why is Java more suitable than other languages?
2. (*a*) Discuss with example how a client application read a file from server through URL Connection.
   (*b*) Describe how a server sends data to client?
3. (*a*) Write a program to connect to a database query and display the results.
   (*b*) What is the usage of PreparedStatement? Discuss with an example.
4. (*a*) What is distributed programming? How is it accomplished in java? Explain with suitable example.
   (*b*) Explain method call with SOAP?
5. (*a*) What are AWT components? How image manipulation is accomplished using AWT components?
   (*b*) How printing is done using AWT components in Java? Explain with the help of an example.
6. What are beans? Explain the bean writing process in details with the help of an example.
7. What are component organizers? How the Swing components differ from AWT components?
8. Write short notes on
   (*a*) Encryption
   (*b*) Lists, Trees, Styled Text Components

B.E.
8TH SEMESTER EXAMINATION MAY, 2009
(CSE-404-E)

Note : *Attempt any five questions. All questions carry equal marks.*

1. Explain the concept of arrays in Java with suitable example. Also explain inheritance and exception handling with example.
2. Explain the concept of E-mail server.
3. Explain the following :
   (*a*) Metadata
   (*b*) LDAP
4. What is remote method call with SOAP? Explain.
5. How styled components are implemented in Java? Also, explain the usage of tables with suitable programs.
6. How images and shapes are manipulated in AWT? Write a program implementing Drag and Drop in AWT.
7. Explain the bean writing process. Explain building an application with beans.
8. What are class loaders? Explain encryption techniques and digital signatures.

B.E.
8TH SEMESTER EXAMINATION DECEMBER, 2009
(CSE-404-E)

Note : *Attempt any five questions. All questions carry equal marks.*

1. What is multithreading? Explain the life cycle of a thread.
2. Describe LDAP and Metadata with example.
3. What is RMI? Differentiate between stub and skeletons with their usage.
4. Explain Java IDL.
5. Explain Transparency, composition and clipboard.
6. What are beans? Explain the bean writing process in details with the help of an example.
7. Write an application to implements tables and progress indicators.
8. Write short notes on
   (*a*) Digital Signature.
   (*b*) Bytecode verification
   (*c*) Security Manager and permission

B.E.
8TH SEMESTER EXAMINATION MAY, 2011
(CSE-404-E)

Note: Attempt any five questions. All questions carry equal marks.

1. (*a*) What is a Default Constructor? How are an object instance variables initialized if a class has only a class constructor?
   (*b*) Explain the concept of multithreading with example.
2. Explain how to install & configure tomcat server for handling URL connections with example.
3. Discuss the following concepts:
   (*a*) JAVA IDL (*b*) CCRA

4. Explain the following:
   (*a*) Clipping (*b*) Rendering
5. What are Java Swings? How to implement a tree using Java Swings?
6. What is RMI? How do you setup RMI for a web server?
7. What are Java Beans? Explain how Java Beans can be used for reusability of code? Explain with example.
8. Does Java has security constraints? Explain Class Loaders and Digital Signatures.

B.E.

8TH SEMESTER EXAMINATION MAY, 2012

(CSE-404-E)

Note: Attempt any five questions. All questions carry equal marks.

1. (*a*) What are Applets? Explain their usage. How they are different from Java programs?
   (*b*) Explain Inheritance and Exception handling in Java with example.
2. Describe implementation of servers. Also explain advanced socket programming with suitable examples.
3. Explain the following:
   (*a*) Metadata (*b*) Design of JDBC
   (*c*) LDAP
4. How parameters are passed in Remote Methods? Explain various remote method calls with SOAP.
5. What are Component Organizers? Write a program in Java to implement Trees and Tables?
6. (*a*) How coordinate transformation and image manipulation is done in AWT/
   (*b*) Write a program to implement Drag and Drop in AWT.
7. Explain the following:
   (*a*) Bean writing process
   (*b*) Naming patterns for Bean components
8. (*a*) How byte code verification is done? Explain.
   (*b*) Explain code signing and encryption.

❑❑❑

Index

J

L

M

N

T

U

V

W

❑❑❑

Win Prizes

ATTENTION READERS

Win Prizes

We request for your frank assessment regarding various aspects of the book as given below:

ADVANCE JAVA

By : KANIKA LAKHANI

Please Fill up the Sheet in Neat and Clean Writing

(i) What topics of your syllabus, important for your examination point of view, are not covered in this book ?

(ii) What are the chapters wherein the treatment of the subject matter is not systematic or organised or updated?

(iii) Have you come across misprints/mistakes/factual inaccuracies in the book? Please specify the chapters and the page numbers.

(iv) What other books on the same subject have you found/heard better than the present book? Please specify in terms of price and quality

(v) Further suggestions and comments for the improvement of this book.

(PS: If need be please attach a separate sheet to write your views)

OTHER DETAILS

(i) Who recommended you this book? (Please tick (✓) in the box near the option relevant to you.

Teacher ☐ ☐ Friends ☐ Bookseller

....

....

....

....

(ii) Name of the recommending teacher, his designation and address:

....

....

....

....

(iii) Name and address of the bookseller you purchased the book from :

....

....

....

....

(iv) Name and address of your institution (Please mention the University or Board as the case may be)

....

....

....

....

(v) Your name and complete postal address:

....

....

....

....

(vi) Course you are enrolled in:

....

....

....

....

The best assessment will be awarded each month: The award will be in the form of our publications as decided by the Editorial Board.

Please mail the filled up coupon at your earliest to:

S.K. KATARIA & SONS®

4885/109, Prakash Mahal, Dr. Subhash Bhargav Lane,

Opposite Delhi Medical Association, Daryaganj, New Delhi–110002 (INDIA)

Phone: +91-11-43551243, Mobile: +91-9871775858, 7678425209

e-mail: katariabook@gmail.com, katariabooks@yahoo.com

Website: www.skkatariaandsons.com